Wearable Devices and Smart Technology for Educational Teaching Assistance

Shafiq Ul Rehman
Kingdom University, Bahrain

IGI Global
Scientific Publishing
Publishing Tomorrow's Research Today

Vice President of Editorial Melissa Wagner
Managing Editor of Acquisitions Mikaela Felty
Managing Editor of Book Development Jocelynn Hessler
Production Manager Mike Brehm
Cover Design Phillip Shickler

Published in the United States of America by
IGI Global Scientific Publishing
701 East Chocolate Avenue
Hershey, PA, 17033, USA
Tel: 717-533-8845
Fax: 717-533-8661
Website: https://www.igi-global.com E-mail: cust@igi-global.com

Library of Congress Cataloging-in-Publication Data

Names: Ul Rehman, Shafiq, 1981- editor.
Title: Wearable devices and smart technology for educational teaching assistance / Shafiq Ul Rehman, Kingdom University, Bahrain.
Description: Hershey PA : IGI Global Scientific Publishing, Publishing Tomorrow's Research Today, [2025] | Includes bibliographical references and index. | Summary: "This book intends to be a pivotal resource that advances the understanding, implementation, and innovation of wearable technologies in education"-- Provided by publisher.
Identifiers: LCCN 2024054536 (print) | LCCN 2024054537 (ebook) | ISBN 9798369378175 (hardcover) | ISBN 9798369378182 (paperback) | ISBN 9798369378199 (ebook)
Subjects: LCSH: Educational technology. | Wearable technology. | Artificial intelligence--Educational applications. | Machine learning.
Classification: LCC LB1028.3 .W395 2025 (print) | LCC LB1028.3 (ebook) | DDC 371.33--dc23/eng/20250227
LC record available at https://lccn.loc.gov/2024054536
LC ebook record available at https://lccn.loc.gov/2024054537

British Cataloguing in Publication Data
A Cataloguing in Publication record for this book is available from the British Library.

Table of Contents

Detailed Table of Contents

The rise of wearable technologies has transformed how we interact with the world around us. This chapter explores the exciting potential of Artificial Intelligence (AI) and Machine Learning (ML) to further enhance wearables for education and teaching. By integrating AI and ML, wearable devices can evolve from simple data collectors to intelligent assistants that personalize learning, improve student engagement, and empower educators. We will delve into the core concepts of AI and ML, their role in wearable technology, and how they can be harnessed to create a more dynamic and effective learning environment. This chapter will explore specific applications of AI-powered wearables in education, analyze the benefits for both teachers and students, and address potential challenges and considerations for responsible implementation.

The field of education has been no exception to the influence of the ever-evolving advancements brought by technology. The implications of smart technology and its inherent tools bestow several benefits and challenges for educational institutions, educators and students alike. This chapter is an attempt to delve deeper into the specific benefits and challenges of integrating wearable technology and Artificial Intelligence (AI) in the teaching-learning process while also offering a snapshot of the concurrent use of both in an educational setting. Substantial research in the field stands evidence of the individual and combined use of wearable technology and AI in education to offer opportunities for improved student engagement, personalized feedback for students and collaborative learning while simultaneously presenting valid discussion points on the potential pitfalls such as data privacy issues, affordances and training needs of educators. Some recommendations on overcoming such challenges have been made and include the active role of educational institutions, educators and learners.

Abstract The prime motive for this chapter is to perform an extensive analysis on smartly how AI along with ML can be incorporated in wearable technologies that will prove to be beneficial for educational practices. This chapter seeks to: Help Improve Understanding, Drive Best Practices, Encourage Inclusivity, Submission of Technical Challenges, Ethical and Customary, Explore Future Directions. Through meeting these objectives, this chapter is intended to help shed light on the broader implications of technology-enhanced learning and offer practical advice for educators, technologists, and policy makers that wish to leverage AI and ML to support more efficient teaching and better learning experiences.

Abstract: This chapter of the book explains the significant role of existing wearable devices on students. Students are basically classified based on what they are placed into such as academics, sports, or challenged. Then under academics, based upon their age and class in which they are studying as per the World Health Organization Standards. And later the sensors which are used for measuring various health like EEG, EEG, Heart rate, etc and psycho -physiological parameters such as stress, Sleeplessness, autism etc are investigated. The wearable devices made by biosensors are attached to the students at different places on the body based on what parameters that are interested to be measured. Theses parameters are used for monitoring student's activities by using recent developed software algorithms including machine learning and Artificial Intelligent algorithms. Some case studies are also presented.

element, and its features significantly contributed to the student's engagement and learning and assessment.

As technology advances, so do its diverse and creative inventions. One of those innovations includes smart wearable technology that goes by different names, such as AI-based smart wearable devices or smart wearables. This paper delves into the multifaceted landscape of smart wearable technology, focusing on its integration and susceptibility in lesser-explored domains such as education, healthcare, and health industrial applications. The paper explores smart wearables' evolution, highlighting their transformative impact on various educational analytics and application cases. It examines the benefits and risks associated with smart wearable technology in healthcare educational settings, shedding light on its widespread adoption's potential opportunities and challenges. By analyzing these aspects, this paper aims to provide insights into the modern-age susceptibility to compact smart wearables and propose measures to address security concerns while harnessing the potential of smart wearable technology for healthcare educational analytics and applications.

This chapter discusses the immense role of wearable health devices and the smart technology in healthcare and medical education. Wearable devices are reshaping the health management system specifically after the COVID-19. People are using these devices to track their vitals such as blood pressure, oxygen saturation, heart rate etc. These health gadgets with smart biosensors allow user to track his health record reducing the hospital visits. Wearable gadgets have come a long way historically from a pacemaker and glucose monitors to the highly sophisticated artificial intelligence

Preface

INTRODUCTION

The intersection of education and technology has long been a space of boundless possibility, but few innovations have shown as much transformative potential as wearable devices. From fitness trackers and smartwatches to AI-powered biosensors and immersive AR/VR headsets, these technologies are redefining how students learn, how teachers teach, and how administrators manage educational outcomes. Wearable devices have evolved beyond simple activity monitors to become integral tools in personalized learning, student engagement, and overall well-being. This book, *Wearable Devices and Smart Technology for Educational Teaching Assistance*, serves as a comprehensive guide to understanding, applying, and advancing the role of wearable technology in education.

The need for personalized learning pathways has never been greater. As educational institutions strive to meet the diverse needs of students, wearable devices powered by Artificial Intelligence (AI) and Machine Learning (ML) present opportunities for more tailored instruction and support. From tracking student engagement to monitoring mental health, these devices provide educators with the insights needed to foster more inclusive, student-centered learning environments. However, alongside these opportunities come significant challenges—issues of data privacy, cybersecurity, and equitable access require thoughtful consideration. This book addresses these concerns while providing a pathway forward for policymakers, educators, and technologists.

This preface offers a glimpse into the structure and purpose of the book. Each chapter dives into a unique aspect of wearable technologies in education, ranging from conceptual frameworks and practical applications to case studies and ethical discussions. Whether you are an educator, researcher, policymaker, or developer, you will find a wealth of information and guidance in these pages. Our goal is to create a collaborative resource that informs, inspires, and ultimately drives the adoption of wearables in education for the betterment of all learners.

CHAPTER OVERVIEW

Chapter 1

This chapter explores how AI and ML are transforming wearables into intelligent educational assistants. It highlights key applications, such as personalized learning, student engagement, and the empowerment of educators. The chapter also addresses implementation challenges and best practices for responsible use.

Chapter 2

Continuing the discussion of AI and ML, this chapter delves into the practical applications of AI-powered wearables for personalized learning. It showcases real-world examples of how wearables support both students and teachers in achieving more dynamic and engaging learning experiences.

Chapter 3

This chapter provides a technical perspective on the integration of AI and ML with wearables. It outlines methodologies for successful integration and offers strategies to address technical and ethical challenges, such as data privacy and algorithmic bias.

Chapter 4

This chapter explores how wearable devices monitor health-related metrics such as heart rate, sleep patterns, and stress levels. It discusses the role of biosensors in collecting student well-being data and highlights the potential for supporting student wellness and academic performance.

Chapter 5

Focusing on Bahrain's innovative approach to education, this chapter analyzes the country's use of AI-powered wearables. It highlights how these tools are being used to enhance personalized learning, promote emotional well-being, and align with global education and workforce trends.

Chapter 6

Gamification is a powerful driver of student motivation and engagement. This chapter discusses the integration of wearable devices into gamified learning systems. It offers insights from empirical studies and statistical analyses, illustrating how wearable tech can positively influence student outcomes.

Chapter 7

This chapter takes a broader perspective, examining how wearable technologies are impacting education, healthcare, and industrial applications. It analyzes the evolution of wearables, their benefits, and the security challenges posed by their widespread adoption.

Chapter 8

Medical education is seeing significant changes with the rise of wearable health devices. This chapter discusses how biosensors and AI-based health trackers are reshaping medical training. It highlights the growing role of augmented reality (AR) and virtual reality (VR) in medical education.

Chapter 9

The mental health of students is an emerging concern in education. This chapter addresses how wearables can support mental health monitoring through biosensors and AI-driven analytics. By tracking stress, anxiety, and sleep patterns, wearable devices provide students with personalized interventions and wellness support.

Chapter 10

Similar to Chapter 9, this chapter takes a deeper look at mental health monitoring. It discusses specific applications of AI-driven wearables, the benefits for students and educators, and strategies for responsible implementation to protect student privacy.

Chapter 11

This chapter presents a case study on EZeCheck, a revolutionary wearable device used to monitor health metrics such as hemoglobin levels. It explores the socioeconomic impact of this device, particularly in supporting the objectives of

"Anaemia Mukt Bharat" (Anemia-Free India), and its potential expansion into educational settings.

Chapter 12

This chapter discusses the broader benefits and challenges of integrating wearables into education. It addresses issues like device comfort, data privacy, and technical support, while also highlighting the opportunities for personalized learning, engagement, and student support.

Chapter 13

Wearable devices pose new cybersecurity challenges in educational environments. This chapter outlines a comprehensive cybersecurity framework for educational institutions, covering device security, data privacy, cyber awareness, and breach management.

CONCLUSION

The rapid adoption of wearable devices in education signifies a profound shift in how students learn and how educators teach. As these devices become more sophisticated with AI and ML integration, they promise a future where learning is more personalized, engaging, and inclusive. Students can benefit from tailored learning paths, while teachers gain deeper insights into student progress and well-being. Wearable devices empower administrators to create data-driven policies that support student growth. However, these advancements come with complex challenges, particularly in the realms of data privacy, cybersecurity, and equitable access. Addressing these concerns requires collaboration across educational, technical, and policy-making communities.

This book emphasizes that wearable technology is not a distant future—it's here, shaping education in real time. The research, case studies, and frameworks presented within these chapters offer clear pathways for stakeholders to navigate this transformative era. As educators, developers, and policymakers work together to embrace wearable devices, they have the opportunity to create more dynamic, inclusive, and student-centered learning environments.

Our journey through this book underscores a central message: wearable technology in education must be guided by principles of equity, ethics, and empathy. The benefits are immense, but only if approached thoughtfully. By fostering an environment where student privacy is protected, technical challenges are addressed, and equitable access is prioritized, we can ensure that wearable devices enhance learning experiences for all students.

We hope this book serves as an essential resource for educators, researchers, and innovators dedicated to shaping the future of education. It invites readers to think critically, act responsibly, and explore new possibilities in wearable educational technology. Our hope is that this book sparks inspiration and paves the way for deeper collaboration, innovation, and the responsible implementation of wearable devices in educational teaching assistance.

This preface sets the tone for the book, guiding the reader to think beyond the technology itself and focus on its transformative potential. It provides an entry point for educators, developers, and policymakers alike, equipping them with the knowledge and insight to navigate the evolving landscape of wearable educational technology.

Shafiq Ul Rehman
Kingdom University, Bahrain

Chapter 1
Significance of AI/ML Wearable Technologies for Education and Teaching

Uddalak Mitra
JIS College of Engineering, India

Shafiq Ul Rehman
https://orcid.org/0000-0003-2266-218X
Kingdom University, Bahrain

ABSTRACT

The rise of wearable technologies has transformed how we interact with the world around us. This chapter explores the exciting potential of Artificial Intelligence (AI) and Machine Learning (ML) to further enhance wearables for education and teaching. By integrating AI and ML, wearable devices can evolve from simple data collectors to intelligent assistants that personalize learning, improve student engagement, and empower educators. We will delve into the core concepts of AI and ML, their role in wearable technology, and how they can be harnessed to create a more dynamic and effective learning environment. This chapter will explore specific applications of AI-powered wearables in education, analyze the benefits for both teachers and students, and address potential challenges and considerations for responsible implementation.

DOI: 10.4018/979-8-3693-7817-5.ch001

1. INTRODUCTION

Artificial Intelligence (AI) and Machine Learning (ML) have brought transformative shifts across many industries(Rao, T. V. N. et al. 2022, Singh, P et al. 2022, Nguyen, D. K et al. 2023, Singh, P. K. et al. 2023, Basnet, S. et al. 2024, Cao, L. et al 2023), but their impact on education is particularly profound(Munir, H. et al. 2022, Sunarya, P. A. et al. 2022, Limna, P. et la. 2022, Shaikh, A. A. et al. 2022, Zafari, M. et al. 2022, Hopcan, S. et al. 2024). As the demand for more personalized and effective learning environments grows, educators and technologists are turning to AI-powered solutions to bridge the gap between traditional educational models and the needs of modern learners (Srinivasa, K. G et al. 2022, Saaida, M. B. et al. 2023, Amer-Yahia, S. et al. 2022, El Asmar, W. et al. 2022, Laak, K. J et al. 2024, Kayyali, M. et al. 2024). Among these innovations, AI-driven wearables are emerging as a key tool for enhancing learning experiences (Aldowah, H. et al. 2017). These devices, combined with sophisticated algorithms, provide real-time insights into students' physical, cognitive, and emotional states, allowing for adaptive, personalized learning.

Wearables such as smartwatches, fitness trackers, and specialized devices are already commonplace in fields like healthcare and fitness, where they track vital signs and physical activity (Sadek, I et al. 2019, Kazanskiy, N. L. et al. 2024, Seneviratne, S et al, 2027). When applied to education, these devices offer educators unprecedented access to student data. They can monitor stress levels, engagement, attention, and even emotions, all in real time. This data can then be analyzed by AI and ML systems, enabling immediate adjustments to teaching strategies and learning materials. Such adaptability allows for truly personalized education, addressing the individual needs of students far more effectively than standardized methods.

AI-powered wearables are not just limited to individual customization (Rehman, S. U et al. 2024). They offer transformative benefits in collaborative learning environments as well, enabling educators to analyze group dynamics, measure participation, and facilitate more balanced, effective teamwork. Furthermore, these technologies are proving invaluable in the support of students with learning disabilities or those in need of emotional and mental health support, with wearables able to detect signs of stress, anxiety, or fatigue and prompt timely interventions.

The potential of AI-powered wearables in education extends beyond the classroom(Zhang, X. et al. 2024, Qasim, S. H. et al. 2024). They are increasingly being integrated into lifelong learning and professional training environments, where workers can receive real-time feedback and coaching, helping them acquires new skills faster and more efficiently. As industries shift towards continuous learning models, wearables are poised to become integral tools in workforce development(Sadek, I., et al. 2022).

However, as with all technological advancements, the integration of AI-powered wearables in education comes with challenges(Aldowah, Hetal. 2019, Mohanty, A. K. et al. 2023). Concerns about bias in AI algorithms, privacy, and data security are becoming more and more prevalent. The long-term viability of these institutions depends on making sure they are equitable, open, and accountable. In order to stop educational disparities from getting worse, it is also becoming more and more important to make sure that these technologies are available and reasonably priced for all students, regardless of their socioeconomic status.

This chapter explores how wearables with AI capabilities can revolutionize education by improving learning outcomes, offering personalized learning pathways, and providing real-time feedback. We will examine the benefits, challenges, and future trends of this emerging technology, offering insights into how wearables can be harnessed to create more efficient, equitable, and effective learning environments. The chapter will also cover the legal and ethical frameworks required to guarantee the proper use of wearables powered by artificial intelligence in classrooms.

Wearables with AI capabilities provide an intriguing new frontier in educational technology at a time when data-driven decision-making is increasingly the standard. These gadgets are helping to mould the future of education by giving teachers useful information on the performance and behavior of their students. This will make learning more human-centered, flexible, and responsive.

2. BACKGROUND

2.1 Wearable Technologies in Education

Wearable devices, such as smart glasses (Google Glass and Microsoft's HoloLens), smartwatches (Apple Watch, Samsung Galaxy Watch), fitness trackers (Fitbit, Garmin Vivosmart), andbrain-computer interfaces (BCIs)like Muse Headband and EmotivEpoc+, are increasingly being adopted in educational environments (**Table 1**). These gadgets have a number of revolutionary advantages that are changing how students engage with and learn from instructional materials.

- Data Collection: Wearable devices gather real-time physiological data such as heart rate, brain activity, attention levels, and physical movement. This provides educators with deep insights into students' cognitive engagement, stress levels, and physical well-being during lessons. For example, BCIs can measure brain waves, allowing teachers to track mental effort, focus, and fatigue (Orban, M. et al. 2022, Nojima, I. et al 2022, Bano, K. S. et al. 2022, Peksa, J. et al 2023). Such data can be used to personalize learning experiences and adjust teaching strategies dynamically, ensuring that students stay focused and avoid cognitive overload.

- Personalized Learning: AI-powered solutions can tailor the learning process to each student's needs by evaluating the data gathered from wearables.Depending on how well students do, adaptive learning systems might change the activities' degree of difficulty or offer different teaching strategies (Gligorea, I et al. 2023, Wang, S. et al. 2023, Minn, S. et al. 2022, Alam, A. et al. 2022, Alam, A. et al. 2023). For instance, if a student's wearable device detects high levels of frustration during a lesson, AI can adjust the content to present simpler examples or offer extra support. Each student may study at their own speed with this individualized method, which improves comprehension and retention.
- Gamification: Wearables offer new opportunities to enhance engagement through gamification. By integrating wearables with gamified learning environments, educators can provide real-time feedback, challenges, and rewards based on students' performance(Nkereuwem, I. V. et al. 2023, Bezzina, S. et al 2023, Chang, Y. et al 2023). For example, smartwatches and fitness trackers can track physical activity and offer rewards like badges or points for completing tasks, while smart glasses can create immersive learning environments where students interact with virtual objects or solve problems in real-time. This interactive learning keeps students motivated and reinforces positive learning behaviors, making education more engaging and effective.
- Collaboration and Accessibility: Wearable devices also enable remote learning and promote inclusive education(Zheng, H. et al 2021, Karagianni, E. et al. 2023)by providing access to students who cannot attend traditional classrooms. Smart glasses allow students to participate in virtual classrooms, where they can interact with teachers and peer in real-time, regardless of their physical location. For students with disabilities, BCIs open new avenues for participation by enabling them to control digital devices using their thoughts, breaking down barriers to learning. This inclusivity fosters a more equitable educational environment, ensuring all students have access to learning opportunities.

Table 1. Transformative role of wearable technology in education

Transformative Aspect	Wearable Devices	Transformative Benefits	Practical Examples	References
Data Collection	Smartwatches, BCIs, Fitness Trackers	- Real-time tracking of physiological data. - Insights into engagement, stress, focus. - Adjust lessons dynamically.	BCIs like Muse track mental effort; educators modify lesson pace to prevent cognitive overload.	Orban et al., 2022; Nojima et al., 2022; Bano et al., 2022; Peksa et al., 2023
Personalized Learning	Smartwatches, BCIs	- Adaptive AI systems - Customized teaching methods based on performance. - Better retention through tailored pacing.	Wearable detects frustration, adjusts lessons to simpler examples, or adds guidance dynamically.	Gligorea et al., 2023; Wang et al., 2023; Minn et al., 2022; Alam et al., 2022, 2023
Gamification	Smart Glasses, Smartwatches	- Real-time feedback. - Engagement via rewards and challenges. - Interactive, immersive learning environments.	Fitness trackers reward active tasks with badges, while smart glasses let students solve virtual 3D puzzles during lessons.	Nkereuwem et al., 2023; Bezzina et al., 2023; Chang et al., 2023
Collaboration & Accessibility	Smart Glasses, BCIs	- Remote learning enabled. - Inclusion for students with disabilities.	Students with BCIs control devices via thought. Smart glasses enable virtual classes for remote learners.	Zheng et al., 2021; Karagianni et al., 2023

2.2 Understanding AI and Machine Learning

1. Artificial-Intelligence-(AI):

Artificial intelligence (AI) describes the ability of robots to carry out activities like thinking, problem-solving, and information adaptation that normally need human intellect. By providing individualised learning experiences, artificial intelligence is revolutionising traditional teaching techniques in the field of education. For example, AI-powered tutoring systems, like Carnegie Learning's "Mika," provide students with tailored lessons that adjust based on their progress and comprehension levels. These systems use AI algorithms to track student performance in real-time, offering personalized feedback and suggesting additional practice in areas where students struggle. Additionally, AI simplifies administrative duties like attendance and grading, giving teachers more time to concentrate on instruction. For example, by identifying trends in student responses, Gradescope and other platforms employ AI to grade assignments more quickly and correctly, particularly in big classrooms. This allows educators to provide quicker feedback and spend more time addressing individual student needs.

2. Machine-Learning-(ML):

A subfield of artificial intelligence called machine learning is concerned with creating algorithms that can learn from data and get better over time. ML is essential to adaptive learning platforms in education, which use information from student interactions to modify the content and level of difficulty of classes in real time. An example is the DreamBox Learning platform, which uses ML algorithms to analyze how students solve math problems and then adjusts the level of difficulty based on their performance. ML can also be used to predict student outcomes by analyzing engagement patterns, test scores, and other data. For example, the AltSchool system uses ML to create detailed profiles of student learning behaviors, identifying students who may be at risk of falling behind. By recognizing these patterns early, educators can intervene with additional resources, personalized learning plans, or tutoring, ensuring that each student stays on track to succeed.

Figure 1. Relation between AI and ML

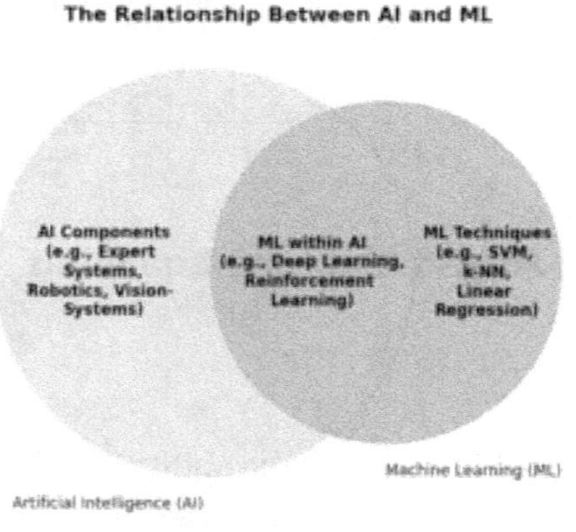

2.3 Integrating AI/ML with Wearables in Education

The integration of Artificial Intelligence (AI) and Machine Learning (ML) with wearable devices is transforming how data is collected, analyzed, and applied in educational settings. Wearable devices, such as smartwatches and fitness trackers, can monitor a wide range of physiological and behavioral metrics, such as heart rate, physical activity, and even attention levels. These wearables, when paired with AI/ML algorithms, are potent tools for improving the educational process by facilitating data-driven decision-making and offering real-time insights.

AI/ML algorithms can be embedded within the wearable devices themselves or operate via cloud-based platforms that analyze data collected from multiple devices. This infrastructure allows for the implementation of advanced features that can significantly improve the learning process:

- **Real-time Feedback**: AI-powered systems provide students with instant feedback on various aspects of their learning, such as performance, engagement levels, or comprehension. For example, if a student's wearable detects declining focus or high levels of stress during a lesson, the AI system can immediately suggest a short break or provide learning materials in a different format to better match the student's current state. This immediate response helps students stay engaged and promotes more effective learning.
- **Adaptive Learning**: ML algorithms analyze the data from wearables to adjust the difficulty, pace, and content of learning materials dynamically. Based on metrics like heart rate variability or focus levels, the system can recommend when to speed up or slow down the presentation of information, ensuring that students are neither overwhelmed nor under-stimulated. For instance, a student showing signs of fatigue may be given simpler tasks or short breaks, while a student exhibiting high engagement may be introduced to more challenging material to maintain interest and foster growth.
- **Early Intervention**: AI systems can detect early signs of struggle, anxiety, or fatigue by analyzing data trends from wearables. If a student repeatedly shows signs of stress or disengagement, the AI system can alert teachers, prompting them to offer personalized support before the student falls too far behind. This proactive approach enables educators to provide targeted interventions, such as additional tutoring, modified lesson plans, or counseling, helping to create a supportive learning environment.

Wearable technology combined with AI and ML can help teachers better understand the learning styles and well-being of their students, resulting in more individualized, responsive, and successful learning environments. This technology-driven strategy aids in bridging the gap between conventional teaching techniques and the data-driven, customized education of the future.

3. APPLICATIONS OF AI-POWERED WEARABLES IN EDUCATION

3.1 Personalized Learning Environments

By adapting educational experiences to each student's unique requirements, preferences, and cognitive states, the integration of cutting-edge technology such as artificial intelligence (AI), wearables, and brain-computer interfaces (BCIs) is transforming personalized learning environments. By providing a more engaging, flexible, and responsive learning environment, these technologies increase the effectiveness and inclusivity of education.

- **Smart Glasses**: Smart glasses represent a revolutionary tool in personalized learning by providing contextual, real-time information directly into a student's field of vision. These glasses can overlay additional learning materials, translations, or real-time subtitles during lectures, making learning more accessible for students with different language skills or hearing impairments. For instance, a student studying history can see supplemental images, video clips, or textual information as they listen to a lecture, enhancing comprehension and retention. Smart glasses also support different learning styles by catering to visual learners through augmented reality (AR) experiences, making abstract concepts easier to grasp through interactive, 3D visualizations.
- **Adaptive Learning Platforms**: In order to generate highly customized learning routes, AI-driven adaptive learning platforms continually examine data from students' interactions with wearables, tests, and instructional materials. These systems dynamically adjust the difficulty, content delivery pace, and learning materials to match each student's skill level, engagement, and progress. For example, if a student is excelling in mathematics, the platform may introduce more advanced problems, while for students struggling with certain concepts, it provides additional practice and customized resources. By preventing students from being disinterested by too-easy or irritated by too-hard information, this method promotes a more interesting and effective learning environment.
- **Brain-Computer Interfaces (BCIs)**: BCIs are an emerging technology that holds the potential to dramatically transform personalized learning environments. By directly measuring brain activity, BCIs can assess a student's level

of focus, engagement, or even cognitive load in real time. These insights can then be used to adjust learning content to the student's current state. For instance, if a BCI detects that a student's attention is waning, the system could switch to a more interactive or engaging format, such as a game-based quiz, to re-capture focus. Conversely, if a student is highly engaged, the system might introduce more complex material. While still in developmental stages, BCIs represent the next frontier in creating truly individualized, responsive learning experiences that adapt in real-time to students' mental and emotional states.

3.2 Enhanced Engagement and Motivation

Through the creation of interactive, customised, and health-conscious learning environments, wearable technologies are becoming more and more important in raising student motivation and engagement. These developments support both cognitive and physical well-being while meeting a variety of learning demands.

- **Wearable Activity Trackers**: Gamified learning environments can be developed using wearable activity trackers that monitor students' movement and physical activity throughout the day. By incorporating game-like elements, such as earning points, badges, or rewards based on participation, schools can promote movement breaks and foster a healthy learning atmosphere. For instance, students may receive notifications to stand up, stretch, or take short walks between lessons, helping to reduce sedentary behavior while maintaining focus. This gamification approach not only motivates students to stay active but also enhances cognitive performance by ensuring regular physical activity during learning sessions.
- **VR/AR Integration**: Immersion learning is being redefined by the combination of wearables with augmented reality (AR) and virtual reality (VR) technology. Wearables such as headsets and motion sensors enable students to engage with interactive, 3D environments that bring subjects like science, history, and mathematics to life. For example, a student learning about ancient civilizations could "walk" through a virtual reconstruction of a historical site, interacting with artifacts and exploring the environment in real-time. This immersion increases engagement and deepens understanding by turning abstract concepts into tangible, interactive experiences, making learning more enjoyable and memorable.
- **Real-time Feedback and Progress Tracking**: Wearables provide students with instant feedback on their academic and physical performance, helping them stay motivated and track their progress. For example, wearable devices

can monitor a student's attention, heart rate, or activity levels during lessons, providing immediate insights into when they are most engaged or distracted. Students can also set goals and milestones, allowing them to visualize their progress over time. This constant feedback loop not only fosters a sense of accomplishment but also helps students self-regulate and make informed decisions about their learning strategies.

3.3 Personalized Support for Students with Special Needs

Wearable technology is very useful for bridging learning gaps and enabling customised interventions, as well as for providing individualised support for kids with special needs. Through the resolution of particular difficulties, improvement of concentration, and encouragement of independence, these assistive technologies empower pupils.

- **Neurofeedback**: Neurofeedback wearables offer real-time monitoring of brain activity, which can be especially beneficial for students with ADHD or attention difficulties. These devices track neural patterns related to focus and relaxation, enabling students to receive feedback on their mental state. For instance, if a student becomes distracted during a lesson, the device can prompt them to engage in mindfulness exercises or take a brief break. This personalized feedback helps students develop greater self-awareness and improves their ability to self-regulate attention, ultimately leading to enhanced academic performance and emotional well-being.
- **Assistive Technologies**: Wearables such as smart glasses, haptic feedback devices, and auditory aids can provide critical support for students with visual, auditory, or learning impairments. For example, smart glasses can translate written text into spoken words for students with visual impairments, while haptic feedback devices can deliver subtle vibrations to guide students with learning disabilities through tasks. Regardless of their skill level, these technologies make it possible for all students to have access to the materials and assistance they require in order to thrive in a varied learning environment. These wearables enable students with special needs to participate more actively in the learning process and become more independent by offering real-time support and feedback.

3.4 Empowering Educators

Wearable technologies, combined with AI and ML, are not only transforming the student learning experience but also empowering educators by providing data-driven insights and tools to improve teaching effectiveness. These innovations help educators to better understand individual student needs, optimize instructional strategies, and provide timely, personalized support.

- **Real-time Student Monitoring**: AI-driven analytics can process data from wearables to give educators real-time insights into student engagement, comprehension levels, and potential areas of difficulty within the classroom. For instance, wearables can monitor physiological indicators such as heart rate and attention levels, allowing teachers to assess whether students are engaged during a lesson or struggling to focus. This continuous monitoring enables teachers to make immediate adjustments in their teaching methods—such as incorporating interactive activities or providing additional clarification on challenging topics—ensuring that every student remains engaged and on track. This real-time data empowers educators to identify struggling students early, allowing for timely interventions and tailored support.
- **Personalized Learning Recommendations**: AI algorithms can analyze the data collected from student wearables, learning platforms, and assessments to recommend personalized learning pathways for each student. These recommendations may include targeted resources, specific activities, or differentiated instruction methods that align with the student's current skill level, learning pace, and engagement patterns. By automating this analysis, AI reduces the burden on teachers, allowing them to focus more on instruction and interaction rather than data processing. For instance, if a student consistently demonstrates high engagement in visual learning tasks but struggles with text-based materials, AI can suggest more visual resources to enhance comprehension, thereby helping teachers to implement more personalized, effective teaching strategies.
- **Formative Assessment and Feedback**: Wearable technologies enable the creation of more dynamic and comprehensive formative assessments that go beyond traditional tests and quizzes. AI can analyze data from wearables—such as attention patterns, cognitive load, and emotional responses during lessons—to assess not only what students know but how they are learning. This deeper analysis allows educators to identify areas where students excel or need additional support, guiding future lesson planning and improving instructional design. For example, if data shows that students consistently lose focus during a specific type of activity, teachers can modify their les-

son plans to incorporate more engaging methods. Additionally, AI-generated feedback based on wearable data provides students with more constructive, personalized guidance, helping them understand their learning strengths and weaknesses in real time.

4. BENEFITS OF AI-POWERED WEARABLES IN EDUCATION

AI-powered wearables bring numerous benefits to both students and educators, revolutionizing the way education is delivered and experienced. These technologies enhance learning outcomes, foster greater engagement, and create more inclusive and personalized educational environments.

- **Improved Learning Outcomes**: Wearable technology can adjust to the individual requirements, preferences, and learning pace of every student by using AI to personalise learning experiences. More focused interventions are made possible by this tailored approach, guaranteeing that students get the assistance they need when they need it. Students' comprehension of difficult ideas deepens as a consequence, and their general academic performance rises. To assist students learn the information more efficiently, a student who is having trouble with a specific topic, for instance, might be given extra resources or practice problems that are suited to their learning preferences.
- **Enhanced Student Engagement**: AI-powered features, such as gamification, real-time feedback, and interactive learning environments, make education more engaging and motivating for students. Wearables can track a student's progress and provide instant feedback, turning learning into a more dynamic and enjoyable experience. Gamification elements, such as earning points, badges, or rewards for completing tasks, help maintain student interest and encourage active participation. For instance, a wearable could notify a student when they've reached a learning milestone, boosting their sense of accomplishment and motivating them to continue.
- **Empowered Educators**: Wearables powered by AI give teachers useful, data-driven insights about the performance, engagement, and problem areas of their students. Teachers may lessen the one-size-fits-all method of education by using these insights to better customize their lessons to each student's requirements. Teachers may make quick adjustments to their teaching methods and provide more individualized help when necessary using real-time data on students' understanding and attentiveness. The educator's capacity to deliver focused interventions is improved by this data-driven strategy, which eventually improves student results.

- **Early Identification of Learning Difficulties**: The capacity of AI-powered wearables to quickly identify pupils who are having difficulty with particular concepts or activities is one of its most potent features. AI can identify trends that point to learning challenges or comprehension gaps by continually observing student performance and engagement. Teachers can step in before issues worsen thanks to this early detection, providing pupils with individualized guidance and tools to help them overcome their obstacles. It has been demonstrated that early intervention greatly enhances long-term academic achievement and keeps pupils from falling behind.
- **Accessibility and Inclusivity**: AI-powered wearables also play a crucial role in making education more accessible and inclusive for students with special needs. These devices can offer assistive technologies, such as real-time translation, speech-to-text capabilities, or haptic feedback, to help students with disabilities engage more fully in the learning process. For example, smart glasses can display subtitles for students with hearing impairments, while haptic feedback devices can guide students with visual impairments. By providing these tailored tools, AI-powered wearables foster a more inclusive educational environment where every student has the opportunity to succeed.

5. CHALLENGES AND CONSIDERATIONS

While AI-powered wearables offer transformative potential for education, their implementation must be carefully managed to address key considerations related to privacy, equity, support, and ethics. Ensuring these aspects are thoughtfully addressed is essential for realizing the full benefits of these technologies while safeguarding students' interests and promoting fair access.

- **Data Privacy and Security**: Protecting student data is paramount when integrating AI-powered wearables into educational settings. Robust security measures and stringent data privacy policies must be established to guard against unauthorized access and data breaches Manickam, S et al. 2023. This includes encrypting data, implementing secure authentication methods, and regularly auditing systems for vulnerabilities. Clear and transparent data privacy policies should inform students, parents, and educators about how data is collected, stored, and used, ensuring that consent is obtained and privacy rights are respected (Aldowah, H et al. 2021). For instance, a school implementing wearables should provide detailed information on how student data will be protected and used, and ensure that any data sharing complies with relevant regulations such as GDPR or FERPA.

- **Equity and Accessibility**: To avoid exacerbating educational inequalities, it is crucial that AI-powered wearables are designed to be affordable and accessible to all students. This involves addressing issues of cost, availability, and infrastructure to ensure that no student is left behind due to financial or logistical barriers. Schools and policymakers must work to provide equitable access to these technologies, potentially through subsidized programs, grants, or partnerships with technology providers. For example, initiatives could be established to provide low-cost or free wearables to underfunded schools or students from disadvantaged backgrounds, ensuring that all students benefit from the enhanced learning opportunities these devices offer.
- **Teacher Training and Support**: Effective integration of AI-powered wearables into the classroom requires comprehensive teacher training and ongoing support. Educators need to be equipped with the skills to use these technologies effectively, interpret the data insights they provide, and integrate them into their teaching practices. Professional development programs should focus on practical applications of wearables, data analysis, and how to leverage insights to enhance teaching and learning. For example, training sessions could include workshops on using data from wearables to personalize instruction, workshops on interpreting engagement metrics, and strategies for incorporating wearable data into lesson planning.
- **Ethical Considerations and Algorithmic Bias**: Wearable AI algorithms need to be well planned and closely watched to avoid biases that can harm particular student populations. Making sure algorithms are fair, transparent, and intended to provide equal results for all students is one aspect of ethics. To find and address any biases in the algorithms, regular audits and upgrades should be carried out. To make sure algorithms don't disproportionately affect students from particular demographic groupings, for example, they should be evaluated on a variety of student demographics. Furthermore, addressing any biases and guaranteeing that the technology benefits all students equally may be achieved by incorporating a variety of stakeholders in the development and assessment of AI systems.

6. CASE STUDY: DEMYSTIFYING MATH WITH AI-POWERED PERSONALIZED LEARNING IN A HIGH SCHOOL CLASSROOM

Background

Let a hypothetical character Mr. Thompson, a high school math teacher, faces a multifaceted challenge in his classroom. His students represent a wide spectrum of learning abilities and needs. Some students quickly grasp mathematical concepts, while others struggle with visualization, conceptual understanding, or find themselves overwhelmed by anxiety. Traditional instructional methods have proven inadequate in providing the individualized support necessary to address these varied learning needs, resulting in frustration and disengagement among many students.

To tackle these issues, Mr. Thompson decides to implement a cutting-edge pilot program that integrates AI-powered technologies with wearable devices to create a more personalized and supportive learning environment.

Intervention

1. Smart Glasses (e.g., Google Glass):

Mr. Thompson introduces smart glasses into the classroom, equipped with advanced AI algorithms. These glasses offer several key features:

- **AI-Generated Visual Aids:** The glasses overlay contextual information directly onto the students' view. For example, if a student is struggling with understanding a geometric concept, the glasses can display animated diagrams or real-world examples that visually demonstrate the concept.
- **Real-Time Feedback:** As students work through math problems, the smart glasses provide immediate, personalized feedback. If a student is struggling with a specific problem, the glasses might offer hints or suggest alternative problem-solving strategies based on the student's current approach and progress.

2. Adaptive Learning Platform Powered by AI and Machine Learning (ML):

The adaptive learning platform analyzes a wide range of student data to provide tailored learning experiences:

- **Data Collection:** The platform gathers data from various sources, including quiz and test scores, smart glasses usage patterns, and BCI (Brain-Computer

Interface) data. For instance, it monitors how much time students spend on different problem-solving steps and how they interact with the smart glasses.

- **Personalized Learning Paths:** The platform uses this information to suggest practice questions, video lessons, and other materials that are tailored to the individual learning requirements of each learner. It makes sure that pupils are neither overburdened nor under challenged by dynamically modifying the tasks' degree of complexity to correspond with their performance.
- **Adaptive Resources:** The platform continuously adapts learning materials to optimize the educational experience for each student. For example, if a student demonstrates difficulty with algebraic concepts, the platform might provide additional exercises and explanations focused on that area.

3. Brain-Computer Interfaces (BCIs) (e.g., Muse Headband):

BCIs are employed to monitor students' focus levels and emotional states:

- **Focus and Emotional Monitoring:** The Muse Headband tracks brain activity to assess how focused or anxious a student is during lessons. This real-time data helps the AI system gauge whether a student is struggling with concentration or experiencing stress.
- **Tailored Instruction:** Based on BCI data, the AI system can adjust instructional methods. For instance, if a student exhibits signs of anxiety, the system might suggest calming techniques or modify the pace of the lesson to reduce stress.

Technology and Approach

1. Smart Glasses Usage:
    - **In-Class Integration:** Students wear smart glasses during lectures, individual practice sessions, and group activities. The glasses provide real-time support by displaying visual aids and feedback as students work on problems.
    - **Behavior Analysis:** The AI algorithms in the glasses analyze how students engage with the learning material, offering insights into their problem-solving approaches and identifying areas where they need additional help.
2. Adaptive Learning Platform:
    - **Data Aggregation:** The platform consolidates data from quizzes, smart glasses interactions, and BCI measurements to create a comprehensive profile of each student's learning needs.

- **Personalized Recommendations:** It uses this data to suggest customized learning paths, practice problems, and instructional resources tailored to each student's unique requirements.

3. BCI Integration:
    - **Focus and Anxiety Assessment:** BCIs provide data on students' cognitive states, helping to identify those who may be struggling with focus or experiencing anxiety.
    - **Instructional Adjustments:** The AI system uses BCI data to modify teaching strategies, such as adjusting lesson pacing or providing additional support, to address identified issues.

Implementation

- **Daily Routine Integration:** Mr. Thompson incorporates the smart glasses and adaptive learning platform into his daily teaching routine. Students use the glasses during lessons and practice sessions, while the platform provides recommendations and resources outside of class.
- **Personalized Learning Paths:** The adaptive platform generates individualized learning paths, allowing students to work on materials suited to their needs and progress independently.
- **Progress Monitoring:** Mr. Thompson monitors student progress through the platform's dashboard, which provides visualizations of performance and BCI data. This enables him to identify students who require additional support, adjust his teaching approach, and address any anxiety-related issues.

Outcomes

- **Improved Math Comprehension:** Students using the AI-powered system showed significant improvements in understanding complex mathematical concepts. The personalized support provided by the smart glasses and adaptive platform helped clarify challenging topics and reinforce learning.
- **Reduced Math Anxiety:** By leveraging BCI data, Mr. Thompson was able to identify students experiencing anxiety and adjust his teaching methods accordingly. This led to a reduction in anxiety levels and a more positive attitude towards math.
- **Enhanced Engagement:** The real-time feedback and interactive features of the smart glasses, combined with the adaptive learning paths, increased student engagement and motivation. Students were more actively involved in their learning and more enthusiastic about solving math problems.

- **Personalized Learning and Growth:** The AI system facilitated tailored learning experiences for each student, ensuring they were appropriately challenged and supported. This personalized approach contributed to continued academic growth and development.

Challenges

- **Initial Learning Curve:** Students required a period of adjustment to become proficient with using the smart glasses and BCIs. This initial learning curve affected the overall effectiveness of the technology in the early stages.
- **Data Privacy Concerns:** Addressing concerns regarding the collection and ethical use of student data, especially from BCIs, was crucial. Transparent communication with students and parents about data privacy policies and security measures was essential.
- **Responsible AI Development:** Ensuring that the AI algorithms used in the smart glasses and adaptive learning platform were unbiased and accurate required ongoing monitoring and refinement. Regular updates and audits were necessary to maintain the integrity and fairness of the AI systems.

Figure 2. Workflow of the proposed strategy

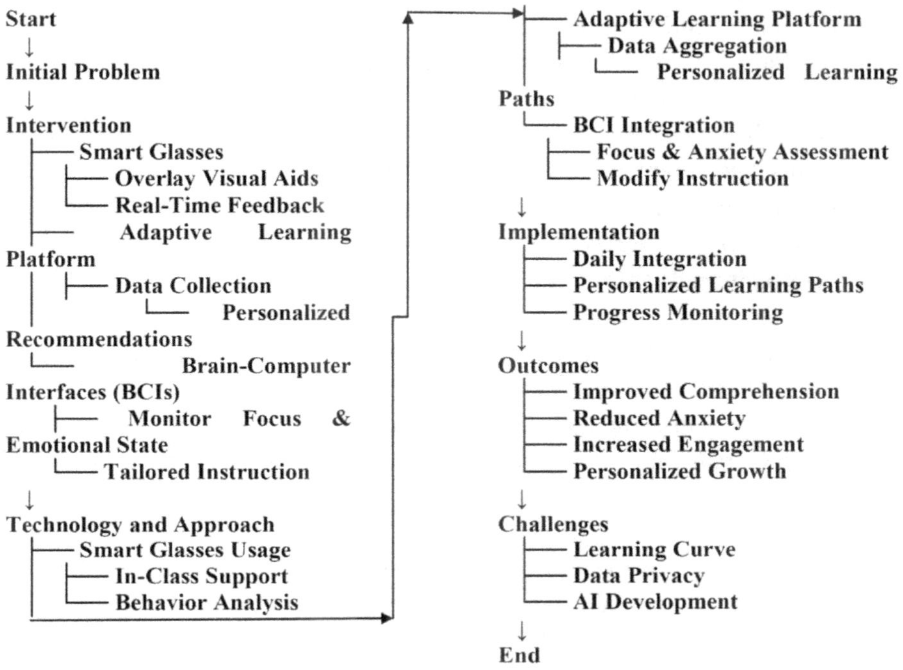

7. CONCLUSION

Wearable technology combined with artificial intelligence (AI) and machine learning (ML) has the potential to revolutionize the educational environment (Table 2). Teachers may create highly customized learning environments that meet the individual requirements and preferences of every student by utilizing this cutting-edge technology. Wearables with AI capabilities can dynamically modify course contents in response to real-time data, creating a more dynamic and interesting learning environment.

Creating Personalized Learning Environments: Wearable technologies equipped with AI can analyze a student's learning style, pace, and comprehension in real-time. This allows for the customization of educational content and methods, ensuring that each student receives instruction tailored to their specific needs. For example, adaptive learning systems modify the level of difficulty and suggest resources depending on individual performance data, and smart glasses may overlay contextual information and real-time feedback.

Enhancing Student Engagement: AI-driven wearables can make learning more interactive and motivating through gamification and immersive experiences. By integrating with virtual reality (VR) and augmented reality (AR), these technologies can create engaging, interactive scenarios that captivate students and maintain their interest. Real-time feedback mechanisms also help students track their progress and stay motivated by providing immediate insights into their performance.

Empowering Educators: Educators gain valuable insights through AI-powered wearables, which can analyze data from various sources to provide a comprehensive view of student engagement and understanding. This data enables teachers to tailor their instruction more effectively, identify students who need additional support, and optimize their teaching strategies. Additionally, formative assessments created through wearable data offer a richer understanding of student progress beyond traditional testing methods.

However, to fully realize these benefits, several challenges must be addressed. Data Privacyand Security are paramount; robust measures and clear policies are necessary to protect sensitive student information from unauthorized access. Equity and Accessibility are also critical considerations; ensuring that AI-powered wearables are affordable and accessible to all students is essential to prevent exacerbating existing educational inequalities. Moreover, TeacherTraining and Support are crucial for effective integration, as educators need proper training to leverage these technologies and interpret the data they generate.

Ethical Considerations and Algorithmic Bias must also be carefully managed. Ensuring that AI algorithms are designed to be fair and unbiased is essential to avoid disadvantaging certain groups of students. As AI and wearable technologies continue to advance, ongoing monitoring and refinement will be necessary to uphold ethical standards and maximize the positive impact on education.

In summary, AI and wearable technologies have the potential to revolutionize education by offering personalized, engaging, and data-driven learning experiences. Addressing the associated challenges will be crucial for ensuring their responsible and equitable implementation. As these technologies evolve, they promise to bring even more innovative solutions that will further enhance the teaching and learning landscape.

Table 2. AI-Enhanced wearable benefits in education

Aspect	Benefit	Example
Personalization	Adaptive learning based on real-time data	AI adjusts lesson difficulty based on student engagement levels
Real-Time Feedback	Immediate interventions based on cognitive/emotional state	Wearables suggest breaks during high-stress moments
Inclusion for Special Needs	Assistive technology for disabilities	Smart glasses provide visual/auditory support
Engagement	Gamified experiences keep students motivated	Fitness trackers reward physical activity with learning incentives

REFERENCES

Alam, A. (2022). Employing adaptive learning and intelligent tutoring robots for virtual classrooms and smart campuses: Reforming education in the age of artificial intelligence. In *Advanced computing and intelligent technologies* [Singapore: Springer Nature Singapore.]. *Proceedings of ICACIT*, 2022, 395–406.

Alam, A. (2023). Harnessing the Power of AI to Create Intelligent Tutoring Systems for Enhanced Classroom Experience and Improved Learning Outcomes. In *Intelligent Communication Technologies and Virtual Mobile Networks* (pp. 571–591). Springer Nature Singapore. DOI: 10.1007/978-981-99-1767-9_42

Aldowah, H., Rehman, S. U., Ghazal, S., & Umar, I. N. (2017, September). Internet of Things in higher education: A study on future learning. [). IOP Publishing.]. *Journal of Physics: Conference Series*, 892(1), 012017. DOI: 10.1088/1742-6596/892/1/012017

Aldowah, H., Ul Rehman, S., & Umar, I. (2019). Security in internet of things: issues, challenges and solutions. In *Recent Trends in Data Science and Soft Computing:Proceedings of the 3rd International Conference of Reliable Information and Communication Technology (IRICT 2018)* (pp. 396-405). Springer International Publishing. DOI: 10.1007/978-3-319-99007-1_38

Aldowah, H., Ul Rehman, S., & Umar, I. (2021). Trust in iot systems: a vision on the current issues, challenges, and recommended solutions. *Advances on Smart and Soft Computing:Proceedings of ICACIn 2020*, 329-339.

Amer-Yahia, S. (2022). Towards AI-powered data-driven education. *Proceedings of the VLDB Endowment International Conference on Very Large Data Bases*, 15(12), 3798–3806. DOI: 10.14778/3554821.3554900

Bano, K. S., Bhuyan, P., & Ray, A. (2022, December). EEG-based brain computer interface for emotion recognition. In *2022 5th International Conference on Computational Intelligence and Networks (CINE)* (pp. 1-6). IEEE. DOI: 10.1109/CINE56307.2022.10037255

Basnet, S. (2024). Artificial Intelligence and machine learning in human resource management: Prospect and future trends. *International Journal of Research Publication and Reviews*, 5(1), 281–287. DOI: 10.55248/gengpi.5.0124.0107

Bezzina, S., & Dingli, A. (2023, July). Rethinking gamification through artificial intelligence. In *International Conference on Human-Computer Interaction* (pp. 252-263). Cham: Springer Nature Switzerland.

Cao, L. (2023). Trans-AI/DS: Transformative, transdisciplinary and translational artificial intelligence and data science. *International Journal of Data Science and Analytics*, •••, 1–13.

Chang, Y., Lee, S., Wong, S. F., & Jeong, S. P. (2022). AI-powered learning application use and gratification: An integrative model. *Information Technology & People*, 35(7), 2115–2139. DOI: 10.1108/ITP-09-2020-0632

El Asmar, W. (2022). *The Effectiveness of AI-Powered Digital Educational Platforms: Students' Attainment and Teachers' Teaching strategies in a private high school in Dubai* (Master's thesis, The British University in Dubai).

Gligorea, I., Cioca, M., Oancea, R., Gorski, A. T., Gorski, H., & Tudorache, P. (2023). Adaptive learning using artificial intelligence in e-learning: A literature review. *Education Sciences*, 13(12), 1216. DOI: 10.3390/educsci13121216

Hopcan, S., Türkmen, G., & Polat, E. (2024). Exploring the artificial intelligence anxiety and machine learning attitudes of teacher candidates. *Education and Information Technologies*, 29(6), 7281–7301. DOI: 10.1007/s10639-023-12086-9

Karagianni, E., & Drigas, A. (2023). New Technologies for Inclusive Learning for Students with Special Educational Needs. *International Journal of Online & Biomedical Engineering*, 19(5), 4–21. DOI: 10.3991/ijoe.v19i05.36417

Kayyali, M. (2024). Future possibilities and challenges of AI in education. In *Transforming education with generative AI: Prompt engineering and synthetic content creation* (pp. 118–137). IGI Global. DOI: 10.4018/979-8-3693-1351-0.ch006

Kazanskiy, N. L., Khonina, S. N., & Butt, M. A. (2024). A review on flexible wearables-Recent developments in non-invasive continuous health monitoring. *Sensors and Actuators. A, Physical*, 366, 114993. DOI: 10.1016/j.sna.2023.114993

Laak, K. J., & Aru, J. (2024). AI and personalized learning: bridging the gap with modern educational goals. *arXiv preprint arXiv:2404.02798.*

Limna, P., Jakwatanatham, S., Siripipattanakul, S., Kaewpuang, P., & Sriboonruang, P. (2022). A review of artificial intelligence (AI) in education during the digital era. *Advance Knowledge for Executives*, 1(1), 1–9.

Manickam, S., & Rehman, S. U. (2023). Secured-KDS: Secret key distribution and authentication scheme for resource-constrained devices. *IET Networks*, 12(1), 1–8. DOI: 10.1049/ntw2.12054

Minn, S. (2022). AI-assisted knowledge assessment techniques for adaptive learning environments. *Computers and Education: Artificial Intelligence*, 3, 100050.

Mohanty, A. K., Ahamed, S., Kamra, R., & Junnarkar, A. A. (2023). Challenges and Future Prospects of IoT and AI Integration in Education. *Progress in Language, Literature and Education Research*, 94.

Munir, H., Vogel, B., & Jacobsson, A. (2022). Artificial intelligence and machine learning approaches in digital education: A systematic revision. *Information (Basel)*, 13(4), 203. DOI: 10.3390/info13040203

Nguyen, D. K., Sermpinis, G., & Stasinakis, C. (2023). Big data, artificial intelligence and machine learning: A transformative symbiosis in favour of financial technology. *European Financial Management*, 29(2), 517–548. DOI: 10.1111/eufm.12365

Nkereuwem, I. V., Obi, N. C., &Oyovwikoghene, S. (2023). Relationship Between Chatbots and AI-Powered Gamification on Chemistry Students' Motivation.

Nojima, I., Sugata, H., Takeuchi, H., & Mima, T. (2022). Brain–computer interface training based on brain activity can induce motor recovery in patients with stroke: A meta-analysis. *Neurorehabilitation and Neural Repair*, 36(2), 83–96. DOI: 10.1177/15459683211062895 PMID: 34958261

Orban, M., Elsamanty, M., Guo, K., Zhang, S., & Yang, H. (2022). A review of brain activity and EEG-based brain–computer interfaces for rehabilitation application. *Bioengineering (Basel, Switzerland)*, 9(12), 768. DOI: 10.3390/bioengineering9120768 PMID: 36550974

Peksa, J., & Mamchur, D. (2023). State-of-the-art on brain-computer interface technology. *Sensors (Basel)*, 23(13), 6001. DOI: 10.3390/s23136001 PMID: 37447849

Qasim, S. H. (2024). Beyond the Classroom: Emerging Technologies to Enhance Learning.

Rao, T. V. N., Gaddam, A., Kurni, M., & Saritha, K. (2022). Reliance on artificial intelligence, machine learning and deep learning in the era of industry 4.0. *Smart healthcare system design: security and privacy aspects*, 281-299.

Rehman, S. U., Sadek, I., Huang, B., Manickam, S., & Mahmoud, L. N. (2024). IoT-based emergency cardiac death risk rescue alert system. *MethodsX*, 13, 102834. DOI: 10.1016/j.mex.2024.102834 PMID: 39071997

Saaida, M. B. (2023). AI-Driven transformations in higher education: Opportunities and challenges. *International Journal of Educational Research and Studies*, 5(1), 29–36.

Sadek, I., Codjo, J., Rehman, S. U., & Abdulrazak, B. (2022). Security and privacy in the Internet of Things healthcare systems: Toward a robust solution in real-life deployment. *Computer Methods and Programs in Biomedicine Update*, 2, 100071. DOI: 10.1016/j.cmpbup.2022.100071

Sadek, I., Rehman, S. U., Codjo, J., & Abdulrazak, B. (2019). Privacy and security of IoT based healthcare systems: concerns, solutions, and recommendations. In *How AI Impacts Urban Living and Public Health: 17th International Conference, ICOST 2019, New York City, NY, USA, October 14-16, 2019* [Springer International Publishing.]. *Proceedings*, 17, 3–17.

Seneviratne, S., Hu, Y., Nguyen, T., Lan, G., Khalifa, S., Thilakarathna, K., Hassan, M., & Seneviratne, A. (2017). A survey of wearable devices and challenges. *IEEE Communications Surveys and Tutorials*, 19(4), 2573–2620. DOI: 10.1109/COMST.2017.2731979

Shaikh, A. A., Kumar, A., Jani, K., Mitra, S., García-Tadeo, D. A., & Devarajan, A. (2022). The role of machine learning and artificial intelligence for making a digital classroom and its sustainable impact on education during COVID-19. *Materials Today: Proceedings*, 56, 3211–3215. DOI: 10.1016/j.matpr.2021.09.368 PMID: 35464152

Singh, P. (2022). Navigating the digital wave: Transformative strategies in modern business with the integration of artificial intelligence (ai) and machine learning (ml). *International Journal of Computing and Artificial Intelligence*, 3(2), 78–85. DOI: 10.33545/27076571.2022.v3.i2a.77

Singh, P. K. (2023). Digital transformation in supply chain management: Artificial Intelligence (AI) and Machine Learning (ML) as Catalysts for Value Creation. *International Journal of Supply Chain Management*, 12(6), 57–63. DOI: 10.59160/ijscm.v12i6.6216

Srinivasa, K. G., Kurni, M., & Saritha, K. (2022). Harnessing the Power of AI to Education. In *Learning, teaching, and assessment methods for contemporary learners: pedagogy for the digital generation* (pp. 311–342). Springer Nature Singapore. DOI: 10.1007/978-981-19-6734-4_13

Sunarya, P. A. (2022). Machine learning and artificial intelligence as educational games. *International Transactions on Artificial Intelligence*, 1(1), 129–138.

Wang, S., Christensen, C., Cui, W., Tong, R., Yarnall, L., Shear, L., & Feng, M. (2023). When adaptive learning is effective learning: Comparison of an adaptive learning system to teacher-led instruction. *Interactive Learning Environments*, 31(2), 793–803. DOI: 10.1080/10494820.2020.1808794

Zafari, M., Bazargani, J. S., Sadeghi-Niaraki, A., & Choi, S. M. (2022). Artificial intelligence applications in K-12 education: A systematic literature review. *IEEE Access : Practical Innovations, Open Solutions*, 10, 61905–61921. DOI: 10.1109/ACCESS.2022.3179356

Zhang, X., Ding, Y., Huang, X., Li, W., Long, L., & Ding, S. (2024). Smart classrooms: How sensors and ai are shaping educational paradigms. *Sensors (Basel)*, 24(17), 5487. DOI: 10.3390/s24175487 PMID: 39275398

Zheng, H. (2021). *Assisting Young Adults with Neurodiversity to Engage in Learning Activities: Smartwatch Applications in Inclusive Education* (Doctoral dissertation, George Mason University).

Chapter 2

Wearables and AI-Smart Technologies: Transforming the Education Sphere

Satya Ratnakaram
https://orcid.org/0000-0002-8908-302X
Bahrain Polytechnic, Bahrain

Durga Madhab Mahapatra
https://orcid.org/0000-0003-1272-7285
North-Eastern Hill University, India

Devi Manikeswari
https://orcid.org/0000-0002-5014-9423
Bahrain Polytechnic, Bahrain

Bhabani Shankar Mohanty
https://orcid.org/0000-0001-9130-6221
Central University of Tamil Nadu, India

Zakir Hossen Shaikh
https://orcid.org/0000-0003-4733-4166
Kingdom University, Bahrain

Anand Soni
https://orcid.org/0009-0001-4896-0639
Bahrain Polytechnic, Bahrain

ABSTRACT

The field of education has been no exception to the influence of the ever-evolving advancements brought by technology. The implications of smart technology and its inherent tools bestow several benefits and challenges for educational institutions, educators and students alike. This chapter is an attempt to delve deeper into the specific benefits and challenges of integrating wearable technology and Artificial Intelligence (AI) in the teaching-learning process while also offering a snapshot of the concurrent use of both in an educational setting. Substantial research in the field stands evidence of the individual and combined use of wearable technology and AI in education to offer opportunities for improved student engagement, personalized feedback for students and collaborative learning while simultaneously presenting valid discussion points on the potential pitfalls such as data privacy issues, affor-

DOI: 10.4018/979-8-3693-7817-5.ch002

dances and training needs of educators. Some recommendations on overcoming such challenges have been made and include the active role of educational institutions, educators and learners.

BACKGROUND

The dawn of technological innovation has transformed almost all walks of human life, and education is no exclusion. In the past ten years, technology has significantly influenced how education is structured globally. Educators, researchers, and government officials are paying close attention to how digital tools can improve education. The recent advancements in digital technology across various industries have inspired researchers to explore how these tools can enhance the quality of education (Qureshi et al., 2021). The increasing favor towards a more learner-centric classroom from the conventional teacher-centric approach is now sought after by educational institutions around the world. The emphasis on technology being an important factor in transforming education is backed by substantial research. In recent times, the assimilation of wearables and artificial intelligence (AI) into educational ecosystems has opened new prospects for individualized, attractive and immensely effective learning experiences.

The education sector has seen a raising prominence on assuming student-centric learning and transferring away from teacher-centric learning. From the conservative instruction strategies where teachers serve as the main source of knowledge and authority in the classroom, there has been an unparalleled interest in active involvement of students in the learning arena and allowing them to take responsibility of their own learning. This shift brings about challenges that often involve changes in organizational culture, teachers' beliefs, values and adjustments that are on par with the structure of the institution. This significant shift also found its place in the Education 4.0 Taxonomy put forward by the World Economic Framework (Manikeswari et al.,2023). This, in turn, is a consequence of the industry dynamics that constantly undergo changes and thereby require graduates and all prospective employees to be work-ready and prepared to adapt to new developments on a continuous basis, The inherent recruitment practices are structured in a manner that they employ a range of recruitment and selection techniques (that include traditional interviews, skill tests and psychometric tests) to screen applicants and find the best fit for the organization (Manikeswari et al., 2024). The purpose being to ensure that organizations are successful in screening, scanning, filtering and recruiting work-ready graduates that are prepared to embrace the challenges of the real-time work environment and to do so, technological literacy is among the most sought-after skills that they would need to have mastery over.

INTRODUCTION

Smart Technology in Education: A Pathway to Tech-Savvy Learning

The integration of smart technology into the field of education has piloted into a new era of learning, thereby altering the nature of the traditional classrooms into highly dynamic and interactive environments. This paradigm shift is illustrated by the extensive adoption of digital tools, techniques, devices and platforms for enhancing the teaching, learning, and assessment processes (Mishra & Koehler, 2006). Smart technology, in education, include devices and systems that are connected to the learning process. More so, Shoikova et al., (2017) define smart learning environments, also referred to as smart education, as a "a new trend in the system of education, including an effective and efficient interaction of pedagogics, technology and their combination towards the advancement of learning processes". In the educational context, smart technology is known to facilitate enhanced learning by providing tailored content, real-time feedback and interactive experiences. For instance, smart classrooms equipped with interactive whiteboards, tablets, and sensors can create immersive and collaborative learning environments. Additionally, smart textbooks and adaptive learning platforms can offer ideal contents in their educational process to individual learners as per their needs and learning styles, promoting deeper understanding and higher engagement (Liu et al., 2020). On another note, Selwyn (2011) argues that even though technology has the ability to transform education, its tangible impact is often constrained by factors such as lack of adequate teacher training, access to technology and educational policies both internal and external to the institutions, some of which, he proposes could be overcome by ensuring that reliable systems are introduced and set in place to ensure that teachers are empowered through training and support, provided access to technology while also consistent with security measures to safeguard both students' and teachers' data privacy. The advent of smart technology in education corresponds to a valuable opportunity for enhancement of learning outcomes and prepare students for the 21st century.

Wearables and AI in Education: An Introduction to a New Paradigm in Smart Technology

Today, students of all ages, from elementary school to college education, are all comfortable with using digital devices like smartphones, tablets, and laptops. While they often use these devices for socializing and fun, there is a growing need for them to be used for learning and education (Qureshi et al., 2021). Some of the most sought-after smart technologies that have gained international favor are e-learning platforms, educational apps, Virtual Reality (VR), Augmented Reality (AR), Artificial Intelligence (AI), Machine Learning, Gamification, Game-based Learning, Interactive whiteboards etc. (Vincent-Lancrin, 2021). While each of them have distinct features and applications, when used judiciously, teachers can reap the benefits of the overall enhancement that becomes inadvertent in the teaching-learning process. In parallel to the application of all available smart technologies in education, Wearable technology, Internet of Things (IoT) and Blockchain technology have also emerged as sought-after advancements that could improve educational outcomes. While the research on their contributions is ongoing, there have been notable benefits that they confer in the classroom. These technologies present distinctive opportunities to augment student experiences, improve engagement and offer significant means to overcome challenges pertaining to data security and privacy (Wang & Liu, 2023) (Al-Tamimi & Al-Smadi, 2022).

The term "Artificial Intelligence" was coined by John McCarthy in 1956 where a group of researchers were invited to a workshop held in Dartmouth College to discuss the feasibility and challenges thereof of developing machines with the ability to think (Tang, 2024).

AI has unique usage of cases that are available to all big enterprises, making it a valuable ally in this era of digital acceleration. Artificial intelligence (AI) is renovating traditional paradigms in different sectors of the economy, including supply chain management, robotics, healthcare, public safety, retail, HR services, financial services, and financial inclusion and governance. (Shaik, et.al 2024). More recently, Artificial intelligence (AI), with its capacity to process large amounts of data and make intelligent decisions, can significantly augment educational outcomes. AI-powered tools can be employed to personalize learning by adapting to students' individual strengths and limitations, for presenting pointed feedback and recommending relevant supporting resources. AI has substantially changed the teaching and learning arena, easing an alteration from teacher-centric to student-centric learning (Tang, 2024).

This chapter explores the current extent of use and imminent potential of these smart technologies in education while simultaneously discussing their benefits, challenges, and future implications. The specific objectives that this chapter attempts to address are as follows:

- To identify the potential benefits that smart technologies can offer to education.
- To analyze the challenges and limitations linked with the implementation of smart technologies in education, including issues related to privacy, equity, and technical difficulties.
- To propose strategies for mitigating challenges and maximizing the merits of smart technologies in education.

The significance of this current chapter examines into the probable benefits and challenges of integrating wearable technology and AI into education to highlight the pertinent benefits to the learning process and other stakeholders. Based on the common knowledge of every coin having two sides it also addresses concerns like data privacy, affordability, ethical considerations and the need for educator training. Providing an important take on potentially convergent literature available on the subject and its practical implications forms the crux of the current chapter.

TRANSFORMATIVE LEARNING THEORY: A MEZIROWIAN PERSPECTIVE AND ITS IMPLICATIONS FOR SMART EDUCATION

As the field of integrating smart technologies, wearables and AI into the stream of education evolves, it is crucial to examine the theoretical underpinnings that guide the development and implementation of these technologies for enhanced outcomes.

While theories such as constructivism, connectivism, cognitive load theory and the universal design for learning offer significant guidance on the same, the transformative learning theory, as theorized by Mezirow (1997), focuses more on a fundamental shift in the perspective or worldview. Mezirow's theory explains how adult learners understand and interpret their experiences. It also explores how societal factors and other influences shape how learners perceive these experiences (Christie et al., 2015). Additionally, the theory discusses how learners' understanding can change when they realize that their current perspective is not working effectively. Smart technologies can facilitate the transformation of learning by allowing learners access to diverse perspectives and challenging their prevailing beliefs. This, in turn, opens opportunities for the introduction of new technologies and innovative tools in the workflow of education for both learners and the faculty.

Transformative learning theory, as proposed by Jack Mezirow in 1997, provides an outline for understanding how individual learners go through significant shifts in their perspectives and beliefs (Taylor & Cranton, 2017). This theory posits that transformative learning appears when individual students analytically reflect on their assumptions, frames of reference, and worldviews, leading to an essential change in their understanding of themselves and the situation and occurrences around them.

Key Components of Transformative Learning Theory

1. **Dissonance:** The process starts with a dilemma that is disoriented, a situation or a circumstance where an individual's prevailing beliefs or assumptions are challenged. Such dissonance can arise from various sources, such as personal experiences, social interactions, or new information.
2. **Critical Reflection:** When confronted with dissonance, individuals engage in critical reflection, observing their assumptions, beliefs, and values. This process involves cross-questioning the validity of their own existing perspective of the world and exploring alternate perspectives.
3. **Reconstruction:** Based on such critical reflection, individuals may then rebuild their view of the world. This involves assuming new beliefs, values, and perspectives that are more compatible with their understanding of the state.
4. **Transformation:** The ultimate effect of transformative learning is a substantial change in the individual's view of the world. This transformation can steer transformations in behavior, attitudes, and relationships. (Mezirow, 1978)

Figure 1. Transformative learning theory

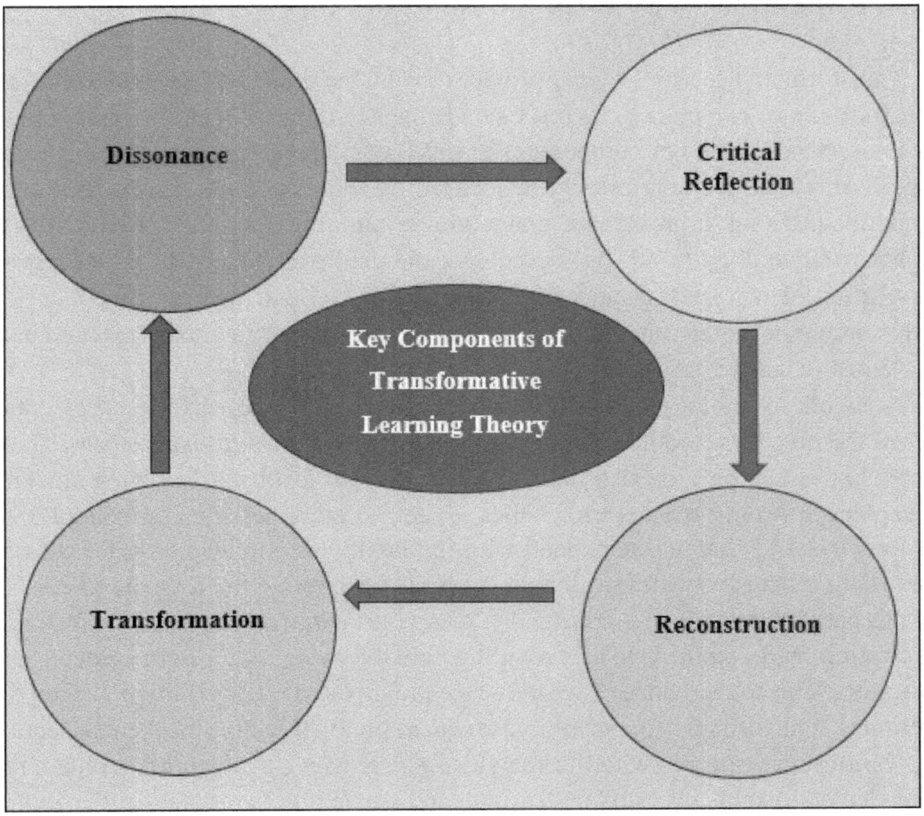

Source: Author's own creation

Transformative learning theory has significant implications for education. By fostering critical reflection and challenging students' existing beliefs, educators and the teaching staff can design learning environments that promote and nurture deep understanding and personal growth. Mezirow's theory has had a deep impact on adult education, presenting a framework for understanding how adults' learning can become more autonomous, critical thinkers, and agents of change. It has also swayed the design of adult education programs and the enhancement of effective teaching strategies (Christie et al., 2015).

Transformative Learning Theory, Wearables and AI in Education: A Synergistic Relationship

Transformative learning theory provides a valuable framework for understanding how individuals can undergo significant changes in their perspectives and beliefs. By understanding the key components of this theory, educators can create learning experiences that promote critical reflection, meaning reconstruction, and transformative learning. It provides a foundation for building upon the benefits offered by smart technology to education wherein the challenged faced in the course of application of new and advanced technology can be dealt with in a manner that fully complements the outcomes of the experiences of learning experience to both teachers and learners alike.

Wearable technology and artificial intelligence (AI) will perform an essential role in assisting for smooth facilitation of transformative learning experiences. These technologies can provide rich data and personalized feedback, allowing students to explore new perspectives and challenge their existing beliefs. For example, the tutoring systems that are embedded with the intelligence of AI can offer tailored feedback and thus providing valuable guidance, prompting the learners to consider alternative viewpoints. Additionally, wearables can track student engagement, motivation, and cognitive load, providing valuable insights for both students and educators. The convergence of wearable technology and artificial intelligence (AI) presents a transformative opportunity for education. By blending the data-collection capabilities of wearables with the analytical power of AI, educators can obtain unprecedented insights into student learning and create highly tailored and engaging learning experiences.

A study by Lee and Chen (2019) explored the use of microlearning modules in professional development. They found that participants who engaged with microlearning modules reported a shift in their professional practices and a deeper understanding of their field, suggesting that microlearning can facilitate transformative learning.

Transformative learning theory and smart education share a common goal: to facilitate deep, meaningful and lasting learning experiences that optimize both current and lifelong learning opportunities. Discussed below are the commonalities that the transformative learning theory and smart education share that create the path for the greater good of education in the era of technological innovation.

Critical Reflection and Challenge: Both smart technology in education and transformative learning theory lay emphasis on the importance of critical reflection for challenging current beliefs and assumptions. In smart education, this could possibly involve the integration of AI to facilitate the provision of offering personalized feedback to students to for them to think critically analyse their understanding.

Meaningful Learning: Transformative learning theory suggests that learning is most effective when it is both meaningful and establishes connections to the learner's personal experiences. As a complementary feature, smart education can employ AI to offer tailored content and activities that match the individual needs and interests of the students interests and needs.

Experiential Learning: Due credit to experiential learning is evident in the foundations of smart technology and transformative learning theory. While smart education can leverage technology to create immersive, interactive learning experiences, transformative learning emphasizes on undergoing reform after critical reflection of experiences.

Collaboration and Community: Transformative learning often fosters a collaborative setting and smart education facilitates effective collaboration opportunities through online platforms and tools. This truly allows crucial settings for gaining an international perspective through collaboration on a global level.

Lifelong Learning: In total assimilation, opportunities for lifelong learning form the crux of both smart education and the transformative learning theory. While preparing learners to have an open mind to learning, an inherent zeal for learning and gaining expertise in new avenues is a fundamental element of the modus operandi.

In summation, transformative learning theory and smart education are complementary approaches that, when used in conjunction, create effective, engaging, and meaningful learning experiences that are on par with the higher order thinking skills taxonomy. By promoting critical reflection, meaningful, experiential, lifelong learning and collaboration these theories can help to empower students to overcome the challenges and make the best of the opportunities of the 21st century.

A NEW ERA OF EDUCATION WITH WEARABLES AND AI: PAVING THE PATH TO OPTIMIZED LEARNING EXPERIENCES

Wearables: Learning's New Look

Evolution of the education field has always been evident in the innumerous teaching-learning tools that have found their place in the classroom. Educators and researchers, through their passion for the field, have brought forward several educational tools and technologies that serve as benchmarked practices in the field.

This section attempts to provide an overview of the manifold benefits of smart technologies, specifically, wearables and AI to the field of education.

The use of wearables in education has seen a considerable rise since the first documented wearable appeared in the 1960s. This mounting interest is driven by continuing research into their effectiveness in augmenting the teaching and learning processes (Havard & Podsiad, 2017 as cited by Havard & Podsiad, 2020).

Wearables are electronic devices that are worn on the body. They often include sensors, processors, and connectivity features that allow them to collect and analyze data, communicate with other devices, and perform various tasks. Wearable technology has evolved significantly from bulky backpack computers to more discreet clothing-based devices and even head-mounted wearables that can interpret brain activity for communication purposes (Mann, 1996; Nanjappan et al. 2017; Dehzangi & Farooq, 2018 as cited by Havard & Podsiad, 2020).

Now a days wearables have shown popularity ever more due to their power to provide personalized information and enhance convenience. The popularity of wearable devices has grown significantly in recent years due to their many features and benefits. Some studies have explored the potential of these devices for educational purposes, including the use of Google Glass. Similar to how people use fitness tracking apps for sports, students could use a set of apps to track their studying hours, achievements, and performance, and how these factors relate to stress, sleep, or anxiety. This information can be crucial for supporting self-regulated learning approaches (Arriba-Perez et al, 2017 Arroyo et al. (2017), created the learning environment where smartwatches or smartphones, strapped in as armbands, were employed for providing students with instructions in multiplayer-embodied games that aimed at assisting them in their learning of Math concepts.

In recent years, people have started using smartwatches and fitness trackers much more often. Wearable technology has made significant advances in recent decades. Researchers and educators are actively exploring approaches to integrate wearables into educational settings to develop the learning experience (Havard & Podsiad, 2023). Wearable technology has transformed education by making learning more interactive and student-centered. Students can now access information and understand concepts faster with less effort. This new approach differs significantly from traditional learning methods, that demand the attendance of the learners in classes at specific times and places (Attallah & Ilagure, 2018).

Researchers have studied how these devices can be used in education. But not all wearables are the same (Shroeder et al., 2023). Smart watches, Fitness trackers, AR glasses, wearable devices etc are as a part of smart technology, have also gained favor in the education arena as they facilitate the transformation of education by putting forth valuable data and insights into students' learning behaviors and physiological states. VR headsets are known to be among the first to be used in the case of wearables in education, more specifically in the teaching-learning of subjects of an abstract nature such as mathematics or geometry (Khosravi et al.,2021). For instance, fitness trackers can be used to monitor students' activity levels, sleep patterns and other health factors, thereby aiding educators to identify potential hurdles to learning. AR glasses, on the other hand, can intersect digital information onto the real world, thus enhancing the learning experience of students extensively making the process more interactive. Through the collection and analysis of mentality data from wearable devices, educators can gain a better identification of students' individual needs and tailor their instruction process accordingly (Huang et al., 2021). Additionally, wearables have been used in education for several other purposes, like comparing teaching methods, understanding how physical data relates to learning, analyzing student data, and trying out new teaching techniques. Because of this, there have been a lot of studies about using wearables in education recently (Shroeder et al., 2023).

Wearables have been used in education with a variety of people, learning methods, and settings. For instance, some studies used ankle wearables (Zhou et al., 2018 as cited by Havard & Podsiad, 2020) to test older adults' thinking skills, while others used gloves to help younger children learn about beekeeping. Wearables have been used in different ways, like in tasks, projects, and games. They've also been tested in places like the military, hospitals, businesses, and schools (Havard & Podsiad, 2020).

Substantial research on the use of wearables indicates that they can be used in education to guide learning activities, collect data to inform learning, teach about technology, make knowledge visible, assist teachers and guide student behavior in addition to providing visual instructions, prompts, or frameworks to guide learning activities, and can also collect data to adjust learning content. They allow hands-free access to information, which opens up new possibilities for mobile learning. Additionally, wearables can be used to teach students about technology, such as programming or computer science, and to make knowledge explicit through direct interaction. Wearables can also help teachers by providing information about student attention levels and can guide students' own behavior by providing them with data about their participation levels (Chu et al., 2023).

Bower et al. (2017), studied the possibility of using this innovative technology in the filed higher education. They surveyed 202 university professors and found that wearable devices could be beneficial for teaching and learning. However, the professors were not very interested in using 4 different devices suggested by the researchers in their classrooms.

Within the field of education, experimenting with wearable technology has gained traction across different levels as well as specializations. GoPro cameras can be used to record lectures in addition to student performance, thereby granting teachers a way to observe the learning environment and assess student work. These cameras capture actions and can help to record embodied skills, sensory, kinesthetic and emotional knowledge (Hyndman, 2017).

VR devices like the Oculus Rift propose a new way to learn by presenting immersive experiences that are often challenging or impossible to achieve in the real world. This technology allows students to actively take part in their learning, unlike conventional classroom methods that bank on passive reading or watching. The Oculus Rift is a comfortable and affordable VR headset that provides a wide field of view for an immersive experience (Hussein & Nätterdal, 2015).

Kim and Lee's 2017 study investigated the effectiveness of VR simulations in teaching chemistry. Their research discovered that students who used VR simulations demonstrated a deeper understanding of chemical concepts. This suggests that VR can present a more immersive and engaging learning experience, leading to improved comprehension and retention of complex scientific information. Gamification and VR complement various teaching theories and methods. When combined for use in education, they can augment traditional learning and teaching. Students and teachers have responded positively to their integration. Gamification elements have substantially impacted learner's accomplishments.

Gamified VR systems in the learning environment has profound affect as compared to traditional practices of learning while enriching the learners and motivating them, to have interactive and personalized collaborative engaged in the enhanced learning chances. The realistic and interactive experiences in virtual reality can improve students' immersion, social presence, knowledge acquisition, and comprehension.

Vibrant amendments in the behavioral attitudes of students, physical and mental appearances, rational and social-emotional enhancement have been detected. Combining virtual reality and gamification in the learning practices and methods can upsurge students' learning outcomes, motivation, engagement, self-efficacy, better achievements in academics. It also increases their involvements to more active way thus providing more satisfaction by improving interest, thoughts, concentration, attention, skills, and competences. Gamified virtual reality has demonstrated the ability to be an interactive and efficient tool of education for advancing learning of varied subjects at multi-levels, and varied contexts (Lampropoulos & Kinshuk, 2024).

Given the prominence of emotions in education, researchers have focused on developing approaches to identify and watch emotions in an educational setting. In recent years, many studies have been conducted on this topic, and numerous literature reviews have summarized the findings. Ba & Hu (2023) conducted a systematic literature review of 50 such journal articles (peer-reviewed and published across 15 years between January 2008 and December 2022), to study the inherent aspects of using wearable technology to measure emotions in education. Their findings pointed towards the fact that research on wearable technology in education has altered towards studying younger learners in real-world settings. Researchers are interested in understanding the relationship between emotions and learning with usage of wearable devices. Electrodermal activity and heart rate are commonly measured, and emotions like engagement, positivity, and anxiety are often studied. Machine Learning and statistical analysis are used to examine the connection between physiological signals and emotions in educational contexts.

AI: A Catalyst for Educational Transformation

To provide a conclusive definition of artificial intelligence (AI) precisely is a challenge even for experts, due to two core reasons. Firstly, AI is continually evolving, and numerous advanced AI technologies are used in everyday applications without being labeled as such. Secondly, AI is an interdisciplinary field with researchers from areas ranging between neuroscience, psychology, and linguistics contributing their own perspectives, knowledge, and terminology. It may be appropriate to regard AI as a system's ability to comprehend and learn from data thereby using that knowledge for accomplishing specific goals. While there is no universally accepted definition of AI, it's clear that it involves not only computer science but also fields like philosophy, neuroscience, mathematics, information science, linguistics, psychology and much more. (Kaplan and Haenlein (2019 as cited by Chen et al., 2020).

AI may be regarded as a system's ability to accurately understand and learn from given data inputs to advance and employ the learning towards the achievement of a specific goal. Though it is commonly agreed that there is no availability generally recognized definition of AI, but it is clearly evident that AI is merged and used in many other disciplines such Psychology, mathematics, linguistics, philosophy, medical scinces apart from the discipline of computers and informatics.

AI-powered algorithms can be employed to analyze learners' data, including the learning styles and preferences, their interests apart from their performance in assessments, and this in turn could aid in creating personalized learning pathways. This can be valuable for addressing and accommodating the diverse learning needs of students in a manner that both content and all inherent resources are tailor-made to make learning more accessible for all. For instance, to modify and modify the difficulty level of assessments considering the progress of learners, AI-powered adaptive learning platforms show enormous role by preventing frustration and boredom (Huang & Wang, 2019). More specific applications of AI into the teaching-learning process are discussed below:

Intelligent Tutoring Systems

Intelligent tutoring systems (ITS) are another powerful application of AI in education. These systems can provide students with individualized instruction and feedback, simulating the role of a human tutor. ITS can adapt to a student's pace of learning, identify areas of weakness, and offer targeted support through personalized guidance to students, helping them overcome learning challenges and develop critical thinking skills. By providing students with immediate and personalized feedback, ITS can help improve learning outcomes and increase student motivation (Huang & Wang, 2019). Additionally, the grading mechanism that is embedded with AI powered tools, will be able to provide free time to the tutors and the teaching faculty with its automated grading system. This will offer a focused way the trainers an opportunity in guiding and providing instructions to the needs of the learners at individual level. (Wang et al., 2019).

Automated Grading

AI can also be used to convert the grading system into an automated task, releasing the teachers' time and helps to focus on more meaningful tasks. These AI-powered evaluation and grading mechanisms will be able to assess the student's work more accurately thus by eliminating the chances and possibilities of human error, including essays, multiple-choice questions, and coding assignments. While there are certain concerns and issues exist in the accurateness and objectivity of automated grading system, advancements in AI technology are making it increasingly reliable (Huang & Wang, 2019).

Computer-Supported Collaborated Learning (CSCL)

AI has also introduced new methods to learn the courses of educational, business, data analytics in computer-supported collaborative learning (CSCL) (Ouyang & Zhang, 2024). CSCL is the study of how technology can help groups learn jointly, both in person and remotely. It focuses on individual learning processes, group interactions, institutional changes, and the development of new educational models. CSCL targets to create new types of knowledge and learning approaches that are relevant to today's knowledge-based society. It follows the principles of scaffolding, where teachers progressively give students more responsibility for their learning, and mediation, where technology is seen as a tool that improves teaching and learning. CSCL highlights the development of specialized knowledge, metacommunication, metacognition, and the ability to rethink tasks through collaboration and the blend of different skills and knowledge. AI-powered learning analytics can provide real-time, personalized, and detailed feedback on collaborative learning processes and group interactions, which can considerably improve collaboration. Unlike traditional manual methods of analyzing group learning after activities, AI can better analyze the large amount of data generated in collaborative learning environments, promoting students' collaboration and technological skills. Compared to the traditional methods that are involved in analyzing the group data manually, the introduction of CSCL activities, AI-driven learning analytics are in a meaningful and better position in analyzing the huge volumes and quantum of learning data within CSCL contexts (Ouyang & Zhang, 2024).

Personalized Learning

The ability of AI in education to tailor the learning experiences of students differently is most beneficial to individual students. (Wang et al., 2019). AI-powered algorithms can analyze student data, such as performance on assessments, learning styles, and interests, to create learning paths that cater at individual level and personal level. This can help address the diverse needs of learners and ensure that students receive instruction that is relevant, engaging, and effective. AI-powered applications and tools, like intelligent robots and adaptive learning systems, are becoming more popular in both K-12 and university education. AI allows for personalized learning, which can cater to the needs of individual learners. Traditional teaching methods struggle to meet the varying demands of all students, but AI can tailor instruction to each learner (Della Ventura, 2017). This can lead to increased motivation, engagement, and independence in learning. AI also offers opportunities to support students with learning disabilities (Della Ventura, 2017; Wang, 2017).

Administrative Tasks

In addition to its educational applications, AI can streamline administrative tasks in schools. For example, AI is capable in the usage of automated activities and tasks such as scheduling, attendance tracking, and communication with parents. This can reduce the administrative burden on teachers and staff, allowing them to devote more time to teaching and student support instruction (Wang et al., 2019).

AI technologies are becoming more widespread in teaching and learning, ridding instructors from repetitive tasks and allowing them to respond to students promptly. This advances adaptive and personalized teaching. Hardware advancements, like high-speed Graphics Processing Units (GPUs) that enhance the speed of Gaming and accessible software and also fueled the growth of AI, particularly due to the progress in Deep Learning (DL) and data analysis. The future of education is intently tied to the development of AI. As intelligent machines become more advanced, education will continue to grow and evolve (Chan & Gary, 2019 as cited by Chen et al., 2020)

The convergence of wearable technology and artificial intelligence (AI) presents a transformative opportunity for education. By combining the data-collection capabilities of wearables with the analytical power of AI, educators can create personalized, adaptive, and engaging learning experiences.

Wearables equipped with sensors can collect a vast array of data on students' behaviors, physiological responses, and learning environments. This data can be analyzed by AI algorithms to provide personalized feedback, identify areas for improvement, and foster critical reflection. For instance, a wearable device that tracks heart rate and brain activity can detect when a student is experiencing cognitive overload, allowing AI to adjust the learning pace or provide alternative strategies.

AI-powered platforms can leverage wearable data to create adaptive learning environments. By analyzing student performance and engagement metrics, these platforms can vigorously adapt the difficulty and pace of learning content to meet needs of individual learners. This customized approach can improve and enrich the student's motivation and enhance the learning outcomes.

Wearables and AI can also be employed to design immersive and engaging learning experiences. Augmented reality (AR) and virtual reality (VR) applications powered by AI can provide students an experience of moving to different historical periods, scientific laboratories, or even other planets. These experiences can challenge students' existing beliefs and persuade them to imagine significantly about the world around them.

However, the combination of wearables and AI in the system of education is involving a lot of challenging issues. Technological limitations, privacy concerns, and equity issues must be thoroughly addressed ensuring that these technologies are used ethically and effectively.

WEARABLES AND AI IN EDUCATION: STEERING THE SHIP IN UNCHARTERED WATERS

Challenges and Considerations

Among the challenges that are imminent in the usage of wearable devices and AI powered tools in education include issues such as battery life, continuity in connectivity issues, security threats to personal data and inconsistencies between different devices and data formats. These issues make it difficult to implement practical applications beyond research experiments. Additionally, privacy concerns and the unclear relevance of certain indicators for education pose further challenges. Despite these problems, the exploration and experimentation of these smart technologies is essential due to their potential value in addition to education (Arriba-Perez et al, 2017).

With all due consideration for the manifold benefits of wearable technology for education, it is crucial to acknowledge the possibility that cheating and other malpractices in exams has increased owing to the initiation of wearables in education. These devices helps to share the valuable information, which is great noting that some or most of which are miniature in size as compared to other devices are more difficult to recognize, has made invigilation of exams difficult. Major tech companies like Apple, Google and Samsung etc. are developing such wearable devices that could be beneficial in the field of education. But numerous reasons exist why people may not be eager to buy them, some of which include high cost, unattractive design, short battery life and overheating issues among others. Additionally, wearable devices often require a nearby smartphone to function at a capacity to facilitate learning. (Attallah & Ilagure, 2018). This could lead to the students getting distracted from the learning activities to other engagements such as the social media calling.

The widespread use of wearable devices raises concerns about privacy, data security, reliance on external parties for the purpose of storing the data and analysis, and accessibility for all students. These devices can be used to secretly take photos and videos without permission, leading to potential legal issues. Educational institutions need clear policies to address these concerns and regulate the use of wearable technology (Sturman & Bower, 2015; Borthwick et al., 2015).

Slimi (2023), based on the findings of a study conducted to analyse the influence of AI in higher education system recommends that the use of AI in higher education is a requirement that is inevitable for all institutions. Nevertheless, AI integration must also be supported by providing training opportunities for academic staff. Simultaneously, ethics and humanity must not be sidelined in the process of integrating AI. Besides, due significance to privacy, dignity and protection of regulations, both local and international laws must be safeguarded. Existing tools predominantly use communication, behavior, and evaluation data to construct results and visualizations. While these tools present different types of feedback, there is the lack of specific design guidelines for their development. Also, though AI has been used to present statistical information, they often lack alerts or suggestions. The research conducted by Ouyang & Zhang (2024) found that while these tools positively influence collaborative learning, they do not offer adequate support for instructional interventions.

There is also the imminent requirement for revised ethical guidelines to address the advanced data collection capabilities of wearable devices. Wearable devices could also be employed to gauge physiological signals and can provide insights into the complex ways emotions affect learning. However, because there are many different types of devices available that can be worn, signalling the physiological and emotional feelings in the educational settings, there is no agreement on the affordance and challenges of using them to measure emotions in education. (Ba & Hu, 2023).

Using or allowing the use of more than one type of wearable device in the classroom is not favored by educators (Bower et al.,2017). On a different note, there are also three other main challenges to using VR wearable devices in education pertaining to the amount of cost involved, serviceability and fear of/resistance to the use technology. In addition, students' attitudes and readiness to use virtual reality are important factors to consider. Some VR headsets, like Google Cardboard, use the smartphone's accelerometer for head tracking, which can lead to lag, motion sickness, or headaches for some users (Hussein & Nätterdal, 2015).

Future Implications and Trends

The future of education is likely to be shaped by the continued development and integration of smart technology, wearables, and AI. As these technologies evolve, more innovative and personalized learning experiences and future trends could significantly reform the concurrent educational practices, some of which are discussed below:

As wearable technology and AI continue to advance, more innovative applications in education are expected. Some potential future trends include:

Augmented Reality (AR) and Virtual Reality (VR): Wearables equipped with AR and VR capabilities can create immersive and interactive learning experiences. AI can personalize these experiences based on individual student preferences and learning styles.

Natural Language Processing (NLP): AI-powered NLP can enable students and learners to interact the contents of their educational courses using natural language, creating learning more accessible and engaging.

Predictive Analytics: AI can explore and analyze data from wearables to forecast student's functioning and detect potential risks of academic failure. This information can be used to provide early intervention and support.

Support for Special Needs Students: Several applications, AI tools and wearable technologies with inherent features that support individuals with special needs are already available in the consumer market. Educators could employ such tools to further support learners with different needs within their classroom to a better standard.

Integrated wearables and AI offer exciting possibilities for transforming education. By leveraging the data-collection capabilities of wearables and the systematic power of AI that has analytical power, educators will make and establish personalized, adaptive, and engaging experiences in the learning that foster student success.

Despite the significant benefits of the smart technology inherent wearables and AI in education, there are also noteworthy challenges and considerations to address:

- **Privacy and security:** The gathering and use of individual data from students' pose worries and issues about their privacy and security.
- **Digital divide:** The availability and affordability of these technologies may exacerbate existing digital divides.
- **Teacher training:** Educators may need additional training to effectively integrate these technologies into their classrooms.
- **Ethical implications:** The usage of AI in educational field elevates ethical questions about bias, accountability, and the potential for job displacement. (Cornejo-Plaza & Cippitani, 2023).

By addressing such challenges and harnessing the potential benefits, educators can establish more personalized, participating, efficient learning experiences to learners making the whole learning process effective.

The Ethical Imperative: Balancing Innovation and Risk

Ethics play a critical role in the advancement and implementation of wearable technology and AI in education. It is a key to ensure that these technologies are used responsibly and in a way that truly benefits students without compromising their rights or well-being. Some recommendations are herewith made for policy makers, educational institutions, educators and learners to build a strong framework or body of guidelines on the responsible use of wearables and AI in education:

Privacy and Data Security: Ethical guidelines ensure that student data is collected and used responsibly, respecting their privacy and protecting their personal information.
Consent and Autonomy: Ethics emphasizes the importance of obtaining informed consent from students and their guardians, ensuring they have a choice in the use of technology and understanding the potential implications.
Equity and Accessibility: Ethical considerations help to address the digital divide and ensure that wearable technology and AI are accessible to all students, regardless of their socioeconomic background or abilities.

Educational Benefits: Ethics guides the use of technology to ensure it aligns with educational goals and objectives, avoiding excessive reliance on technology and maintaining a balance with traditional teaching methods.
Teacher Professional Development: Ethical guidelines help to ensure that teachers are equipped with the knowledge and skills to use wearable technology and AI responsibly and effectively.
Continuous Monitoring and Evaluation: Ethics requires ongoing evaluation of the ethical implications of technology to ensure it is used in a responsible and beneficial manner. (Simbeck, 2024).

By prioritizing ethical considerations, educational institutions and all potential stakeholders can harness the potential of wearable technology and AI while minimizing risks and ensuring that these technologies are used for the betterment of students.

Wearables, and AI have the ability to transform education by providing tailored, engaging, and real learning experiences. By carefully considering the benefits, challenges, and ethical implications of these technologies, educators can harness their power to create a brighter future for education. Educational institutions must work towards educating teachers and students about wearable technology, provide training programs and create clear policies to encourage the use of these devices in academic activities, teaching and learning, student engagement, interaction, assessment, evaluation, and feedback (Yanamandra, 2024).

CONCLUSION

Like two sides to every coin and similar to all advancements, the application of this wearable technology and AI in education brings to the table both benefits and challenges. The identifiable limitations vary depending on the type of wearable device, extent of AI use and the educational context, but there are notable circumstances where using wearable technology in teaching and learning can have negative consequences. A huge responsibility of restricting, if not fully eliminating, such undesirable consequences fall on the shoulders of the educational institutions and the educators. While the emphasis on creating and fostering opportunities for integrating technology into the classroom setting to thereby enhance the learners' digital literacy cannot be sidelined, it is quintessential that the pitfalls are not overlooked. Through policy, quality audits, training opportunities for teachers, ethical guidelines for all parties involved and rigorous control systems to deal with privacy, security, affordance, ethics and other issues, educational institutions could optimize the opportunity bestowed by wearables and AI for education.

REFERENCES

Al-Tamimi, A., & Al-Smadi, O. (2022). The Impact of Internet of Things (IoT) on Education: A Review. *International Journal of Advanced Research in Computer Science and Engineering*, 11(12), 106–111.

Arroyo, I., Micciollo, M., Casano, J., Ottmar, E., Hulse, T., & Rodrigo, M. M. (2017). *"Wearable learning: multiplayer embodied games for math," in Proceedings of the Annual Symposium on Computer-Human Interaction in Play*. ACM., https://dl.acm.org/doi/pdf/10.1145/3116595.3116637 DOI: 10.1145/3116595.3116637

Attallah, B., & Il-agure, Z. (2018). Wearable Technology: Facilitating or Complexing Education? *International Journal of Information and Education Technology (IJIET)*, 8(6), 433–436. DOI: 10.18178/ijiet.2018.8.6.1077

Ba, S., & Hu, X. (2023). Measuring emotions in education using wearable devices: A systematic review. *Computers & Education*, 200, 104797. Advance online publication. DOI: 10.1016/j.compedu.2023.104797

Borthwick, A. C., Anderson, C. L., Finsness, E. L., & Foulger, T. S. (2015). Special article personal wearable technologies in education: Value or villain? *Journal of Digital Learning in Teacher Education*, 31(3), 85–92. DOI: 10.1080/21532974.2015.1021982

Bower, M., Sturman, D., & Alvarez, V. (2016), "Perceived utility and feasibility of wearable technologies in higher education," in Proc. *15th World Conference on Mobile and Contextual Learning, mLearn*, Sydney.

Chen, X., Xie, H., Zou, D., & Hwang, G. (2020). Application and theory gaps during the rise of Artificial Intelligence in Education, *Computers and Education. Artificial Intelligence*, 1, 100002. Advance online publication. DOI: 10.1016/j.caeai.2020.100002

Christie, M., Carey, M., Robertson, A., & Grainger, P. (2015). Putting transformative learning theory into practice. *Australian Journal of Adult Learning*, 55(1).

Chu, S. L., Garcia, B. M., & Rani, N. (2023). Research on wearable technologies for learning: a systematic review. In Frontiers in Education (Vol. 8). https://doi.org/ DOI: 10.3389/feduc.2023.1270389

Cornejo-Plaza, I., & Cippitani, R. (2023). Ethical and Legal Considerations of Artificial Intelligence in Higher Education: Challenges and Prospects. *Revista de Educacion y Derecho*, 28. Advance online publication. DOI: 10.1344/REYD2023.28.43935 de Arriba-Pérez, F., Caeiro-Rodríguez, M., & Santos-Gago, J. M. (2017). "Towards the use of commercial wrist wearables in education," 4th Experiment@International Conference (exp.at'17), Faro, Portugal, 2017, pp. 323-328, Doi: DOI: 10.1109/EXPAT.2017.7984354

Della Ventura, M. (2017). Creating inspiring learning environments by means of digital technologies: A case study of the effectiveness of WhatsApp in music education. *E-Learning, E-Education (Eds.), Online training*, Springer (2017), pp. 36-45. http://dx.doi.org/DOI: 10.4108/eai.26-7-2017.152906

Havard, B., & Podsiad, M. (2020). A meta-analysis of wearables research in educational settings published 2016–2019. *Educational Technology Research and Development*, 68(4), 1829–1854. Advance online publication. DOI: 10.1007/s11423-020-09789-y

Huang, L., & Wang, F. (2019). The impact of artificial intelligence on education: A systematic review. *International Journal of Educational Technology in Higher Education*, 16(1), 1–23.

Huang, Y., Chen, Y., & Liu, Y. (2021). Wearable technology in education: A systematic review. *Journal of Educational Technology*, 48(1), 1–22.

Hussein, M., & Nätterdal, C. (2015). *The benefits of virtual reality in education*. Department of Computer Science and Engineering, Chalmers University of Technology, University of Gothenburg.

Hyndman, B. P. (2017). A simulation pedagogical approach to engaging generalist pre-service teachers in physical education online: The GoPro trial 1.0. *The Australian Journal of Teacher Education*, 42(1), 84–102. DOI: 10.14221/ajte.2017v42n1.6

Khosravi, S., Bailey, S. G., Parvizi, H., & Ghannam, R. (2022). Wearable Sensors for Learning Enhancement in Higher Education. *In Sensors* (Vol. 22, Issue 19). https://doi.org/DOI: 10.3390/s22197633

Kim, H., & Lee, J. (2017). The impact of virtual reality on student learning outcomes in chemistry. *Journal of Educational Technology*, 8(2), 123–142.

Lampropoulos, G., & Kinshuk, . (2024). Virtual reality and gamification in education: A systematic review. *Educational Technology Research and Development*, 72(3), 1691–1785. DOI: 10.1007/s11423-024-10351-3

Lee, J., & Chen, C. (2019). The impact of microlearning modules on professional development: A transformative learning perspective. *Journal of Educational Technology*, 10(2), 123–142.

Liu, Y., Wang, H., & Xu, J. (2020). Smart classroom: A review of its concept, application, and challenges. *International Journal of Educational Technology in Higher Education*, 7(4), 1–16.

Manikeswari, D., Ratnakaram, S., Harris, S. B., & Saseendran, H. (2023), A literature review on the effectiveness of Problem-based learning (PBL) approach in Higher education with reference to the Education 4.0 Taxonomy, *Proceedings of Global conference on Emerging technologies, Retrieved from*: https://www.researchgate.net/publication/384184489_Conf_Proceedings_Dec'23

Manikeswari, D., Ratnakaram, S., Shaikh, Z. H., Alawi, S. H., & Kumar, A. (2024). PBL in Higher Education: Empowering Students for Industry Dynamics. *European Economic Letters*, 14(3), 1137–1140. DOI: 10.52783/eel.v14i3.1873

Manikeswari, G. D., Kumar, R. S. P., & Sirisetti, S. (2024). Psychometric testing - the game changer: A 360-degree view of psychometric testing and its applicability in HR functions. *International Journal of Business Excellence*, 33(1), 48–75. DOI: 10.1504/IJBEX.2024.139112

Mezirow, J. (1978). Transformative dimensions of adult learning. *Adult Education Quarterly*, 28(4), 230–244.

Mezirow, J. (1997). *Transformative dimensions of adult learning*. Jossey-Bass.

Mishra, P., & Koehler, M. J. (2006). Technological pedagogical content knowledge: A framework for teacher knowledge. *Journal of Educational Computing Research*, 35(1), 60–75.

OECD. (2024). *The Future of Education and Skills: A Global Perspective.* OECD Publishing. https://www.oecd.org/en/about/projects/future-of-education-and-skills-2030.html

Ouyang, F., & Zhang, L. (2024). AI-driven learning analytics applications and tools in computer-supported collaborative learning: A systematic review, *Educational Research Review*, Volume 44, 2024, 100616, ISSN 1747-938X, https://doi.org/DOI: 10.1016/j.edurev.2024.100616

Qureshi, M. I., Khan, N., Raza, H., Imran, A., & Ismail, F. (2021). Digital Technologies in Education 4.0. Does it Enhance the Effectiveness of Learning? *International Journal of Interactive Mobile Technologies*, 15(4), 31. Advance online publication. DOI: 10.3991/ijim.v15i04.20291

Schroeder, N. L., Romine, W. L., & Kemp, S. E. (2023). A scoping review of wrist-worn wearables in education. *Computers and Education Open*, 5, 100154. Advance online publication. DOI: 10.1016/j.caeo.2023.100154

Selwyn, N. (2011). Is technology really transforming education? A critical perspective. *Educational Researcher*, 40(2), 75–80.

Shaikh, Z. H., Irfan, M., Nomran, N. M., & Ratnakaram, S. P. (2024). *Unlocking the Potential of AI for Efficient Governance: Innovative Approaches of Bahrain., AI-Driven Decentralized Finance and the Future of Finance*. IGI Global., DOI: 10.4018/979-8-3693-6321-8.ch005

Shoikova, E., Nikolov, R., & Kovatcheva, E. (2017). Conceptualising of smart education. E+E, 52(3–4), 29–37.

Simbeck, K. (2024). They shall be fair, transparent, and robust: Auditing learning analytics systems. *AI and Ethics*, 4(2), 555–571. Advance online publication. DOI: 10.1007/s43681-023-00292-7

Sturman, D., & Bower, M. (2015). What are the educational affordances of wearable technologies? *Computers & Education*, 88, 343–353. DOI: 10.1016/j.compedu.2015.07.013

Tang, K. H. D. (2024). Implications of Artificial Intelligence for Teaching and Learning. *Acta Pedagogia Asiana*, 3(2), 65–79. DOI: 10.53623/apga.v3i2.404

Taylor, E. W., & Cranton, P. (2017). TRANSFORMATIVE LEARNING THEORY. In *Mapping the Field of Adult and Continuing Education: an International Compendium* (Vol. 1). Adult Learners., DOI: 10.4324/9781003445913-9

Vincent-Lancrin, S. (2021). Smart Education Technology: How It Might Transform Teaching (and Learning). *New England Journal of Public Policy*.

Wang, C., He, Y., & Yang, Y. (2019). The role of artificial intelligence in education: A systematic review. *International Journal of Artificial Intelligence in Education*, 29(3), 397–422.

Wang, F. H. (2017). An exploration of online behaviour engagement and achievement in flipped classroom supported by learning management system. *Computers & Education*, 114, 79–91. DOI: 10.1016/j.compedu.2017.06.012

Wang, X., & Jiang, J. (2021). The potential of wearable devices and AI for personalized learning. *International Journal of Educational Technology in Higher Education*, 18(1), 1–15.

Wang, Y., & Liu, J. (2023). Blockchain Technology in Education: A Review. *IEEE Access : Practical Innovations, Open Solutions*, 11, 11439–11453.

Yanamandra, R. (2024). Developing an integrated framework of application of wearable devices in education. *International Journal of Business Excellence*, 32(1), 85–104. Advance online publication. DOI: 10.1504/IJBEX.2024.135923

Chapter 3
Harnessing AI and Machine Learning to Elevate Educational Wearable Technology

Marwah Lateef
Kingdom University, Bahrain

ABSTRACT

Abstract The prime motive for this chapter is to perform an extensive analysis on smartly how AI along with ML can be incorporated in wearable technologies that will prove to be beneficial for educational practices. This chapter seeks to: Help Improve Understanding, Drive Best Practices, Encourage Inclusivity, Submission of Technical Challenges, Ethical and Customary, Explore Future Directions. Through meeting these objectives, this chapter is intended to help shed light on the broader implications of technology-enhanced learning and offer practical advice for educators, technologists, and policy makers that wish to leverage AI and ML to support more efficient teaching and better learning experiences.

INTRODUCTION

The integration of Artificial Intelligence (AI), Machine Learning (ML), and wearable technologies provides alluring avenues for revolutionizing the traditional educational environments. The potential to alter the way that teachers teach, and students learn is becoming more apparent as these technologies mature. The traditional education system, in which teaching methods were the same for all students,

DOI: 10.4018/979-8-3693-7817-5.ch003

is now replaced by tailored and personalized experience with the evolution of AI and ML built into Wearable devices.

By harnessing wearable technology, AI and ML has transformed the data collection and analysis in real-time, which will help educators gain insights into student performance, behavior, engagement etc. This approach of data-driven feedback leads to the creation of personalized teaching strategies according to student learning styles, competences and preferences. In addition, wearables infused with AI and ML can also monitor holistic metrics like cognitive engagement, emotional well-being, physical activity etc., further strengthening the inventory of tools available to engineer a complete learning landscape that makes way for both academic excellence and personal well-being.

The mixing of wearable technology with AI and ML in the context of education is not new, but it really gives a boost to its potential. Historically, devices like fitness trackers and smartwatches were built to track physical activity like step counts, heart rates, or sleep quality. In our AI/ML era, wearables have taken a new face from much more fleshed out sensory input/output like fine-grained cognitive and emotional tracking, to direct connections into how a student is performing.

These technologies provide educators in the classroom today the ability to see what their students are learning as well as how they are learning. An AI-powered wearable might be able to detect that a student is struggling with something say, heart rate spikes or eye tracking studies, or how they are seated. The AI can then suggest interventions or adapt the learning material to improve satisfaction for a particular student.

Key Terms and Definitions

Artificial Intelligence (AI)

Artificial intelligence (AI) is defined as the ability of a machine to imitate human intelligence processes. These processes include learning (the obtaining of data and rules for using the data), reasoning (the use of the rules to reach approximate or definite conclusions), and self-correction. AI allows automation, adaptive learning, intelligent tutoring systems, etc., in education, replacing traditional teaching methods in making education smarter.

Machine Learning (ML)

Machine Learning (ML) is a subset of AI and refers to using algorithms and statistical models to allow computer systems to improve at performing a given task with experience, reduction of accurateness, and without being programmed to perform by a human being. Within education, ML enables the analysis of data, tailored delivery of content, and forecasting of performance, creating a better-aligned learning experience.

Wearable Devices

Wearable devices are those technology-enabled accessories that may be strapped to the body in the form of smartwatches, fitness trackers, and augmented reality glasses. In education, they enable real-time collection of data, encourage interactive learning experiences, and aid accessibility for student populations with different needs.

Augmented Reality (AR)

Augmented reality (AR) expands the real-world experience by adding elements from a digital world like images, sounds, other sensory enhancements through a device, like a smartphone or AR glasses. AR engages learners by delivering interactive simulations, visualizations, and immersive experiences that connect theoretical knowledge with practical applications.

Virtual Reality (VR)

VR puts users in a completely virtual environment via headsets and similar devices. While AR allows you to experience an augmented reality, VR replaces the physical world entirely with a virtual environment. VR brings experiential learning possibilities in the shape of virtual field trips, lab simulations, and skill-based training within educational mindsets.

Personalized Learning

Personalized learning approaches to education involve customizing and personalizing learning strategies for everyone according to their specific needs, interests, and learning speed. AI and machine learning are used to adjust content and assessment dynamically as students' progress through personalized learning systems to help them engage and succeed.

BACKGROUND

Artificial intelligence (AI): Is the simulated ability of machines to mimic human intelligence, allowing them to perform tasks that typically require human understanding and a cognitive process such as learning, reasoning, problem-solving and decision making. These AI systems are able to do all data processing to recognize patterns and take informed decisions based on that information.

Machine Learning (ML): One of the most important areas within AI, machine learning involves creating algorithms that allow machines to learn from data and improve their performance over time without being explicitly programmed (Russell & Norvig, 2021).

Using wearable devices for data collection, analysis and interpretation of performance and behavior relevant student activities in education. This type of technology provides instant feedback to both teachers and students, which allows for an even greater personalized learning experience. AI wearables, holographic systems that take students on virtual field trips or let them dissect digital cadavers, are other examples of education technologies already envisioned. Armed with these details, AI algorithms will improve to suggest means of modifying the learning environment by slowing the rate of lessons down or offering extra assistance, for example, and bringing in new educational materials (Holmes, 2021).

On the other hand, ML can be used to study long-term trends, of a student's performance and behavior which helps educators in prediction analysis, of future challenges before they get out of hand. As an example, an ML model could look at performance data for a student over time and predict that based on this history the student is likely to start failing in English. The system can recommend appropriate interventions, which can be targeted at the problem, before it affects the student as whole to clear his/her course (Goodfellow, Bengio & Courville, 2016).

Table 1 showcase specific wearable devices, their applications, and outcomes as a comparative analysis of AI and ML capabilities in wearables, as we can see different wearable devices can be utilized in the educational process by integrating the AI and ML technologies to increase the students' engagement and enhance the educational outcomes.

Table 1. Comparative table of AI & ML capabilities in educational wearables

Device Type	Main AI/ML Capabilities	Educational Applications	Data Collection Types	Outcomes
Smartwatch	Real-time Feedback	Tracking engagement levels and cognitive load	Heart rate & movement	Personalized pacing
AR/VR Headset	3D Interactive Learning	Virtual field trips & 3D modeling	Visual input & motion	Immersive learning experiences
Biometric Tracker	Emotional Monitoring	Tracking stress and emotional states	Physiological signals	Enhanced emotional well-being
Smart Glasses	Environmental Sensing	Monitoring environmental factors	Temperature & noise	Optimized learning environment

A Short History and Evolution of Wearable Tech in Education

While wearable technology has been used for education purposes over several years, its capability increases exponentially with the new addition of AI along with machine learning capabilities. You might remember a few years ago when most of the early wearable devices (such as fitness trackers and smartwatches) just tracked your motion or provided health stats like steps, calories burned or sleep quality. Although somewhat obsolete when it came to education, these devices were a boon in physical health.

Some more recent wearables have also expanded far beyond simple fitness tracking to including biometric monitoring, environmental sensing, interactive learning applications and upper-body gesture control-based user experiences. Many devices like smart glasses, AR/VR headsets and gesture-control wearables have allowed students to interact with digital content in unique ways, providing more immersive and higher quality learning experiences (Bower & Sturman, 2020). For example, virtual field trips, simulated experiments, and 3D visualizations of difficult-to-understand concepts are already possible using AR and VR wearables.

These new generations of wearables on the market are also now incorporating machine learning and AI to enable them capturing real-time data and pairing it directly with insights that can help instructional strategies, feedback responses, and unpredictable patterns learning environments. Ultimately, these technologies help educators go further than traditional categorical levels of comprehension and take a more holistic approach in student performance by encompassing cognitive and emotional data as part of the process to inform about it.

In addition to this, AI-powered wearables can help make learning spaces more accessible for all students through offering individualized assistance to those with different learning requirements. Thus, as seen herein for wearables tracking learning engagement and progress of students with disabilities from a university, it highlights how educators could tailor their pedagogical practices to adapt their instructions to the differentiated needs of each student within the education spectrum (Hernandez & Patel, 2021). Wearables can do the same for students with physical disabilities as well.

AI, ML, and Wearable Tech Leading to One Common Intersection

When all these components work together, we see the disease and potential cure, simply put that is what AI, ML devices should do in education to enhance the edtech IoT user experience from a teacher, student aspect. AI and ML driven wearables can collect a lot of data in respect to student behavior as:

- **Cognitive Engagement:** Using biometric data such as HRV, and eye tracking to track how engaged a student is with the learning material. For example, if we have a student that shows indicators of lacking focus on his or her biometric data (such as HR detected feeling stressed), the AI may suggest changes in the pace with which they are going through a lesson and provide extra aids to reengage the student.
- **Emotional States:** Columbus based behavioral wearable tracks stress, anxiety and emotional states using physiological signs. For instance, if a test or presentation takes place and the student's heart rate rises during this time, cognitive algorithms can identify it as a stress indicator and offer solutions such as mind-body exercises at the moment or take a short break (Liu & Zhang, 2020).
- **Physical Activity:** Wearables can track movement patterns of physical activity and exercise levels with AI providing insight on how different activities impact the well-being of students. Another case in point, an AI wearable might suggest a student to have more physical activity during the day to help improve attentiveness and focus on learning contexts (cf., Park & Lee, 2021).

In the realm of physical education classes, adaptive AI in fitness trackers has been used to increase student participation (2022). Students physical and learning achievement was enhanced through bespoke target setting derived from the individual baseline data points entered the system for each student. It was corrected on the fly how active students should be in terms of pushing them into participating sometimes but without overwhelming them (Lopez et al., 2022).

- **Environmental Interactions:** AI-enhanced wearables can track environmental factors including noise levels, temperature, or lighting conditions and provide insights on how they may influence student learning. If there is too much noise in a classroom, for example, the AI could recommend moving to a different part of the room or putting on noise-cancelling headphones to help focus.

Using these data points, AI and ML models can give real-time feedback to both teachers as well students thereby personalizing learning as a result. Also, this technology is capable of recognizing patterns in students' behavior and performance over time so that educators can begin using predictive analytics to become more proactive before any problems occur.

CURRENT TRENDS IN AI AND ML-ENHANCED EDUCATIONAL WEARABLES

The AI and ML Trends in Wearable Tech Industry

Wearable technology has also started as an education trend, we are already seeing a handful of devices that harness AI or ML for better educational practices. The biggest & most meaningful AI & ML-based wearables trends include:

1. **Augmented and Virtual Reality Devices (e.g., Microsoft HoloLens, Oculus Rift):** Smart glasses or AR/VR headsets used to build classroom experiences where students can interact with educational content in novel ways. By having a 3D model based on an AI-driven simulation, AR/VR applications are able to mimic real-world situations where students can directly interact with the models and gain an experience-based knowledge of complex ideas. As an example: a biology major could use AR glasses to virtually dissect a frog, offering them the option to inspect body structure in fine detail beyond traditional classroom constraints (Johnson et al., 2021).
2. **The applications of smartwatches powered with AI:** Smartwatches supported by modern-day technology can track a student's state both physically and cognitively and even provide real-time feedback and personalized learning strategies. In an exam, for instance, a smartwatch could track the heart rate of a student and add information that this might be how stress is affecting their results. From there, the AI system could recommend stress management strategies such as deep breathing exercises to help the student calm down and do better on test day (Holmes, 2021)

3. **Biometric wearables:** include high-end fitness trackers and biometric devices that can collect health and physiological records such as heart rate variability, skin temperature or electrodermal activity. The third is the AI algorithms that can analyze this information and then provide insights into student stress levels, engagement, and concentration for providing them with teaching strategies accordingly. For instance, if a biometric wearable shows that over time a student experiences high levels of stress when engaging in certain activities, they could suggest altering the learning environment or offer more resources and assistance to help with stress management (Perez & Kumar, 2021).
4. **Interactive Learning Devices:** Wearables like Smart pens, as well as watches and other gadgets are now designed for educational purposes that get interactive among students that engage them dynamically with digital content. The way that this operates on a practical level is that AI and machine learning are able to personalize the content delivery based on how students interact in real time. As an example, a smart pen could analyze the quality of a student's handwriting in real-time during a lesson and then give the student tips on how to perfect his or her penmanship while adapting lesson content to better fit the unique learning style of that person (Bower & Sturman, 2020).

Figure 1 illustrates key features and current trends in AI and ML technologies with wearable devices to improve the educational outcomes and increase students' engagement.

Figure 1. Key features in AI-Enhanced wearables for education

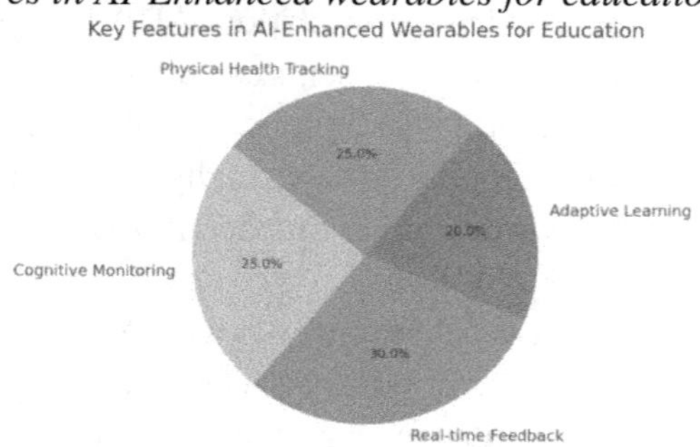

Wearable Data-Driven Adaptive Learning Systems

Driven by AI and machine learning, adaptive learning systems revolutionize the way educators teach students, delivering instruction that is tailored to each student on a personal level. We used these systems to track student work engagement in real-time during the baseline period and provide the students with personalized learning experiences which were updated in real time based on each student's own progress and preferences using data from wearable devices (Murray et al., 2020).

Another example is if an adaptive learning system powered by AI may take the data from a wearable device of, say a student, while in a class to assess their engagement levels. If the system captures that a student does not seem to understand a certain concept, it can adjust the speed of instruction, deliver additional materials or recommend different resources to aid in better understanding for the student. In contrast, it releases difficult content if the system recognizes that the student is benefited with a subject to challenge their learning abilities and motivation (Liu & Zhang, 2020).

Adaptive learning systems become useful especially in the settings where personalized education is delivered to students who have various styles of learning, capabilities and preferences. By leveraging AI and machine learning, these systems are able to create personalized learning pathways for each student, providing them with the assistance they need to succeed. Adaptive learning systems also give immediate feedback to teachers, empowering them to keep track of student progress and informed in their curricular decision making.

Figure 2 gives a visual perspective on where wearable devices adoption is highest. There are various adoption rates at different levels of education starting from schools to professional training. The average rates of AI wearables across various educational stages lies in: K-12 (30%), Higher Education (50%), and Professional Training (20%).

Figure 2. AI in wearables adoption in different educational levels

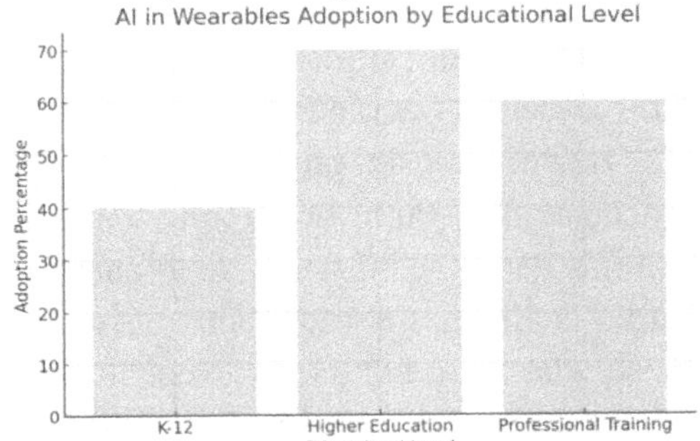

USE CASES OF AI/ML IN EDUCATION

The possibilities for AI and ML to change the way we learn with wearables have not even been fully realized yet, but some of them are useful now. Those aspects span to personalized learning, gamification, immersive experience, real-time feedback and so much more in the student education technology landscape. Below, we delve into some of the most promising use cases for AI and ML in education.

Personalized Learning Experiences with Wearables: Case Studies

Personalized experiential learning that meets the unique requirement for every student is one of the greatest advantages of AI and ML enabled wearables. The key benefit of this approach is that it allows teachers to move beyond the traditional, "one-size-fits-all" method of teaching by offering personalized instruction tailored to the unique learning style, preferences, and capabilities of each student.

1. Case Study 1: AI-powered wearables to improve literacy Skills

Third graders in a case study that used smartwatches with AI-powered functions to improve literacy. The smartwatches kept track of students' reading behaviors, such as their reading speed and comprehension levels, as well as how engaged they were when tackling those tasks. They analyzed the data through machine learning algorithms to provide students with reading recommendations including age-appropriate books and mapping areas where the child lacked practice. As a result, students who utilized smartwatches exhibited significantly more raised degrees of constraints compared to the students who don't have access to this technology (Smith & Jones, 2020).

2. Case Study 2: AI Based Math Tutoring using Wearables

Further in research, middle school students used an AI-enabled math tutoring program equipped with biometric wearables to monitor their cognitive engagement during lessons. The devices tracked physiological signals such as Heart Rate Variability, Skin Conductance, and Pupil Dilation to check in on the focus levels of students. Armed with this data, the AI was able to change math problem levels in real-time, aiming harder challenges at more active learners and providing extra practice support for those who were not as involved. In this study, students who participated in the AI tutoring program scored significantly higher on their math assessments during the semester compared to a control group (Murray et al., 2020).

The case studies above demonstrate how AI and ML-powered wearables can ultimately drive a more personalized future of learning that enriches student success. The ability to analyze data in real-time means feedback is personalized, instruction can be adjusted, and support is differentiated for each student based on their specific needs.

Wearable Devices Integrated With AI, Interactive Learning Through Gamification

Among all methods introduced to education, gamification is one of the most used tools, with its ability to motivate learners through game elements designed into non-game environment. Combining gamification with AI and wearable technology can result in increased engagement and an enriched learning experience that allows learners to be more active participants and achieve a better understanding.

- **Gamified Learning Platforms:** Wearables powered by AI can be combined with gamified learning platforms that track student progress and reward achievements in the form of points, badges or any other incentives. Wearable fitness trackers could be used, for example, to measure a level of physical activity in a PE class, and award points for those who have reached daily exercise goals. As a result, the AI can adjust the challenges according to individual student fitness level so that exercises are always of adequate difficulty while allowing students to see progression (Johnson et al., 2021).
- **Interactive Learning Games:** Wearables also allow for interactive learning games, which promote cooperation and problem-solving. As an example, high school students leveraged AI-enhanced AR headsets to play a science-based game together. This game asked players to gather up and solve science problems like running a virtual experiment or exploring a digital ecosystem. The AI system watched closely for individual students' participation in the collaborative task and provided immediate input on their teamwork, communication, critical thinking skills (Holmes 2021). This new, gamified method of learning not only made the experience of learning more interesting for students, but also assisted them in forming social and cognitive skills crucial for development.
- **Learning through Physical Activities:** Using gamification in education, this can help provide not just learning from a screen but getting active during learning Using gamification as an example, a wearable fitness tracker offers the advantage of students seeing their daily goals transformed into physical activities. Students can earn points by fulfilling specific tasks, e.g. taking a set number of steps or participating in group fitness challenges, making

the physical education more fun and encourages healthy habits (Park & Lee, 2021)

Gamification with AI wearables is the real fun part which makes learning interactive and easily consumable. Moreover, it provides students with personalized feedback and incentives that encourage them to learn better.

AR/VR Wearables for Immersive Learning Experiences

Augmented reality (AR) and virtual reality (VR) wearables Enable immersive learning, whereby students get greater exposure to educational content through hands-on and interactive experiential ways. These technologies can transport students to virtual environments outside of their present framing, allowing them to investigate new concepts in an applied manner with practical experience.

- **Virtual Field Trips:** Using AI-powered AR & VR headsets helps students to take an immersive virtual field trip to various locations around the world. So, for example a history course can travel through ancient Rome using VR headsets that would help students to learn by visiting historical places and meeting digital representations of historical personalities. AI algorithms can even shape the content to better match individual student interests and comprehension, keeping them informed as well as engaged (Johnson et al., 2021).
- **Simulated Science Labs:** Through the integration of AR and VR wearables in science education, simulated lab environments let students perform experiments without the need for real-world substances. As an example, a student learning about chemistry could wear AR glasses, combine virtual chemicals and observe chemical reactions occurring in real-time. This platform allows the AI to offer real-time feedback during the student's performance, give clues or corrections, and adjust the experiment according to the progress of each student (Bower & Sturman, 2020).
- **VR wearables in the teaching of technical material:** As students need practical time for skill acquisition and development, from K12s to vocation, technical education. For example, aspiring electricians or automotive technicians may be able to wear VR headsets so that they can practice their skills in virtual environments that resemble real-world scenarios. AI can track how students are progressing, identify where improvement is needed, and provide personalized feedback for them to practice required skills before entering a real-world environment (Liu & Zhang, 2020).

Williams and Carter (2024) explore simulations enabled with VR in science classrooms, finding that students who used virtual reality labs showed an improvement in retention and understanding over those who engaged in non-VR methods. AI can be used in virtual reality wearables to identify when a student is struggling through a concept, suggesting appropriate resources that enable a more interactive learning environment (Williams & Carter, 2024).

Students, equipped with wearable AI and machine learning enhanced devices, navigate more interactive forms of educational content to help students better understand complex topics which leads to experiential comprehension and active learning.

Real-Time Feedback for In-Class Student Engagement

One of the other key benefits that AI and ML-powered wearables provide is feedback in real time for educators, as much as students. The feedback is instant, and teachers can adjust their methods quickly to ensure that students are listening, taking in the content and meeting learning goals.

According to a study by Kim and Hwang (2023), the implementation of AI feedback systems enabled through wearable devices has led to an increase in engagement levels among university students. These systems utilize biometric data and adapt the teaching material accordingly as per cognitive load while providing live recommendations about pace adjustments, resulting in students with better academic performance and overall satisfaction (Kim & Hwang, 2023).

AI probably will tell the different signs of student engagement such as where one is looking most often and for how long, their apt which can be decoded based on body posture, eye movements, facial expressions etc. For example, when a wearable worn by a student sees distract signs (such as seeing away from the screen or showing distracting behaviors), the AI system can indicate it to teacher and suggest some strategies on how to re-engage, e.g., proposing more interactive activity or adjusting pacing in instructing task (Perez & Kumar, 2021).

- **Cognitive Load Assessment:** AI-supported wearables, in live with tracking engagement can also evaluate cognitive load, how much mental effort it takes to perform a task. Based on biometric data such as heart rate variability, eye tracking and brainwaves. AI can detect cognitive overload or boredom in students. Using this information, the system should be able to adjust the difficulty of the learning material so that is neither too challenging nor too easy for a student based on what they know, without overwhelming them (Goodfellow, Bengio, & Courville, 2016).
- **Instant Feedback for Students:** In the classrooms, wearables can instantly provide students with feedback that helps them stay focused and improve

their performance. For example, a student could receive an alert from their wearable that they are not paying attention because of the way they are standing or that their heart rate is up, and they may be feeling more anxious during an exam. The AI system can then provide personalized suggestions for the person to pause briefly and relax techniques making it easier for the student to manage stress better and focus on its work (Holmes, 2021).

AI-powered wearables provide on-the-fly notifications of engagement or cognitive load helping students stay engaged and ultimately improving learning outcomes.

Social and Ethical Cost

While AI and ML based wearable technologies are very exciting for Education, the same holds true with facing some serious social and other ethical challenges as well. As with any technology that collects and analyzes personal data, it is important to approach privacy, security and fairness concerns head on. Additionally, the use of AI and ML-powered wearables should be guided by ethical standards that ensure their implementation promotes equity, transparency, and accountability.

Compliance and Security Standards for AI Controlled Wearables Making School Safe with Students Data Privacy

One of the biggest concerns with artificial intelligence powered wearables in education is acquiring and securing specific student data. As a result, they end up collecting huge amounts of data in the form of biometric or psychographic data or academic performance metrics. While that data can help improve custom learning experiences, it also raises important questions about its utilization, access to it and the means of protecting it.

1. **Encryption and Secure Storage:** The educational institutions need to adopt security measures like data encryption and secure storage to protect the student confidential information. According to (Luca, 2020), encryption is important in keeping the data complementing wearables away from unauthorized access, followed with a secure storage system that allows for less chances of data breach.
2. **Informed Consent:** making sure students (and ideally their parents) fully understand what data from wearables is collected, how it is used and who else might have access to this data. Educational institutions should have informed consent from students and their parents before wearable technologies are used at school (Campbell et al., 2020).

3. **Data Minimization:** Schools should implement a data minimization approach to collect the necessary information for educational purposes only. In doing so, the schools do not have to collect unnecessary data hence they can reduce privacy breach risks and ensure students information is not used in ways that it was unintended (Liu & Zhang, 2020).
4. **Access Controls:** Proper access controls shall be made in order that most effective designated college individuals can get entry to scholar-related records. Role-based permissions limit what data a person and his/her applications can interact with based on their role and need to know, thereby reducing unauthorized access or exposure of sensitive information (Perez, Kumar, 2021).

Educators can support the responsible and ethical application of AI wearables by emphasizing data privacy and security.

Privacy and Ethical Issues in AI and ML with its Applications in Educational Wearables

When it comes to the AI and ML in wearable technologies, it makes sense not just data privacy problems but also some paramount ethical concerns regarding fairness, bias, transparency of decisions, susceptibility from getting addicted to this tech. This means we need to make sure education with these technologies also respects the rights of students and represents ethically sound decision making.

A recent study by Sato et al. states that there is increasing necessity for ethical standards in educational wearables based on encryption and transparency of data. There findings imply that AI wearables should embrace a data management system centered on the students themselves, in an approach aligning with student autonomy and providing clear explanations about how their data is used (the latter of which has been proven beneficial to gaining trust from students and teachers toward the technology itself. (Sato et al., 2023).

1. Bias in AI Algorithms

An ethical issue at the forefront of AI and machine learning in education would be the potential for bias in algorithmic decision-making. So much in fact, that the quality of the data is one of our most potent weapons for debiasing AI systems. In other words, if the training data includes biases, then that will lead to biased outputs of the AI model as well. An AI system embedded in a wearable device that tracks students' performance, for example, and is informed by perpetuating stereotypes with respect to such data could only likely cement or exacerbate current inequities (Campbell et al., 2020).

To solve this problem, developers need to ensure that AI models are trained on a large and varied set of datasets with respect to classes. Additionally, the AI systems should be audited and tested on a regular basis for early identification of biases that may arise. As such, fairness should be programmed as prerequisite in AI systems to ensure that every single student regardless of race or economic background is treated fairly and given the same opportunities.

2. Transparency and Explainability.

AI and machine learning systems, especially those employing sophisticated algorithms, often defy human understanding. This opacity can lead to a "black box" effect where the people impacted by AI outcomes (students, teachers, and parents) do not understand why those decisions happened in the first place. For example, if a wearable device makes a recommendation for a student to get additional help because of their health data, it might not be clear how or on what basis the AI reached that conclusion (Russell & Norvig, 2021).

It is important for AI systems to be transparent in their decision-making for building trust in its usage. Educators who know how these systems work and understand their recommendations, a collaboration between schools and developers to create accessible AI algorithms on top of that, students and parents need to be informed on how these AI-powered wearables are operated, as well as what data gets stored and how that data affects educational determinations.

3. Human Oversight and Accountability

While AI and machine learning can provide important input data on how students are performing or behaving, it is critical to remember that they are there to assist human judgment. In this light, educators need to be able to have the final say and see AI recommendations as a support tool that empowers their expertise rather than replace them. This is where human oversight is crucial to ensure that the models are used in a responsible and ethical manner while also considering recommendations made from this type of model within the broader picture of what each student needs (Holmes, 2021).

This also comes from accountability. It should also be defined when decisions resulting from AI-generated insights are made by whom in the schools. However, there needs to be a process if wearables powered by AI are incorrectly reading the data or providing improper recommendations and steps to mitigate harm against students.

4. Consent and Autonomy

Ethical concerns also extend to the data rights and informed consent of students, and their parents where appropriate, related to wearable technology use within educational settings. Students and parents should also have the option to opt out of the collection of wearable data, as well as some control over how it is used (Luca, 2020).

Also, students should always feel a sense of autonomy that they oversee their own learning journey. Like the human tutors of yesteryear, AI and ML systems should not dictate learning experiences rather they ought to be designed so educators can remain in control. It means building systems that are flexible, open and responsive to individual choices and desires.

5. Designing Inclusive and Adaptive Wearable Technology Solutions

Along with the ethical implications of AI / Machine Learning in educational settings, we must also consider the principles and initiatives to design for inclusiveness and adaptability when designing wearable devices. These can provide tailor made learning options for a wide range of needs, though this is only possible if the whole design processes are by nature inclusive.

Wearable Technology Initiatives and Scaling Strategies

One of the biggest challenges for wearable technology education initiatives is moving them beyond the limited pilots that often fund these programs and into regular use by all students here and around the country as more schools are expected to welcome devices like health trackers onto their campuses. This involves managing infrastructure, price tags, as well as employee training

1. Infrastructure and Technology Readiness

The success of wearable tech initiatives depends on the infrastructure that schools have access to. It implies internet connectivity, sufficient bandwidth and required hardware (wearable devices). The absence of these key elements makes an effective integration of wearables in learning difficult (Ramirez & Thomas, 2021).

Also, the school may have to invest in specialized software platforms that can process and analyze the data recorded by wearables. Cloud-based systems are one of many ways to solve the problem of elite-scale data storage and analysis. Colleges need these resources so that their plans to roll out wearables work.

2. Cost and Resource Allocation

Wearable devices and the software platforms that accompany them are expensive. The reality is they have no reach in most underfunded schools or areas that do not possess the resources. For wearable technology to be more accessible, educational institutions may consider funding through government grants or collaborations with the private sector (Martinez et al., 2021). Also, schools may require spending money for maintaining and upgrading the wearable devices as well as providing tech support. Inclusion in terms of devices. All students must be on the same page when it comes to having a device, no one gets left behind.

3. Educator Training and Professional Development

Wearable technology initiatives will be in vain if educators are unable to capitalize on the devices. For this to happen, schools must invest in professional development programming whereby teachers are learning how to better integrate wearable technologies into their classrooms (Holmes, 2021).

This training should cover interpreting the data gathered from wearables, how to adjust pedagogical practices according to this information, and how to address any technical issues which may present themselves. In the long run, schools should provide continued support and training so that educators have what they need to make use of wearable technology in order to amplify student learning.

Designing Inclusive Solutions for Diversity of Learners in the Classroom

Wearable technologies have an immense potential to benefit every kind of student, those with disabilities, Head Start kids learning English for the first time, and a learner with autism. Paired with AI and ML software, these wearables can be used to develop personalized feelings around the learning style of every individual.

1. Accommodated Technologies for Special Education Students

Without a doubt, student wearable technology can be personalized to address assistive solutions for students with both physical and cognitive issues. A wearable can be equipped with haptic feedback to assist in directional guidance for a visually impaired learner, an AI-driven device may provide live speech-to-text transcription for those learners that are Deaf or Hard of Hearing (Hernandez & Patel, 2021).

They can also tailor learning materials to everyone. A student with dyslexia might use a wearable that visually alters the text, changing the size and color to improve readability. Similarly, students with attention deficit could employ wearables to be regulated reminders or resting and break suggestions leading them the way for staying on task during lessons (Liu & Zhang, 2020).

A 2022 study by Nguyen & Lee, found that AI-wearables have enhanced the accessibility to learning of students with disabilities like providing voice-to-text support for the hearing impaired. These applications of wearables, the study states, narrow educational inequities and provide every student with essential tools for success (Nguyen & Lee, 2022).

2. Multilingual Support for English Language Learners

AI-empowered wearables can help English Language Learners (ELLs) with immediate translation and personalized feedback. For example, a wearable might interpret classroom instructions to the native language of students, so they are able to participate more easily with their lesson. The AI system also had elements of personalized language practice which could help to advance their English language development (Campbell et al., 2020).

3. Personalized Learning for Diverse Learning Styles

Because each student learns differently, wearable technology allows for customized learning solutions targeted at specific learning modalities. Students excelling in visual method of learning would especially be served by means of AR wearables to interact with 3D models or simulations, and students who do better when they are listening may benefit from a wearable device where information is given audibly (Johnson et al., 2021).

Lesson plans can be automatically adjusted by AI and machine learning for the purpose of observing how students engage with various sorts of learning materials. This model helps to ensure that each learner receives the right support and resources he or she needs to succeed.

CHALLENGES AND FUTURE DIRECTIONS

AI and ML driven wearables have the highest potential to revolutionize education but face lots of challenges in their adoptions on a larger scale. Understanding and addressing the concerns above will be critical in enabling wearables to continue maturing into a valuable tool for education.

Technical Challenges

1. Data Accuracy and Reliability

How well do those wearables sense, anyway. Providing incorrect recommendations or feedback due to wrong, or not enough, data can also be a common problem of AI and ML systems which prevents delivering the best possible learning experience. Securing the inclusion of accurate sensors in wearable devices and maintaining their accuracy throughout data collection is a big technical task (Perez & Kumar, 2021).

2. Integration with Current Education Systems

It is important to integrate wearable devices seamlessly within the existing educational ecosystem (LMS, SIS & other digital platforms). Achieving such a level of integration presents technical challenges, particularly in schools that implement a mix-and-match strategy when it comes to the use of software and hardware solutions (Luca, 2020).

3. Empirical Data Set translates into data overload and Interpretation difficulties.

How do we integrate these new, AI and ML enhanced wearable techs in educational environments. Wearables Tech provides in-depth data on students' movements, brain engagement and emotion states. While this data is invaluable for understanding student performance, the wealth of data can be overwhelming for teachers and school administrators. To process the data in an insightful manner, strong AI systems are required to sort out and interpret useful things from the data collected.

Different people need to do different stuff (educators need tools and training to make sense of data). Misinterpreting data can easily lead to misuse or misinterpretation, and in turn to faulty conclusions about a student's progress or behaviors. Dashboards should be constructed in such a way that only relevant data is displayed, and minimal information which could clue the educator with informative recommendations on what to do next (Goodfellow, Bengio & Courville, 2016).

4. Cross-platform Interoperability

Two nontrivial technical challenges are to ensure interoperability of diverse wearable devices and education platforms. Most schools use a variety of digital services like Learning Management Systems (LMS), Student Information Systems (SIS), and digital assessment platforms. These wearable devices need to communicate well with these systems and have seamless data sharing to provide a holistic picture of student performance.

The biggest hurdle to making this vision a reality, the researchers said, is establishing common standards and APIs (Application Programming Interfaces) that enable wearables data to easily be integrated with existing educational platforms. In the worst case, without interoperability, schools will have fragmented systems with wearable data siloed off unable to contribute to more general educational strategies (Ramirez & Thomas, 2021). As we look to the future, it will be crucial for wearable makers to collaborate with software developers and educational institutions to integrate them in a meaningful way.

5. Consequences on Data Security and Ethics

The more that students are tracked by wearables, the greater the need for data security. Education institutes have a responsibility to protect wearable information and secure the data from breaches. This includes implementation of encryption related protocols, secured data storage mechanism and strong access controls to avoid any potential leakage of data.

Moreover, this raises questions from an ethical point of view: who can use the data and to what capacity? For example, who gets access to the data from wearables educational technology companies or governments? What would be the consequences for student privacy if the biometric data of students got abused or misused by an outside entity? These are moral questions that schools must weigh carefully, as they already collect other data as part of the duty of care, and we need clear governance set out (Luca, 2020).

6. Inclusive Access to Wearable Technologies

The single largest barrier to the adoption of wearables in education is to ensure that every student has equitable access to these tools. Every student does not have the same access to necessary resources (wearable devices, internet connectivity and supporting technological ecosystem) The consequences of this digital divide can exacerbate preexisting educational disparities, particularly for low income or rural students (Martinez et al., 2021).

This information suggests that schools and policymakers will have to work together to address the opportunity gaps by committing financial resources to make it possible for every student to have access to the wearable technologies. The solutions may be government funded programs, tech industry partnerships, or strategic efforts to provide devices directly to underserved locations, schools need to establish what is acceptable when it comes to these cutting-edge wearable technologies while at the same time ensure that their students have a right of access independent of their background.

Financial Constraints and Time to Implement

Although the benefits of AI and ML enabled wearables in education are huge, the cost to adopt such technologies becomes a burden for many schools. Wearable technology, the kind our software and infrastructure support accessibility to, is a rather pricey investment, especially for an educational institution with an already tight budget. In addition, to move forward, the cost of keeping all these devices updated and maintained with technical support may be more than schools can afford in later years.

1. Investment in Hardware and Software

The major obstacle to the widespread adoption of wearable technologies is the financial outlay required by students and their teachers in the form of an initial investment in wearable devices. While one could argue that certain wearable technology like fitness trackers are budget friendly or more affordable compared to AR/VR headsets or advanced biometric monitors, a lot of wearables still come with a hefty price tag. A way for schools to do this is through buying the necessary software platforms in which they can gather the data from all these products (Park & Lee, 2021).

Some of these expenses could be lower by leveraging wearable device leasing agreements, applying for government allocations, or partnerships with manufacturers that give educational grade tech at a cut rate price. Some are piloting them; others are using it with everyone.

2. Ongoing Maintenance and Support

The initial investment is not the only cost, schools also must factor in upkeep for wearable devices, and that includes tech support. These devices may periodically need to be updated, fixed or replaced, necessitating schools with IT personnel to deal with technical issues. In addition, wearable technology use with existing school systems might be an additional software development or IT expense (Ramirez & Thomas, 2021).

Recall that the cost of technology does not begin and end at the purchase of new hardware/software, schools should also add these ongoing expenses to their overall tech budget and make them bring their most. For example, a school might select a centralized device management system which allows IT staff to see and manage many devices from afar, so that there is less need for one-to-one troubleshooting.

3. Funding and Resources Allocation

Although wearable technologies are increasingly popular, justifying costs is always the concern, and schools want to be smart with their expenditures. That could look like finding the right wearables for students in certain topics or grade levels where they could have the most impact. For example, stem classes have the potential to integrate AR/VR wearables to provide engaging learning in laboratory investigations (Martinez et al., 2021).

Schools might also investigate new financing methods, such as related local enterprises or neighborhood groups that support educational technology programs. Schools can reduce the financial burden of implementing wearable technologies while still improving student benefits using external funding sources.

Future AI and ML in Educational Wearables

However, as AI and wearable technologies continue to evolve in tandem, a slew of new features are likely to impact the future of education. These changes have the potential to mark a major step forward in personalized learning, improved student outcomes, and more immersive and engaging educational experiences.

1. AI-Powered Personalized Learning Assistants and Virtual Tutors

There is a potential of creating a new category, AI-powered virtual tutors that could support tailored, personalized instruction and learning assistance outside of the classroom. Virtual tutors might rely on the bracelet, which uses data to modify how it helps a student based on their progress. One example is that a virtual tutor would assess the stress of students during a math lesson by detecting their biometric data using wearable devices and decide whether the difficulties of problems can be adjusted (Martinez et al., 2021).

Apart from providing tailored assistance, AI-based virtual tutors might even evaluate student assignments instantly, suggest additional study material and keep a tab on academic progress. The answer is a massive level of individualized learning that would allow students to learn at their own pace, circling back in areas where they struggle and accelerating through subjects of strength (Liu & Zhang, 2020).

2. Smart Classroom Ecosystems

Zeng et al. (2023) describes AI-enabled smart classroom ecosystems that relay behavioral data from wearables to dynamically change the environment, such as reducing light and noise, to ensure students maximize focus and lower cognitive load. The system has been proven to positively impact the overall engagement and attentiveness in the class and provides insights to improve smart classroom effectiveness (Zeng et al., 2023).

Educational wearables are paving the way for a future in which AI-enabled devices and systems work together to create a smart classroom ecosystem that augments students learning environment. As an example, wearable technologies might interact with intelligent whiteboards, interactive displays, and smart luminaires to foster a more flexible and responsive teaching/learning environment (Johnson et al., 2021).

AI algorithms could assess data from wearables to alter aspects of the learning environment, such as lighting, temperature or noise levels, things that could make it optimal for students to learn in this smart classroom. In turn, these systems could provide teachers with live feedback about the engagement status of their students, allowing them to adjust the way they teach accordingly.

3. Collaborative Learning with AI-enabled Classroom Projects

Collaborative Learning through Wearable Technologies AI-based wearables could track how groups interact, their communication patterns and the individual contributions of each member when it comes to working on joint projects, giving them a roadmap to work more cohesively within a group (Perez & Kumar, 2021).

AI systems that capture data from wearables could determine which students participate in group discussions or need some assistance for motivation. Armed with information such as this, teachers could more effectively ensure that all students were participating in group work fairly.

4. Intelligent Wearables in an Emotional Perspective

The development of wearables with emotional awareness, i.e. a technological empathic with the emotion, that can sense emotional state and act before on when student need help is one likely future evolution. These tools could be equipped with biometric sensors, such as heart rate variability or skin conductance, to automatically assess an individual's emotional health and provide them with person-specific support where necessary (Campbell et al. 2020).

As an example, if a wearable is alert of high stress being experienced by a student through for example an exam, the AI can suggest ways to calm down like breathing exercises and short breaks. Wearable devices with emotional intelligence could also help teachers detect such students and intervene at the earliest to provide support (Liu & Zhang, 2020).

5. Artificially Intelligent Professional Development for Educators

The potential of AI is not limited to students, however; it can also be a massive boon for improving educators' professional development. Wearable devices containing integrated artificial intelligence can provide teachers with instant feedback on their teaching methods, help improve what they are doing wrong and better handle their students' requirements (Johnson et al., 2021).

For example, they could be used to track a teacher's movements and speech while giving a lesson, providing an understanding that the educator was either actively participating in the class or successfully communicating. Armed with this knowledge, AI systems could start offering personalized suggestions on better ways to handle a classroom, pace a course or deliver instruction.

CONCLUSION

Wearable technologies with AI and ML are expected to revolutionize education by delivering new ways to engage learners, drive better student outcomes, and enable a more personalized form of teaching. Wearable technology gives educators critical information about what students are doing, how they are engaging and performing, all leading to actionable data that is personalized for each student.

These wearable technology devices, fueled by Artificial Intelligence, will facilitate increased engagement and a greater degree of interactivity in the learning environment which can help students absorb knowledge in newer and more innovative perspectives. Whether personalized learning assistants or AI-powered virtual tutors, these techs will give students the support they need to succeed in the fast-paced educational world. Similarly, the use of wearable gadgets can even be customized for group work on assignments that support social-emotional learning as well as adaptive and empathetic learning experiences aligned with each learner's needs.

However, there are some major issues associated with this integration of wearable technologies within the education sector including technical concerns, ethical dilemma and budget constraints. They should also protect privacy protections, address equity of access to technology for all students and provide professional development for teachers when integrating AI-powered wearables.

From now on, the AI and wearable technologies in education have great prospects. As these technologies develop in tandem with each other, the impact they wield on the educational landscape will continue to grow more and more, rendering education to be a much more personalized experience capable of reaching every student. While it will require the right infrastructure, resource allocation and ethical standards in place, AI-run wearables can redefine how education is delivered, offering both a more inclusive approach along with ensuring its effectiveness.

REFERENCES

Bower, M., & Sturman, D. (2020). *Wearable technology and immersive learning*. Journal of Educational Technology.

Campbell, L.. (2020). Ethical considerations in AI for education. *Educational Research Review*.

Goodfellow, I., Bengio, Y., & Courville, A. (2016). *Deep Learning*. MIT Press.

Hernandez, R., & Patel, S. (2021). *Inclusive design in educational wearables*. Journal of Inclusive Education.

Holmes, B. (2021). *Personalized learning through AI wearables*. Technology in Education.

Johnson, R.. (2021). Gamification in the AI-powered classroom. *Interactive Learning Environments*.

Kim, J., & Hwang, S. (2023). Enhancing student engagement through AI-powered real-time feedback in educational wearables. *Journal of Educational Technology Research*, 45(2), 132–147.

Liu, Y., & Zhang, W. (2020). *AI in education: Emerging trends*. International Journal of AI in Education.

Lopez, A., Garcia, T., & Lee, P. (2022). Adaptive learning in physical education: AI-driven fitness trackers and student engagement. *Physical Education and Technology*, 14(1), 58–73.

Luca, J. (2020). Ensuring privacy in wearable tech for education. *Computers & Security*.

Martinez, J.. (2021). Future innovations in AI-driven education. *AI & Society*.

Murray, A.. (2020). Adaptive learning through wearable tech. *Educational Technology Research and Development*.

Nguyen, T., & Lee, K. (2022). AI wearables for accessibility in education: Enhancing learning for students with disabilities. *Journal of Inclusive Educational Technology*, 11(2), 140–155.

Park, S., & Lee, H. (2021). *Wearables and student well-being*. Journal of Student Well-being.

Perez, C., & Kumar, R. (2021). Real-time feedback in AI-enhanced education. *Education in Science*.

Ramirez, M., & Thomas, D. (2021). *Technical and financial challenges in AI for education*. Journal of Education and Technology.

Russell, S., & Norvig, P. (2021). *Artificial Intelligence: A Modern Approach*. Pearson.

Sato, Y., Wang, M., & Chen, L. (2023). Ethical standards in AI-driven wearable technologies for education: Addressing data privacy and transparency. *International Journal of Educational Ethics*, 12(3), 211–225.

Smith, T., & Jones, P. (2020). *The impact of personalized feedback on student engagement*. Learning Technologies Journal.

Williams, D., & Carter, H. (2024). Immersive VR labs and their impact on student comprehension in science education. *Journal of Interactive Learning Environments*, 32(1), 87–102.

Zeng, R., Johnson, E., & Patel, M. (2023). Building smart classrooms: Leveraging AI and wearables for adaptive learning environments. *Smart Learning Environments*, 9(4), 334–349.

Chapter 4
The Role of Wearable Devices on Students

Satyanarayana Sanku
https://orcid.org/0000-0003-0623-9293
Sreenidhi Institute of Science and Technology, India

ABSTRACT

Abstract: This chapter of the book explains the significant role of existing wearable devices on students. Students are basically classified based on what they are placed into such as academics, sports, or challenged. Then under academics, based upon their age and class in which they are studying as per the World Health Organization Standards. And later the sensors which are used for measuring various health like EEG, EEG, Heart rate, etc and psycho -physiological parameters such as stress, Sleeplessness, autism etc are investigated. The wearable devices made by biosensors are attached to the students at different places on the body based on what parameters that are interested to be measured. Theses parameters are used for monitoring student's activities by using recent developed software algorithms including machine learning and Artificial Intelligent algorithms. Some case studies are also presented.

1. INTRODUCTION

The student's life between primary to post higher studies are very crucial as they are exposed to tremendous pressure and stress for their goal achievement. Monitoring the study behavior, metabolic activities, mental health and balancing them is required. Therefore, students need wearable devices for aid in education, to monitor physical and to maintain psychological and mental stability during regular class work and more importantly while examination periods. The students those who choose sport as their life career, also undergo the psychological pressure and physical strain (for

DOI: 10.4018/979-8-3693-7817-5.ch004

outdoor games) during practice or while in the game. These sport students also need the wearable devices to aid them to maintain balance in all aspects.

There are various wearable devices such as wrist bands, fitness bands, smart watches, smart glasses, smart shirts, smart shoes, smart socks, smart pants, smart pant, smart ring, smart finger and smart bracelet. Each of the wearable devices have their own significance and working principle for tracking or monitoring or scanning or tracing some parameter under measurement.

The fitness bands track the calories burned, weight gained, and sleep hours, number of footsteps and distance travelled. It also suggests remedial solutions for out of range parameters. The wrist bands and smart watches monitor health related parameters such as SPO2, heartbeat, pulse count, Blood Pressure levels, sugar level in blood etc. And also provide notifications such as call/ message alerts, wake time alert etc. The smart shoes tracks calories consumed, steps made, posture, fatigue and other health parameters to prevent foot and leg related injuries and pains. The smart cloth devices detect temperature, respiration rate, heart rate and other health-related information of persons who is sensitive to extreme environments and even for military personnel posted at high altitude and extreme climate condition places. The wearable smart glasses are used for security, healthcare, navigation, indoor or outdoor sports activities and hand-free cooking assistance.

Users of the wearable smart gloves can manipulate virtual items by naturally and immediately touching them. It can be used, for example, to snap home appliances, control objects in games, or most importantly to bridge the gap between the general public and members of the deaf or dumb population. They facilitate easy communication between the deaf and dumb by translating sign language into text or speech.

A. Students Classification

There are different groups of students who sometimes need different varieties of wearable devices to cater certain desired parameters. Primarily students need to be classified based on various parameters which are shown in Figure 1.

Figure 1. Classification of students

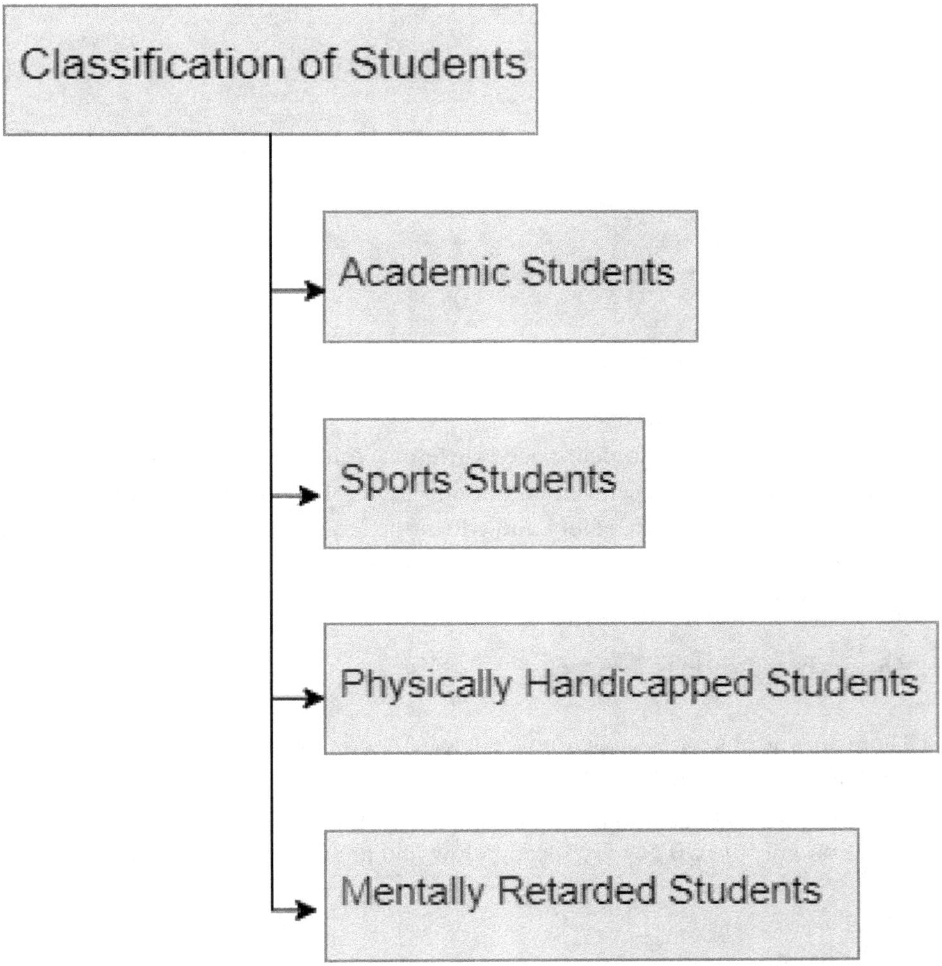

Figure 2. Time line of WBT evolution in educational context from 2004 to 2022

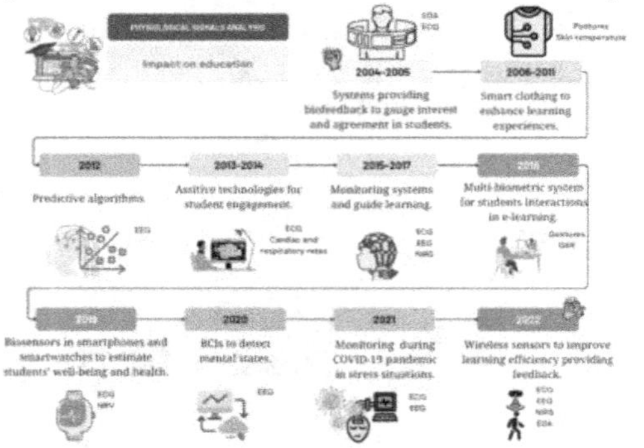

(Lima-Carmona et al., 2024)

Various categories of students need different kinds of sensor. Different types of Wearable Biosensors are invented and are being under use to monitor various parameters related to health, safety and education from past few years. The Figure 2 shows the evolution of various wearable biosensors used for education 2004 to 2022 (Lima-Carmona et al., 2024)

B. Wearable Device Types

There are various types of wearable devices based on where they are worn on the human body which are especially used for students of various categories depending up on the convenience and the required parameter for measurements in the context of education purposes. They are listed below and are also shown in Figure 3

a. Head worn
b. Wrist worn
c. Chest worm

Figure 3. Head-worn wearable devices: (A) Emotiv EPOC EEG system(Tutorials – EMOTIV, n.d.) (B) Google Glasses in (Lindeque et al., 2014) and (C) Wearable wrist-worn like Apple Watch used in(Laverde et al., 2018) and (D) Wearable chest-worn

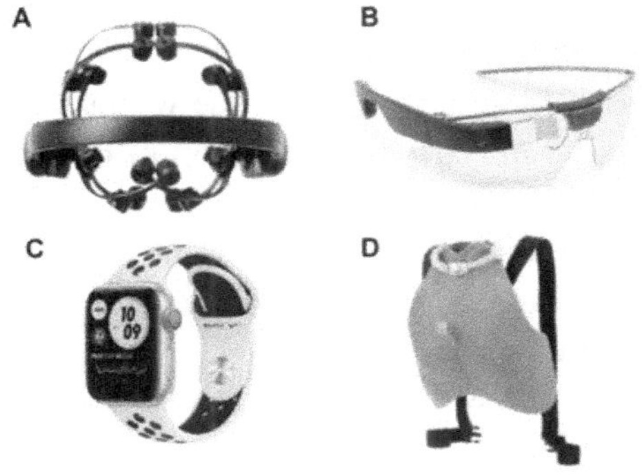

(Cowperthwait et al., 2015)

a. ***Head-Worn Devices***(Khosravi et al., 2022): In 1968, Ivan Sutherland demonstrated the first head-mounted display at Harvard University. During the past four decades, scientists and researchers have been investigating ways of developing full color opaque displays that enable users to see through to the real-world. Examples of commercial devices include Microsoft's Hololens. In terms of their use in higher education, the University of South Australia found evidence to support the impact of real-time information overlay on learning using head-mounted displays. EEG sensors were used in educational design programs, to assess brain activity through the wearable plug and play headset, combined with Oculus Rifts VR to conduct spatial assessments. Emotiv EPOC® EEG head-mounted gaming system Figure 3A has been used in cognitive and brain science to measure brain activity in the Macquire University. The Advantages of Head-worn wearable devices such as head-mounted displays (HMD) being that they offer alternate learning styles with the use of VR, for students who are mainly visual, active and global learner.
b. ***Wrist-Worn Devices***(Laverde et al., 2018): Devices worn on the wrist can be traced back to the beginning of the 20th Century. Currently, due to their unobstructiveness to the wearer, various types of wearable devices have been developed, which have been used for educational purposes. Examples include: Wristbands, Watches ECG and Sensors etc. In, wrist-worn activity trackers equipped with bio-metric sensors were used in higher education regarding eHealth literacy acquisition. Advantages of Wrist-Worn Wearable devices in

Education to collect physiological signals of the human body and are proven beneficial for education purposes.

c. ***Chest-Worn Devices***(Cowperthwait et al., 2015): Examples of chest-worn devices used for educational purposes include the wearable chest plate. In a study, students and faculty from engineering and nursing developed a wearable Tracheotomy Overlay System (TOS), for use with standardized patients. This device was designed to improve the education of health professional students while learning assessment and care of a patient with the tracheostomy in clinical practice. The advantage of Chest-Worn Wearable's in Education which includes different sensors that can collect and store data such as heart rate, heart rate variability, and respiration.

2. MONITORING ACADEMIC STUDENTS USING WEARABLE DEVICES

The students in this category join the academic schools for well education and knowledge. They improve their thinking skills, writing skills and reading skills. They will be able to disseminate social responsibility, possess general awareness, and show intellectual skills towards self, interpersonal, societal, national and global wellbeing. It is the prime responsibility of all human beings to provide basic educational needs to their next generations.

The Academic students can be classified into various categories based on their age and the standard that they are studying in school. The classification is:

a. Play group and Kindergarten students
b. Primary school students
c. Middle school students and
d. Junior Secondary school students

The monitoring of students under these categories using different wearable devices is investigated.

A. Monitoring Play Group and Kindergarten Students (Age Below 5 Years)

The children of this group are very sensitive to surroundings, climate changes, and emotional attachments and exhibit low interpersonal behavior. They learn and built the character from their teachers, parents, siblings and friends. Early stage learning helps them to build good personality, which is crucial for societal and nation welfare. Along with this, reaching the milestones as per the guidelines given by World Health Organization (WHO)(Sommer et al., 2021) is essential for the overall growth of the students under this category.

The following Table 2 summarizes the recommendations suggested for various age group toddlers Health Organization (WHO)(Sommer et al., 2021):

Table 1. Recommendations suggested by the WHO for growth of the toddler

Age group	Activities	List of activities	Duration	Activities not suggested
Less than 1 Year	Physical activity Time	Floor based play including Prone time calculated throughout the awake time	Minimum of 30Min	Excess activities and less sleep times
	Sedentary Screen Time	Engaging in reading and story telling	Not more than 1 Hour	Screen time
	Good quality sleep Time	Sleep including naps	14 to 17 Hours(for 0-3 months of age) 12 t0 16 Hours(for 4-11 months of age)	Disturbances
1 to 3 Years	Physical activity Time	Moderate to vigorous activities spread throughout the day. More is suggested	180Min	Very less time
	Sedentary Screen Time	Engaging in reading and story telling	Not more than 1 Hour	Watching TVs or videos and playing video games
	Good quality sleep Time	Sleep including naps	11-14 Hours	Disturbances
3-4 years	Physical activity Time	Variety of physical activities at any intensity, of which at least 60 minutes is moderate- to vigorous- intensity physical activity, spread throughout the day; more is better	At least 180 Min	Very less sleep times
	Sedentary Screen Time	Prams or sit for extended periods of time. Sedentary; less is better. Engaging in reading and storytelling.	Not be restrained for more than 1 hour at a time	screen time should be no more than 1 hour
	Good quality sleep Time	Good quality sleep, which may include a nap, with regular sleep and wake-up times.	10–13 hours of	Disturbances

Although there are CCTV and monitoring staff/ attendants present, it is not all the time possible to monitor various activities of this age group student. To monitor below 5 age group students at play schools and crèches as per the guidelines specified by WHO physically is tedious. A Wearable Device Technology (WDT) or Wearable devices can be used.

The use of Wearable Devices under different category for this age group shown in [REMOVED REF FIELD]Figure 4 includes:

a. Activity Related Wearable Device Technology
b. Health Related Wearable Device Technology
c. Safety Related Wearable Device Technology

Figure 4. Classification based on what is being monitored

a. ***Activity related Wearable Device Technology***: This section presents the some Wearable Devices or Wearable Device Technologies used for monitoring students.
   i. **Smart Hat**(Hsiung Chang et al., 2020): Smart Hat is a IoT and AI enabled device for studying real object name and their purpose, it consists various modules and tools such a camera sensor, Arduino module, Raspberry pi module, open CV tool, Deep learning module and uses Google cloud server for text to speech translation. The kid wearing this smart hat presses the button standing in front of it, to study about that object. Then the camera sensor captures the object and sends the image captured to the deep learning module which is used to identify the image and provides the details of it. These details are uploaded to the Google cloud server for producing corresponding text related info about that object. And the Google Translate API translates the text to Audio, which is played on the small speaker mounted in the Hat by reading out the name and little details of the object. This enables the student at play school to explore more about the objects surrounding him and learn about them.

Figure 5. Student of age 4-year exploring the details about the object

(Hsiung Chang et al., 2020)

The authors here have experimented by mounting the camera sensor for image capturing at three different places: a) mounting on Hat b) mounting on shoulder c) mounting on Hand. The Figure 5 shows the four year student exploring the details about the object in front of her.

ii. **Language Environment Analysis(LENA)**(Choo et al., 2017): A wearable device with supporting smart phone apps was developed by the authors to monitor the maternal language learning skills from their respective mothers. The parents feedback was also recorded, which states that they feel happy while using the app and gadget places in the kid's pocket.

b. ***Health related*:** The best-known health recommendations are to perform at least 60 min of Moderate-To-Vigorous Physical Activity (MVPA) daily, including at least three days of Vigorous Physical Activity (VPA), as established by the Center for Disease Control and Prevention and U.S. Guidelines(*About Physical Activity | Physical Activity | CDC*, n.d.)(HHS, n.d.) aimed at children aged 6–17. However, U.K. Guidelines and E.U. Physical Activity (PA) Guidelines established that for younger ages under five years, it is recommended that a child perform 60 of MVPA daily(Department of Health, n.d.)(Bhawra et al., 2023) and 180 min of total PA per day according to World Health Organization (WHO), and the National Association for Sport and Physical Education (NASPE) recommends an increase to 120 min of MVPA per day for ages 2–3 years and older from the time they start walking by. Another option is to consider PA in steps volume, with recommendations of between 10,000 and 14,000 steps for these ages.

i. **Monitoring Posture and Movement**(Hendry et al., 2023): The Early five years of a child's life are characterized by substantial and rapid neuro-physiological development. An essential aspect of baby development is the ability to go into different postures and perform a required range of movements. Little ones typically develop rapidly to be able to roll from supine to prone (~3–6 months), sit (~5–8 months), and crawl (~6–11 months)(Group & de Onis, 2006). Movement activities continue to increase throughout the toddler's life from birth to preschooler phases of childhood, including learning walk, climbing stairs, running, and jumping. Achieving these posture and movement abilities signals healthy development of the child. Failing to attain or significant delay in development of these skills slowdowns the participation of child in the society and may pay paths towards physical and mental health issues. Therefore there is much interest from health and education professionals and parents in measuring the posture and movement of young children (0–5 years of age).

Wearable sensors known as accelerometers are commonly used to compute time spent at different physical activity intensities.

A systematic review concludes that accelerometers and accompanying software for physical activity were feasible for all-day assessment in children and can provide a good indication of the total amount of activity and temporal patterns of activity(Bruijns et al., 2020). The commercially available software often uses count-based algorithms that sum the data over pre-set time periods (e.g., 15 s), and then uses thresholds to classify this data into different intensities of movement, such as sedentary, light, moderate, or vigorous-intensity activity (Romanzini et al., 2012). Studies utilizing this technology have been pivotal in understanding the link between childhood physical activity and health. However, categorizing children's movements into energy expenditure intensity categories overlooks potentially important aspects of specific postures and movements such as prone lying, sitting, standing, walking and running.

c. ***Safety Related***: Safety of the students is the primary act of any parent and educational institution. There has been many child abuse cases reported all over the world at various instant of times. To counter this act, various wearable devices and the methods are proposed by numerous researchers. The following are the highlighted works in many research papers.
   i. **Stress Monitoring:** To monitor the stress on the students at this age group authors in (Choi et al., n.d.), proposed a three-step stress detection algorithm which is introduced to provide accurate detection performances. The heart

rate and the audio signal from the attached wearable devices are captured using the hardware modules what they developed and later these two signals are processes to understand the condition of stressfulness. The three step method is: First, raw audio signals are preprocessed to extract meaningful features, and are combined with heart rate. Second, features having high discriminative power for the stress and normal states are selected. Then, a classifier is learned from training data composed of the selected features and stress states of children are determined using the classifiers. The Figure 6 shown below explains the frame work of stress monitoring wearable device. (Choi et al., n.d.)

Figure 6. Framework for real time stress monitoring for child using wearable devices

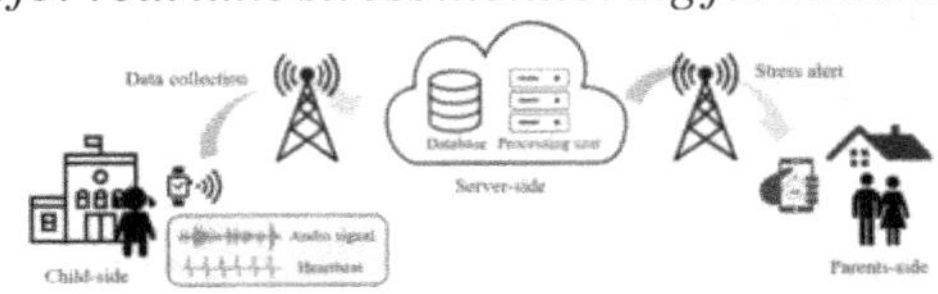

(Choi et al., n.d.)

B. Monitoring Primary School Students (Age Group 5-12 Years)

To monitor some aspects of these age group students, the following activates are identified.

Activities:

a. Interactivity and Attentiveness Related Wearable Device Technology
b. Health Related Wearable Device Technology
c. Safety Related Wearable Device Technology
   a. ***Interactivity and Attentiveness Related Wearable Device Technology:*** Devices used to track the activities are:

Smart Interactive Watch(Liang et al., n.d.): Traditional teacher centric teaching methods are not attracting school students' attention, especially the junior high school students, elementary school students, or even younger students and which is a critical issue for educators. Therefore, authors have developed an intelligent interactive education system that leverages wearable devices (smart watches) to accurately capture hand gestures of school students and respond instantly to teachers so as to increase the interaction and attraction of school students in classrooms. In addition to this, through physical information of school students from their smart watches, it can found out the root points of the learning process according to the deep data analysis. In this way, it can provide teachers to make instant adjustments and suggest school students to achieve multi-learning and innovative thinking. The developed system is mainly consists of three components: (1) smart interactive watch; (2) teacher-side smart application (mobile-app); and (3) cloud-based analysis system. Specifically, the smart interactive watch is responsible for detecting the physical information and interaction results of school students, and then giving feedback to the teachers. The teacher-side app will provide real-time learning suggestions to adjust the teaching pace to avoid learning disability. The cloud-based analysis system provides intelligent learning advices, academic performance prediction and anomaly learning detection. This kind of can potentially enhance teaching and learning processes for both educators and school students. The Figure 7 shows the model proposed by the authors.

Figure 7. System architecture using smart interactive watch

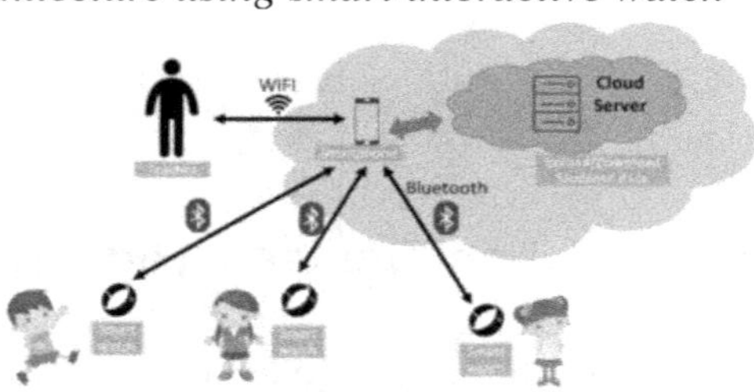

b. ***Health Related Wearable Device Technology:*** This health related wearable devices are similar to the one discussed in section-A
c. ***Safety Related Wearable Device Technology:*** This safety related wearable devices are similar to the one discussed in section-A

C. Monitoring Middle School Students (Age Group 12-14 Years)

The middle school children are one who enter into their teens and most likely to develop changes in the body physically and mentally. Due to these change, they may gain or lose weight and other challenging kind of issues which are highlighted in this section.

a. ***Health Related monitoring*:** The Health related monitoring of these group students needs wearable devices which need to be different than that of the one studied in previous sections as these group of students enter into the teens, where extra monitoring is required.

   i. **Obesity and Overweight Monitoring**(Emilio Standoli et al., 2016): The teen agers in this age group develop over weight and obesity issues due the psychological pressure from their livelihood and biological changes in their body. To assess the effect of obesity and overweight, a project named PEGASO was proposed—an FP7 project funded by the European Union (EU)—tackles prevention, focusing on obesity and related co-morbidities, in which lifestyle coaching plays a very important role by authors in(Emilio Standoli et al., 2016) . This goal is achieved by developing a platform that—uses mobile and ICT technologies and through wearable sensors and mobile apps—supports young students in becoming alert of risks and motivates them in a behavior change path towards healthy lifestyles. PEGASO provides guidance towards inculcating good habits and provides a social platform to stimulate young student's willingness to engage enthusiastically in their health management.

The architecture of PEGASO system is based on a mobile platform that gathers and integrates data coming from physical measures, healthy behaviors as well as environmental responses, which is shown in Figure 8. Three levels of sensors are used to gather this data. Firstly there is a smart phone which remains the central and principal part of the system. Usually teenagers possess and utilize smart phones which have installed PEGASO applications that support its functionalities and enable it offer the minimum amount of information needed in order to allow its operations. A smart wristband can be referred to as second tier PEGASO sensors. This wearable device has an activity tracker capable of measuring one's movements throughout the day, utilizing, three-axis accelerometer while calculating energy expenditure. Users could wear it and all the data acquired are transferred to a dedicated app via Bluetooth. The third level of PEGASO sensor system consists of wearable electronic systems and intelligent clothes. This portable equipment is meant for short-term monitoring of physiological parameters during sports practice. The electronic module collects information about heart activity (with I-lead ECG), body movement (thanks to 3-axis accelerometer) and calorie burn. Attached with two press-studs to garments, this gadget sends analyzed information through wires.

Figure 8. PEGASO sensors system

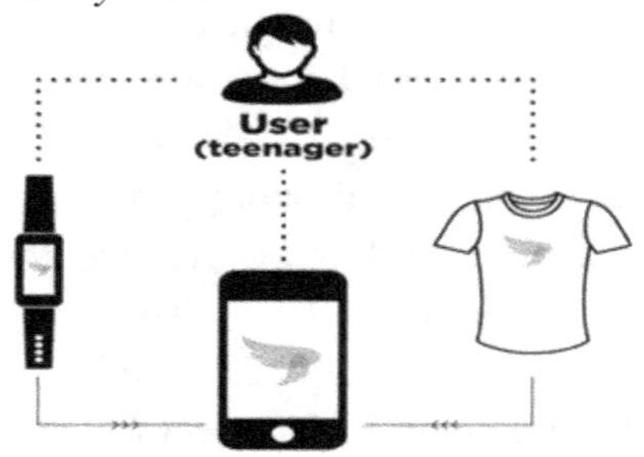

(Emilio Standoli et al., 2016)

The design samples of some PEGASO garments (two sport t-shirts) were shown in Figure 9, to evaluate the pivotal features that the teenager students consider significant to improve the acceptance of the wearable clothes. Multiple samples of garments were developed and presented to the students group: one set of garment was made with the "visible sensors" concept, the other one following the "hidden sensors" vision. .

Figure 9. The contrast of the two male versions of the first set of sensorized t-shirts: (a) the style with technical textile and the visible device to be attached to the two snap buttons; (b) the style with a more fashionable color and the sensing device hidden

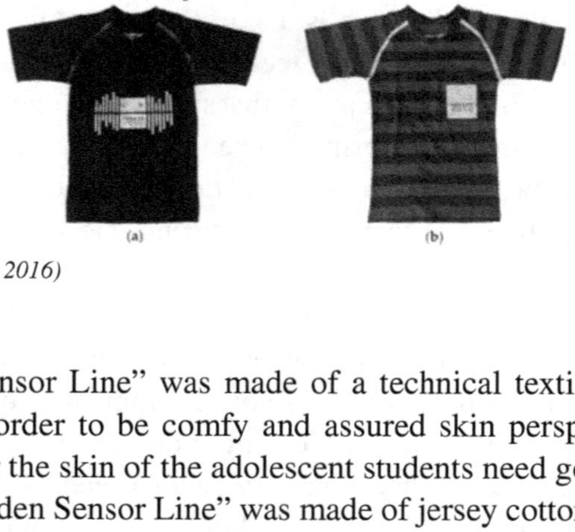

(Emilio Standoli et al., 2016)

The "Visible Sensor Line" was made of a technical textile, used for trained sport activities, in order to be comfy and assured skin perspiration. For proper sensor stability over the skin of the adolescent students need good elasticity of the garments. The "Hidden Sensor Line" was made of jersey cotton, in order to have a look like everyday regular garments. The device is placed hidden in a pocket on the chest. Two textile electrodes are embedded in the smart garments: they sense the recording of the I-leading ECG signal, along with respiration; the sensing device also consists of a three-axis Inertial Monitoring Unit (IMU) to detect and notify activity and movements. The smart garments are later connected with the electronic wearable device by the two snap buttons.

The PEGASO framework was tested upon secondary school students during the pilot test stage. Four studies were carried out involving about 400 students in Italy (Lombardy), Spain (Catalonia) and the United Kingdom (England/Scotland).

b. ***Physical fitness related monitoring***: Physical fitness is most essential for any child for their overall success

   i. **(Smart Bracelet System)**(Guo et al., 2022): The lack in accurate computational models and in-depth data analyses for physical fitness and to solve the challenge faced by the traditional methods for monitoring students at teenage,, the authors in (Guo et al., 2022), propose a novel evaluation model for predicting the physical fitness of teenagers. Initially, the PPGs samples from 1024 teenagers were collected under the supervision of the proposed three-stage running paradigm. After that, the median filter and wavelet transform are applied on them to denoise the original signals and find HR and SpO_2. Then, the Pearson correlation coefficient method was used to finalize the feature set, based on the extracted nine physical features. Finally, a 1D-CNN was built with LSTM model to classify teenagers' physical fitness condition into four levels: excellent, good, medium, and poor. The accuracies obtained are 98.27% and 99.26% for boys' and girls' physical fitness prediction correspondingly. The experimental results have provided evidence and proof for supporting the feasibility of predicting teenagers' physical fitness levels by using their running PPG recordings.

The bulk transmission of PPG recordings from the smart bracelets to the personal computer (PC) was achieved by Self-designed smart bracelet system which is displayed in Figure 10. The PPG voltage value recordings 25 times per second are collected by the smart bracelets.

Figure 10. Smart bracelet system

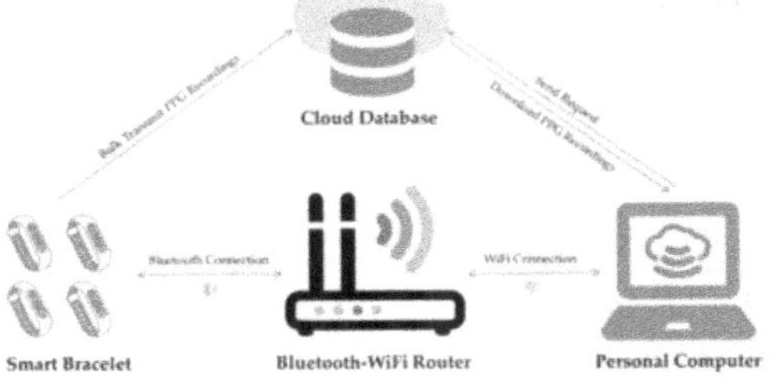

(Guo et al., 2022)

The collected student's PPG recordings with the help of smart bracelets are then transmitted to the self-designed system for further analysis.

Below is an explanation of how the entire system works efficiently

1. The Bluetooth-enabled wrist bands connect wirelessly to the Bluetooth Wi-Fi router and are always on. The personal computers are connected to the same router through Wi-Fi, thus enabling all devices such as bracelets, routers and PCs to be within a single Wi-Fi environment.
2. PPGs values are transmitted automatically in bulk to the cloud database via Bluetooth.
3. When a request is sent from a local computer requesting for some data from cloud database, one receives original signals recorded by all used bracelets at their location.

The teenagers' physical fitness conditions are classified into four levels: excellent, good, medium and poor. The developed framework for teenager physical fitness evaluation system for physiological recordings, are shown in Figure 11. First, the fourteen year old middle school pupils' PPGs were recorded under designed experimental paradigm. Second, MF-WT method was applied to preprocess the collected PPGs. Later, the HR and SpO2 were calculated based on preprocessed signals and extracted nine useful features from it. Finally PCC was used to finalize the feature set. At last, a 1D-CNN along with LSTM machine learning evaluation model was developed to classify the teenagers' physical fitness condition into four different levels.

Figure 11. Framework of the proposed evaluation system

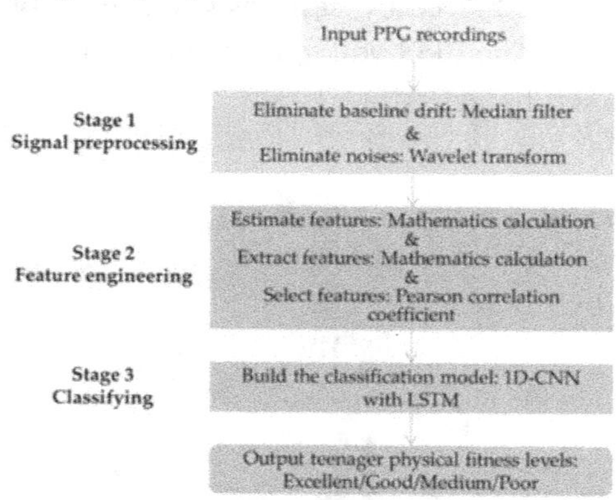

(Guo et al., 2022)

The experimental setup was designed according to the three aspects as per the physical Health standards in china. Volunteer students were allowed to wear smart bracelets and made them to do three-minute warm-up exercises before they are instructed for official run. Later the volunteers were divided into two groups as per their gender, the boys were instructed to run 1000 m while the girls to run 800 m, respectively. Ten Students maximum per experiment was considered. After that the students were suggested to take rest, so as to let their hurried breathing and heartbeat sluggish fall until they were recovered to the normal fitness condition. The flowchart is shown in Figure 12.

Figure 12. Flowchart of the experimental paradigm

(Guo et al., 2022)

The performance measures of the designed model like accuracy, precision, recall and f1-score for every class was calculated for better classification of students from different gender. The physical fitness evaluation accuracy values for boys' and girls' are 98.27% and 99.26%, correspondingly.

c. ***Attentiveness related monitoring***(Ayearst et al., 2023): As per DSM-5 diagnosis for Attention-Deficit/Hyperactivity Disorder (ADHD), there are difficulties in one or both two factors: attention and hyperactivity. Signs such as impulsivity and hyperactivity are more easily visible and noted by the overall behavioral trouble that these symptoms may cause. Students possessing ADHD may appear to be paying attention, but in actuality their mind won't be stable, In other words they appear to be like they are out of the task because they are moving around when, in fact, they are processing what is going on. consequently an individual students with ADHD who suffers attention problems like patients with "Cognitive Disengagement Syndrome" (earlier called "Sluggish Cognitive Tempo"; or elder adults of age >55 years be apt to be under diagnosed.

The majority of the parents, who have trained in psychological treatments for their kids with ADHD, primarily focus on behavior management. Attention deficiencies of psychological interventions are effective when they are administered in classrooms as per some evidences, which also includes attention deficit as it occurs either in the context of combined or inattentive presentations. Recently, several digital wearables were created attempting to train people's attention using digital games. Initially, some of the intervention that appeared promising has ended up failing to show far-transfer effects to real-life outcomes after randomized controlled trials (RCTs). Latest studies conducted have demonstrated how a computerized game can affect outcomes as indicated by a subtest from TOVA – Test of Variables of Attention – an ADHD evaluation test(Llorente et al., 2001). Nevertheless, the computer game unable to show success for ADHD symptoms for complete or functional deficit, compared to control groups. It is hypothesized that one possible cause for this inability to demonstrate any far transfer effects could be due to the fact that interventions are mostly limited to home settings only.

i. **Wearable Watch(Revibe Connect)**(*Attention-Deficit/Hyperactivity Disorder (ADHD)*, n.d.): The Revibe Connect, a wearable watch-like device produces unique vibration signals that create meta-cognitive awareness and self-realization improvises attention capabilities. The wearable vibrates sporadically and signals the child to "tap back" to alert for paying attention. The developed wearable targets the attention and reduces off-task behavior by combining cutting edge technological invention like biosensors embed in wearable device, which are susceptible to physiological parameters, algorithmically-guided adaptive response, and machine learning supported digital assessment and feedback. It learns the student's behavior over and on time, provides a custom-made, adaptive, and modest intrusion at the time of act.

A Revibe Connect wearable provides various functions therapeutically. One of which is "pay attention" remainders from a person to a device. This reduces the cost and burden on stigmatizing teacher or paraprofessional support in classroom by increased on-task behavior by that device, which operates continuously with high frequency at that moment on spot in the classroom. This increases the potential to train attention to ADHD victims in the classroom sitting. The wearable continuously collects data to examine attention and aids the caregivers and teachers to become more aware of the ecological factors of attention symptoms.

D. Monitoring Junior Secondary School Students (Age Group 14-16 Years)

This category of students undergo immense pressure as they need to appear for school board exams and entrance exams for pre-university studies and competitive exams for entry into top universities. They might have spent numerous sleepless nights for engaging themselves into thorough preparations. The investigations suggest them to use wearable devices to monitor their health and fitness alongside their studies.

a. ***Monitoring Optimal sleep***(Wang & Chen, 2023)**:** Sleep has always been an essential part of human life. Everyone needs sleep irrespective of your profession or age or status. The students in present scenarios use of laptops and mobile phones may be for the study purpose, which make them to expose to blue light radiation. Studies exhibit that; extreme exposure to blue light can cause secretion of melatonin that impairs sleep.

Sleep health monitoring is achieved by using equipment which continuously updates physiological information. There were three types of equipment utilized for gathering information on physiological characteristics. They are a).The MP150, device from Biopac Student Lab of United States b).The wGT3X-BT: wrist monitor by ActiGraph shown in Figure 13 and c).The smart watch (Series 5) from Apple Inc., USA, shown in Figure 14. The MP150 senses and stores physiological signals like waves of brain, electrocardiograms waves from heart and electromyography signals from body. AcqKnowledge® 5.0 software interprets these signals. And for monitoring daily activities and sleep-related physiological info the wGT3X-BT wrist monitor is used. The examination and normalized movement patterns and their intensity during wear is done by the ActiLife 6.0 softwar. Sleep cycles, sleep efficiency, swing during sleep and latency are also assessed by it.

Figure 13. The wGT3X-BT: A biophysiological information gathering equipment. (a) An ActiGraph manufactured GT3x-BT wrist-meter host. (b) An ActiGraph: The ActiLife analysis software

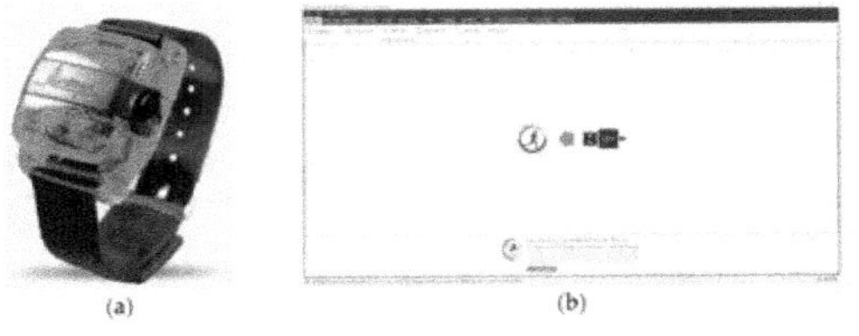

(Wang & Chen, 2023)

A smart watch from Apple (Series 5) commercial graded with multiple features. It possesses advanced technology for heart rate storage using the optical volume variation method from optical science.

Figure 14. Apple Watch 5 by Apple Inc.

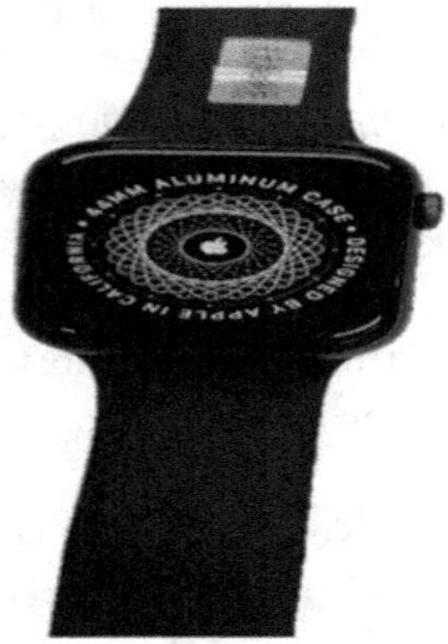

(Wang & Chen, 2023)

3. MONITORING SPORTS STUDENTS USING WEARABLE DEVICES

Wearable technology has become essential for improvising performance in sports by using real-time data investigation and examination. Training efficiencies and completion outputs of the athletes are relied upon wearable sensors. The overall performance optimization, injury prevention and rehabilitation of athletes are measured and scrutinized body wearable sensors.

Figure 15. Layered structure of wearable computing

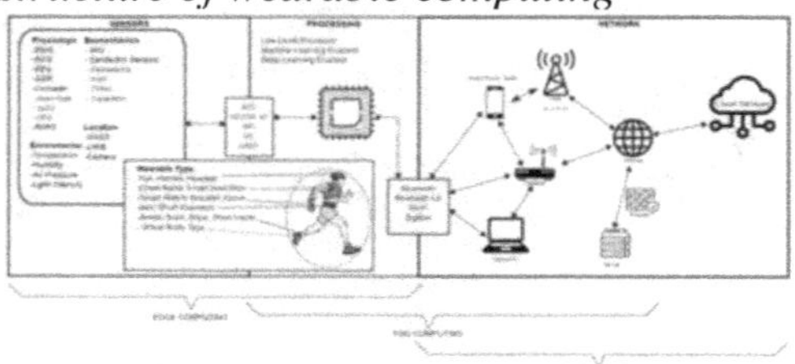

(Wang & Chen, 2023)

A wearable technology device consists of three layers shown in Figure 15: sensor layer, processing layer and network layer. Sensing and processing happens on electronic hardware under the first two layers. And network layer integrates various devices, computational methods, and communication protocols (Seçkin et al., 2023). The Sensor layer consists of Physiological Sensors, Biomechanics Sensors, Location Sensors and Environmental Condition Sensors.

Table 2. Commercially available wearable tracking device

S. No.	Commercially available Device	Sports	Wearing place	Application / purpose
1.	MotusBASEBALL (MOTUS) wearable	Baseball	The athlete has to insert this sensor it into a sleeve positioned at the throwing arms distal end.	It is used to measure various parameters and is useful for monitoring, preventing and rehabilitating, and injuries in baseball players.
2.	KINEXON wearable	Widely used system in handball, football, basketball, and ice hockey.	It is worn on the upper body garments of the athletes	It tracks player position, direction, heart rate and movements. Also measures physical distance and time among players.
3.	ZEPP-Golf wearable	Golf	It is worn on the outside of the leading arm glove	It is useful by providing golf related metrics like club speed, hand plane, downswing, backswing, hip rotation, tempo ratio etc.
4.	Clearsky T6, Optimeye S5, and Optimeye S6 (wearable products of Catapult)	Basketball		It provides current info on player positions, speed, acceleration, distance, and heart rate. Its intention is to avoid injuries while complex sports generating alerts to lessen injuries by interpreting data.
5.	Viper Pod	It is extensively used by top sports clubs, primarily in first-league football and basketball franchises	The devices can be comfortably worn on the player's body using a belt	It measures distance, speed, accelerations, high metabolic load distance, decelerations, heart rate and dynamic stress load.
6.	Lumo Run	Athletics	An embedded system consists of a small sensor that is attached to any pair of shorts or tights working like a personal coach.	It gives real time aural feedback to correct running form, lessen the risk of hurt, and optimistically helps to run faster.

The motor competence is an essential feature of physical activity which must be improved during school years especially for sports aspirants. Wearable devices like smart bands monitor student's imperative health info, rate of heartbeat, using a blend of sensors and machine-learning technology. With regard to school children, smart bands are used to track physical activities levels, patterns of sleep and complete health grade. Classification in machine learning was used to investigate the father data by smart bands to identify patterns that help improvement in health and goodness of sports children.

Researcher in (Sulla-Torres et al., 2024) explored the use of smart band wearable device and the application of machine learning to assess physical activities and motor competencies in educational and sports centers. They examined the levels of motor competence using smart bands, hyper-parameter optimization for machine learning, and the benefits of using them in educational and sports centers. Smart bands using machine-learning for schoolchildren involves numerous steps such as data gathering and interpretation. Smart bands are contains sensors to collect data of physiological parameters like heart rate and physical activity levels. The collected data is interpreted to, classify patterns.

Anthropometric measurements and smart band utilization followed the principles of the Declaration of Helsinki (World Medical Association) concerning ethical standards for human research. The Table 4 below following attributes was used for data collection.

Table 3. Parameters for physical activity measurement

S.No	Physical activities	Measurements of a sports student during activity
1.	Anthropometric info	Age, height, weight, sitting height and waist circumference.
2.	Avg. pace	Walk time per kilometer.
3.	Avg. cadence	Steps per minute.
4.	Steps	Steps taken during the activity
5.	Calories	Calories burned during the activity
6.	Avg. speed	Average speed moved in kilometers/hour
7.	Avg. stride	Average distance in each step
8.	Heart rate	Beats per minute of heart
9.	Max. heart rate	Maximum number of beats per minute

The reasonable use of smart band wearable device could able to collect the necessary information and obviously the machine learning algorithms at the background helped for assessment of motor tests for sports which is shown below in Figure 16.

Figure 16. Modeling of data for sports students

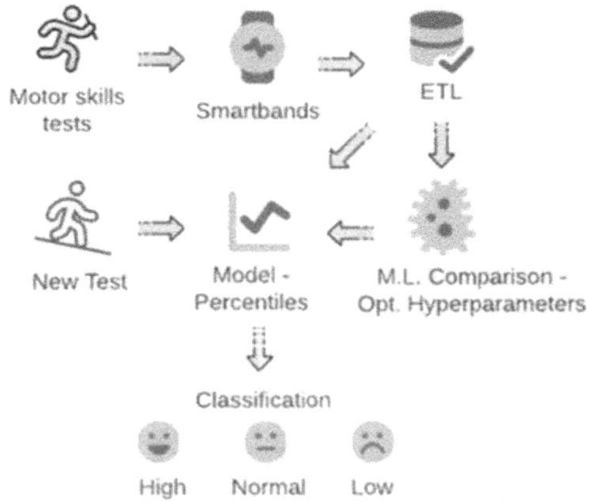

(Wang & Chen, 2023)

4. MONITORING MENTALLY DISABLED STUDENTS USING WEARABLE DEVICES

To improve the learning process of mentally disabled students in special schools, a system which recognizes the brain's current emotional state is essential. From past many years, the teachers at special school commonly use the conventional methods for the purpose of educating mentally disabled students.

A. Computer Aided Education System for Mentally Disabled (Mehmood & Lee, 2017)

A novel computer aided method for instructors (special education teachers) at special schools was proposed, where they teach special students the proposed system using wearable technologies. Students and instructors are the prime users of the system. Teachers directly connect with the system as well as with students. The brain signal device is mounted on the mentally disabled student during the class work. The special education teachers monitor the emotional changes in the disabled students with the help of the wirelessly connected head mount device and the application software. After carefully analysis of the subject under study, the special education teachers record and suggest the treatment procedures when the mental condition of the subject worsens. The central database is used to store all treatment related information which can be accessible on all times.

The Figure 17 demonstrates the Computer Aided Education System (CAES) flow diagram.

- The instructor helps the disabled student in complete process as a system admin.
- The subject's internal brain activity generates a different signal patterns.
- Brain signals are recorded from the scalp of subject's brain.
- The emotional behaviors like happy, calm, sad, and scared are recognized during the test.
- Based on behavior or emotion of the each subject, the corresponding psychological treatment is suggested.
- The instructor then adds or updates the treatment procedure for the every subject in the system, which are finally recorded in the database of the central repository.

Figure 17. Process flow diagram of CAES

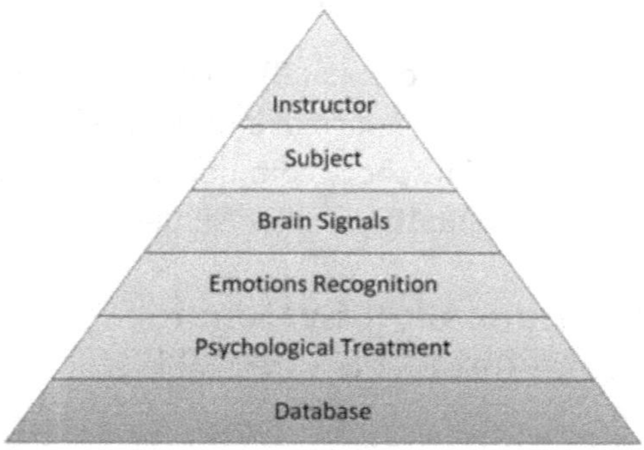

(Mehmood & Lee, 2017)

The system mainly constitutes the important features listed below.

- User (Student) Information Module: This contain existing subject details, also allows creating new subject detail by connecting the brain side device.
- Device Control Module (Head mount): This visualizes the status of all electrodes and displays the brain activity signal out from each electrode on 2D chart.
- Events Module: This is used to record actions of system: device connection, exceptions, training, user log-in, user log-off, etc.

- Expressive Module: This records and displays the current mood of the subject. Also allows the special education teacher to modify the treatment procedure of current subject.

The Figure 18 presents the signal flow of the computer aided education system for mentally disabled.

Figure 18. Computer-aided education system for mentally disabled

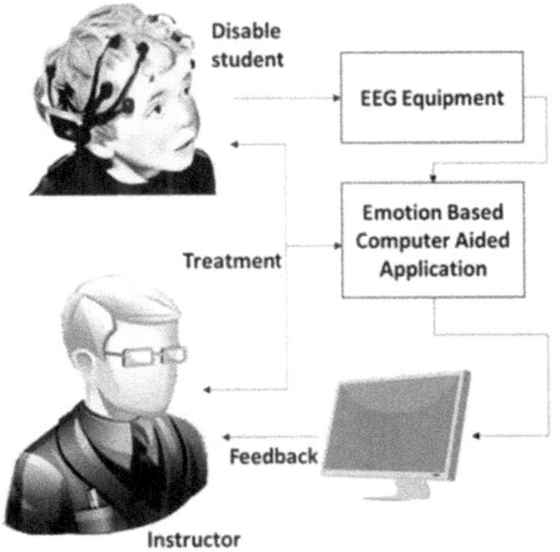

(Mehmood & Lee, 2017)

B. Monitoring Challenging Behaviors Among Children with Autism (Alban et al., 2023)

There is a growing interest in incorporating wearable sensors in autism therapies for the students who have developed autistic behavior either by nature or by circumstances they live in. Research was performed on wearable devices by many researchers so as to cater the autistic problems in students. Wearable technology along with the machine and deep learning techniques are used to interpret and analyze the behavior of the autism students.

The stereotypical motor movements detection in children having autism using three axis accelerometers that were worn on different body parts like wrists and torso was investigated in (Goodwin et al., 2011).

In (Pollreisz & TaheriNejad, 2017) an Empatica E4 wristband smart watch for developing emotion recognition system by collecting various parameters from it such as electro-dermal activity, heart rate, and temperature values from ten young adults was used. The success rate of 65% using decision tree algorithm was recorded for emotion recognition. The study investigated in (Jarraya et al., 2021), on different emotions faced during meltdowns and developed an recognition system for motions via machine learning methods. Neural network based emotion identification was performed by (Lee et al., 2006) using the measured heart rate variability and galvanic skin response. Their results show that the change in the heart rate is an early sign for change in the behavior.

Social robots and regular children's toys were used for analyzing the autistic behavior. The robots used are humanoid Nao robot from SoftBank Robotics, Tokyo, Japan with a white furred robotic seal from PARP robots, Itasca, IL, United States. Figure 19a shows the wrist wearable device, the Figure 19b social robots and the Figure 19c shows the recorded data of various sensors used. The toy robots contain a green spongy rubber ball, brass cymbals, multicolor train and wooden letter blocks placed on a toy truck.

A sensor on wristband wearable watch (Empatica E4, Milano, Italy) was used to obtain the data readings from the children during the experiments. The E4 wearable device sensor contains a real-time clock and records physiological data signals to an internal memory. The physiological signals considered are Acceleration (ACC), Electrodermal Activity (EDA), Temperature (TEMP), Heart Rate (HR) and Blood Volume Pulse (BVP) as shown in Figure 20c

Figure 19. An adopted methodology overview .(a) The wearable device: Empatica E4. (b) Kid interaction with the social robot. (c) Acquired data from wearable device

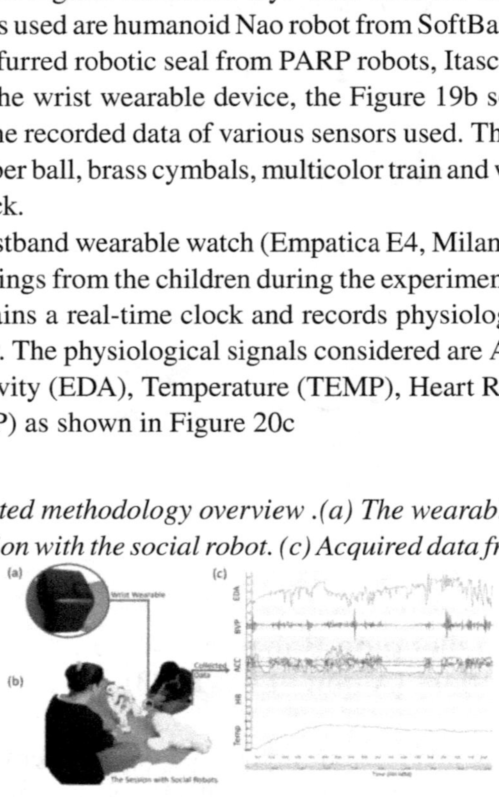

(Alban et al., 2023)

The Heart Rate Variability (HRV) is yet another measurement for measurement of multiple values for sake of health in autism students (K. Li et al., 2023). The HRV based on calculating the Root Mean Square of Successful Differences (RMSSD) was measured to investigate the significance of heart rate parameters. The RMSSD was calculated form inter beat interval signal sampled by 64 Hz frequency. The changes in HRV for one of the children during different states were calculated in Figure 20a-c. During the case of a challenging behavior, the HRV values appear to be highest as in Figure 20.

Figure 20. The HRV changes corresponding to different states. (a) The simulation of the overwhelmed child using bubble gun toy. (b) The rest state of the child. (c) The challenging behavior of child

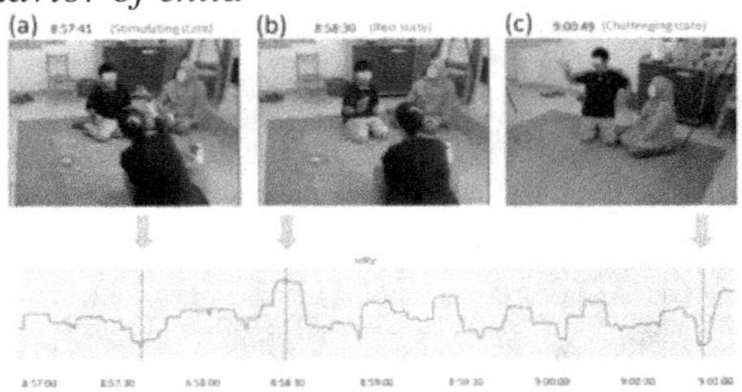

(Alban et al., 2023)

5. SUMMARY

Various categories of students under different age groups use different wearable devices. The Table 4 briefs the devices used for analyzing the different students' various activities:

Table 4. Summary of the wearable devices used for monitoring various category of students

Student category	Monitoring parameter	Device used	Sensors / sensing parameter	Purpose	Remarks
Play group and Kindergarten students	Activity related	Smart Hat	Video camera.	Learning about physical things around the kid	Head mounted
		Smart phone	Smart phone app	Analyses the maternal language learning skills	Kept in Pocket
	Health related	Posture and movement	Accelerometer and required software	Monitoring prone lying, sitting, standing, walking and running.	Body wear
	Safety related	Heart rate and audio signal	Heart rate sensor and audio recorder	Stress monitoring	Wrist wear

continued on following page

Table 4. Continued

Student category	Monitoring parameter	Device used	Sensors / sensing parameter	Purpose	Remarks
Primary school students	Interactivity and Attentiveness Related	Smart watch	hand gestures	Interactivity and attentiveness in the classroom	Wrist wear
	Health related	Similar to the play group	Similar to the play group	Similar to the play group	Similar to the play group
	Safety related	Similar to the play group	Similar to the play group	Similar to the play group	Similar to the play group
Middle school students	Obesity and Overweight Monitoring	Smart T-shirt, Smart watch, 3-axis accelerometer	Heart activity with I-Led ECG	Monitoring physiological parameters during sports practice	Body wear
	Physical fitness related monitoring	Smart bracelet	Heart rate sensor and SPO2 sensor	PPG recordings from bracelet	Wrist wear
	Attentiveness related monitoring	Wearable Watch	Vibration sensor	alert for paying attention	Wrist wear
Junior Secondary school students	Optimal sleep	MP150 device, wGT3X-BT wrist monitor and The smart watch (Series 5) from Apple Inc.	waves of brain, electrocardiograms waves from heart and electromyography signals from body	Assess Sleep cycles, sleep efficiency, swing during sleep and latency	Wrist wear
Monitoring Sports Students	physical activities and motor competencies	smart bands	Physiological Sensors, Biomechanics Sensors, Location Sensors and Environmental Condition Sensors	Various parameters in different sports	Wrist wear, Chest wear, leg wear etc.
Monitoring Mentally Disabled Students	Education System among students	EEG device	electrodes and displays	Analyzing brain activity signal	Head mounted
	Challenging Behaviors among Children	E4 wearable device sensor	Acceleration (ACC), Electro-dermal Activity (EDA), Temperature (TEMP), Heart Rate (HR) and Blood Volume Pulse (BVP)	Heart Rate Variability	Wrist wear

6. CONCLUSION

This chapter has elevated the significance and effectiveness of wearable devices for students of various domains i.e. academic students at academic institutions, sports students at sports academies and sports clubs, and both mentally and physically challenged students at their respective special schools. The students need the aid of wearable devices and its associated technologies like Internet of things, latest communication protocols either wired or wireless for monitoring different kinds of health related, safety related, physical activities related, and psychological activity related parameters for their wellbeing and corrective measures. After collecting various parameters, they are analyzed by using computer aided software and various cutting edge technologies like machine learning algorithms and Artificial Intelligence algorithms to figure out the lacuna or achievement in the studies and learning skills of the corresponding students individually.

In future the use of wearable device on students will help analyze their abilities, capabilities and strengths of students by latest technologies like Artificial Intelligence and Machine Learning algorithms with the data gathered by Internet of Things enabled wearable devices. The data gathered at cloud are processed, analyzed and interpreted by these algorithms to come out with the usable decisions and suggestions. Evaluation of student's intelligence, analysis of emotional behavior of individual students, understanding emotions and skill improvement in mentally ill students may be carried out with the future generation wearable devices. Now the future work on wearable devices uses fine processing of data by using reinforce learning and advanced AI algorithms. Some of the recent research works are presented here which may be useful for understanding the prospectus of wearable devices for students. The integration of IoT in education system enhance the overall teaching and learning process, the work in (Hercog et al., 2023) discusses the use of latest technologies in education system in various aspects like providing remote lab facility, learning with augmented reality, management of students attendance, creating smart environment for learning support. Using deep learning techniques the student attention monitoring is performed in (Trabelsi et al., 2023). The anxiety and depression are most common in college students, the monitoring of these with multi-layer perceptron based wearable devices are proposed in (X. Li et al., 2024). Physical activities in pre-school and primary school are very much lacking due to easy availability of internet connected TV and smart phone, the systematic review is presented in (Sousa et al., 2023) for monitoring and assessment of physical activities in them.

REFERENCES

About Physical Activity | Physical Activity | CDC. (n.d.). Retrieved September 11, 2024, from https://www.cdc.gov/physicalactivity/php/about/index.html?CDC_AAref_Val=https://www.cdc.gov/physicalactivity/basics/pdfs/FrameworkGraphicV9.pdf

Alban, A. Q., Alhaddad, A. Y., Al-Ali, A., So, W. C., Connor, O., Ayesh, M., Ahmed Qidwai, U., & Cabibihan, J. J. (2023). Heart Rate as a Predictor of Challenging Behaviours among Children with Autism from Wearable Sensors in Social Robot Interactions. *Robotics (Basel, Switzerland)*, 12(2), 55. Advance online publication. DOI: 10.3390/robotics12020055

Attention-Deficit/Hyperactivity Disorder (ADHD). (n.d.). Retrieved September 15, 2024, from https://my.clevelandclinic.org/health/diseases/4784-attention-deficithyperactivity-disorder-adhd

Ayearst, L. E., Brancaccio, R., & Weiss, M. D. (2023). An Open-Label Study of a Wearable Device Targeting ADHD, Executive Function, and Academic Performance. *Brain Sciences*, 13(12), 1728. Advance online publication. DOI: 10.3390/brainsci13121728 PMID: 38137176

Bhawra, J., Khadilkar, A., Krishnaveni, G. V., Kumaran, K., & Katapally, T. R. (2023). The 2022 India Report Card on physical activity for children and adolescents. *Journal of Exercise Science and Fitness*, 21(1), 74–82. DOI: 10.1016/j.jesf.2022.10.013 PMID: 36408207

Bruijns, B. A., Truelove, S., Johnson, A. M., Gilliland, J., & Tucker, P. (2020). Infants' and toddlers' physical activity and sedentary time as measured by accelerometry: A systematic review and meta-analysis. *The International Journal of Behavioral Nutrition and Physical Activity*, 17(1), 14. DOI: 10.1186/s12966-020-0912-4 PMID: 32028975

Choi, Y., Jeon, Y., Wang, L., Sensors, K. K.-, & 2017, undefined. (n.d.). A biological signal-based stress monitoring framework for children using wearable devices. *Mdpi. ComY Choi, YM Jeon, L Wang, K KimSensors, 2017•mdpi.Com*. DOI: 10.3390/s17091936

Choo, D., Dettman, S., & Care, R. D.-… and C., & 2017, undefined. (2017). Talking to toddlers: Drawing on mothers' perceptions of using wearable and mobile technology in the home. *Ebooks.Iospress.NlD Choo, S Dettman, R Dowell, R CowanIntegrating and Connecting Care, 2017•ebooks.Iospress.Nl*, 239, 21–27. DOI: 10.3233/978-1-61499-783-2-21

Cowperthwait, A. L., Campagnola, N., Doll, E. J., Downs, R. G., Hott, N. E., Kelly, S. C., Montoya, A., Bucha, A. C., Wang, L., & Buckley, J. M. (2015). Tracheostomy overlay system: An effective learning device using standardized patients. *Clinical Simulation in Nursing*, 11(5), 253–258. DOI: 10.1016/j.ecns.2015.03.001

Department of Health. T. (n.d.). *Start Active, Stay Active publications/physical-activity-guidelines-uk-chief-medical-officers-report.* Retrieved September 11, 2024, from https://www.gov.uk/government/

Emilio Standoli, C., Renata Guarneri, M., Perego, P., Mazzola, M., Mazzola, A., Andreoni, G., & Sp, N. (2016). *A Smart Wearable Sensor System for Counter-Fighting Overweight in Teenagers.* DOI: 10.3390/s16081220

Goodwin, M. S., Intille, S. S., Albinali, F., & Velicer, W. F. (2011). Automated detection of stereotypical motor movements. *Journal of Autism and Developmental Disorders*, 41(6), 770–782. DOI: 10.1007/s10803-010-1102-z PMID: 20839042

Group, W. H. O. M. G. R. S., & de Onis, M. (2006). WHO Motor Development Study: Windows of achievement for six gross motor development milestones. *Acta Paediatrica (Oslo, Norway)*, 95(S450), 86–95. DOI: 10.1111/j.1651-2227.2006.tb02379.x PMID: 16817682

Guo, J., Wan, B., Zheng, S., Song, A., & Huang, W. (2022). A Teenager Physical Fitness Evaluation Model Based on 1D-CNN with LSTM and Wearable Running PPG Recordings. *Biosensors (Basel)*, 12(4), 202. Advance online publication. DOI: 10.3390/bios12040202 PMID: 35448262

Hendry, D., Rohl, A. L., Rasmussen, C. L., Zabatiero, J., Cliff, D. P., Smith, S. S., Mackenzie, J., Pattinson, C. L., Straker, L., & Campbell, A. (2023). Objective Measurement of Posture and Movement in Young Children Using Wearable Sensors and Customised Mathematical Approaches: A Systematic Review. *Sensors (Basel)*, 23(24), 9661. Advance online publication. DOI: 10.3390/s23249661 PMID: 38139507

Hercog, D., Lerher, T., Truntič, M., & Težak, O. (2023). Design and Implementation of ESP32-Based IoT Devices. *Sensors (Basel)*, 23(15), 6739. Advance online publication. DOI: 10.3390/s23156739 PMID: 37571523

HHS. (n.d.). *Physical Activity Guidelines for Americans 2 nd edition.*

Hsiung Chang, I., Keh, H. C., Dande, B., & Roy, D. S. (2020). Smart hat: Design and implementation of a wearable learning device for kids using AI and IoTS techniques. *Journal of Internet Technology*, 21(2), 593–604. DOI: 10.3966/160792642020032102026

Jarraya, S. K., Masmoudi, M., & Hammami, M. (2021). A comparative study of Autistic Children Emotion recognition based on Spatio-Temporal and Deep analysis of facial expressions features during a Meltdown Crisis. *Multimedia Tools and Applications*, 80(1), 83–125. DOI: 10.1007/s11042-020-09451-y

Khosravi, S., Bailey, S. G., Parvizi, H., & Ghannam, R. (2022). Wearable Sensors for Learning Enhancement in Higher Education. *Sensors (Basel)*, 22(19), 1–17. DOI: 10.3390/s22197633 PMID: 36236732

Laverde, R., Rueda, C., Amado, L., Rojas, D., & Altuve, M. (2018). Artificial neural network for laparoscopic skills classification using motion signals from apple watch. *2018 40th Annual International Conference of the Ieee Engineering in Medicine and Biology Society (Embc)*, 5434–5437.

Lee, C., Yoo, S. K., Park, Y., Kim, N., Jeong, K., & Lee, B. (2006). Using neural network to recognize human emotions from heart rate variability and skin resistance. *2005 IEEE Engineering in Medicine and Biology 27th Annual Conference*, 5523–5525.

Li, K., Cardoso, C., Moctezuma-Ramirez, A., Elgalad, A., & Perin, E. (2023). Heart Rate Variability Measurement through a Smart Wearable Device: Another Breakthrough for Personal Health Monitoring? *International Journal of Environmental Research and Public Health*, 20(24), 7146. DOI: 10.3390/ijerph20247146 PMID: 38131698

Li, X., Zou, L., & Li, H. (2024). Multilayer Perceptron-Based Wearable Exercise-Related Heart Rate Variability Predicts Anxiety and Depression in College Students. *Sensors (Basel)*, 24(13), 4203. DOI: 10.3390/s24134203 PMID: 39000984

Liang, J.-M., Su, W.-C., Chen, Y.-L., Wu, S.-L., & Chen, J.-J. (n.d.). *Smart Interactive Education System Based on Wearable Devices*. DOI: 10.3390/s19153260

Hernández-Mustieles, M. A., Lima-Carmona, Y. E., Pacheco-Ramírez, M. A., Mendoza-Armenta, A. A., Romero-Gómez, J. E., Cruz-Gómez, C. F., & Lozoya-Santos, J. D. J. (2024). Wearable Biosensor Technology in Education: A Systematic Review. *Sensors (Basel)*, 24(8), 2437.

Llorente, A. M., Amado, A. J., Voigt, R. G., Berretta, M. C., Fraley, J. K., Jensen, C. L., & Heird, W. C. (2001). Internal consistency, temporal stability, and reproducibility of individual index scores of the Test of Variables of Attention in children with attention-deficit/hyperactivity disorder. *Archives of Clinical Neuropsychology*, 16(6), 535–546. DOI: 10.1093/arclin/16.6.535 PMID: 14590152

Mehmood, R. M., & Lee, H. J. (2017). Towards building a computer aided education system for special students using wearable sensor technologies. *Sensors (Basel)*, 17(2), 317. Advance online publication. DOI: 10.3390/s17020317 PMID: 28208734

Pollreisz, D., & TaheriNejad, N. (2017). A simple algorithm for emotion recognition, using physiological signals of a smart watch. *2017 39th Annual International Conference of the Ieee Engineering in Medicine and Biology Society (EMBC)*, 2353–2356.

Romanzini, M., Petroski, E. L., & Reichert, F. F. (2012). Accelerometers thresholds to estimate physical activity intensity in children and adolescents: A systematic review. *Revista Brasileira de Cineantropometria & Desempenho Humano*, 14, 101–113.

Seçkin, A. Ç., Ateş, B., & Seçkin, M. (2023). Review on Wearable Technology in Sports: Concepts, Challenges and Opportunities. *Applied Sciences 2023, Vol. 13, Page 10399, 13*(18), 10399. DOI: 10.3390/app131810399

Sommer, I., Nußbaumer-Streit, B., & Gartlehner, G. (2021). WHO Guideline: Physical Activity, Sedentary Behavior and Sleep for Children under 5 Years of Age. In *Gesundheitswesen* (Vol. 83, Issue 7). DOI: 10.1055/a-1489-8049

Sousa, A. C., Ferrinho, S. N., & Travassos, B. (2023). The use of wearable technologies in the assessment of physical activity in preschool-and school-age youth: Systematic review. *International Journal of Environmental Research and Public Health*, 20(4), 3402. DOI: 10.3390/ijerph20043402 PMID: 36834100

Sulla-Torres, J., Calla Gamboa, A., Avendaño Llanque, C., Angulo Osorio, J., & Zúñiga Carnero, M. (2024). Classification of Motor Competence in Schoolchildren Using Wearable Technology and Machine Learning with Hyperparameter Optimization. *Applied Sciences (Basel, Switzerland)*, 14(2), 707. Advance online publication. DOI: 10.3390/app14020707

Trabelsi, Z., Alnajjar, F., Parambil, M. M. A., Gochoo, M., & Ali, L. (2023). Real-time attention monitoring system for classroom: A deep learning approach for student's behavior recognition. *Big Data and Cognitive Computing*, 7(1), 48. DOI: 10.3390/bdcc7010048

Chapter 5
Enhancing Education in Bahrain With AI and ML: Through Wearable Technology

Anand Soni
https://orcid.org/0009-0001-4896-0639
Bahrain Polytechnic, Bahrain

Zakir Hossen Shaikh
https://orcid.org/0000-0003-4733-4166
Kingdom University, Bahrain

Mohammad Taqi
https://orcid.org/0000-0002-6701-7267
Bahrain Polytechnic, Bahrain

Vishwanath Bijja
https://orcid.org/0009-0009-7828-6460
UTAS-Ibri, Oman

Satya Pavan Kumar Ratnakaram
https://orcid.org/0000-0002-8908-302X
Bahrain Polytechnic, Bahrain

ABSTRACT

The study explores the transformative potential of integrating AI and ML in wearable technologies to enhance Bahrain's education sector. It emphasizes the personalized learning experiences and holistic student support such technologies can offer, promoting academic and emotional well-being. Additionally, AI-powered wearables can aid teachers in classroom management and professional development. By preparing students for digital-centric work environments, Bahrain aims to align

DOI: 10.4018/979-8-3693-7817-5.ch005

with global workforce trends and Sustainable Development Goals. The chapter discusses integration challenges, benefits, and ethical considerations, advocating for a supportive ecosystem. It reviews Bahrain's educational projects, regulatory frameworks, and policies driving AI and ML adoption in education. Leveraging secondary data, the study underscores the significance of AI and ML in wearables for educational advancement in Bahrain.

1.INTRODUCTION

Bahrain's education sector is going to be transformed because of how artificial intelligence (AI) and machine learning (ML) are getting mixed into wearable technology; these amazing tools can make learning material extremely tailored to what each student needs, so everyone's education becomes better. Personalized learning experiences that these technologies bring to the table are going to make a major difference in how much students can get out of school (Government of Bahrain, 2023). Wearable devices embedded with AI algorithms can analyze student data, provide tailored feedback, and adjust content delivery to cater to individual learning styles (Shishakly et al., 2024). By monitoring various aspects of student performance, including physical and emotional states, these wearables can identify areas for improvement and intervene promptly to support student progress comprehensively (Shajari et al., 2023) and (Kabudi et al., 2021). AI-driven wearables aren't only amazing for students; they may potentially help teachers manage their classrooms better, give feedback immediately, and understand ways to learn new things in a manner that's just for them. But it's vitally important to use these devices to make teaching better without forgetting that teachers need their privacy and to make their own choices too (Seo et al., 2021). In this new era, educators can explore new teaching methods, and students learn differently due to smartwatches, augmented reality glasses, and biometric sensors. These remarkable tools are making a significantly large difference in students' performance, their eagerness to learn, and how lessons can be customized specifically for them, thus opening entirely new opportunities in education (Amos, 2024). The research looks into how AI-powered gadgets that can be worn are being used in educational institutions in Bahrain. It looks at how these spectacular technology pieces don't simply make learning easier to get to--but also give instant feedback on learning and make students more into their studies. What's more, it dives into seeing how items such as virtual reality, technology clothes, and eye-tracking devices can make learning feel like you're right there with everything. It also focuses specifically on what's good and what's tough regarding throwing AI and machine learning into the amalgam with wearables for all levels of education, really hitting the point that it's vitally important to keep students' data safe, make sure their data's secure, and be wholly invested in making it clear how decisions are made by the algorithm.

1.1. Background of the Study

Wearable devices that use AI and ML are changing the tools in education. It is not only just about simulating human items such as learning and reasoning anymore (Forero-Corba & Bennasar,2024). It's greatly improving how people learn. And then you have these wearable biosensors that are not only spectacular but make a big difference by personalizing how everyone learns. The major issue regarding mixing AI, ML, and these gadgets is focused on making education better, more tailored, and kinder to the planet. To keep everything safe while learning with this technology (Amos,2024) also having clear rules regarding data is vitally important. AI in education needs to focus on us, humans, to make things more efficient and cut down on the ominous, or questionable, material (Aggarwal and peers, 2023).

Because of machine learning and AI, the data these wearables collect is extremely accurate, which means everyone can learn in a manner that's just right for them. This combo is doing more than just making classes interesting; it's changing how education works on a major level, building a road for everyone to learn better. The crossroads of technology and education is a glimpse at a hopeful future; but we must be intelligent and informed regarding how we walk down that road. The concrete culmination of this? AI, ML, and wearable technology joining forces could transform education, making it extremely personal and effective for everyone (Shishakly et. al., 2024).

1.2 Need and Scope

This study breaks down AI and ML and how they're used in wearable technology for educational organizations. It has some spectacular ideas that might help create rules and guidelines for putting these technology tools to work in a more effective way: making learning more enjoyable, students improving, and keeping up with Bahrain's major goals for a sustainable future.

1.3 Objectives

The primary objective of this study is to examine the potential of AI and ML-powered wearable technologies in enhancing Bahrain's education sector. The study aims to:

- To examine the effectiveness of AI-powered wearables in monitoring student performance and timely interventions for improved learning outcomes.
- To analyze how the ethical principles of the SDGs in education and the integration of AI and ML in wearable technology correspond within the legal framework.
- To assess the benefits, challenges, and impact of personalized and adaptable learning experiences facilitated by wearable AI and ML technology.

1.4 Research Questions

RQ1: Do wearable technology enhance learning outcomes for students and provide more customized learning opportunities?
RQ2: What are the benefits, challenges, and impact of incorporating AI and ML into wearable technology in an educational system?
RQ3: How do the incorporation of AI and ML in wearables align with ethical considerations within regulatory framework towards SDGs.

2. LITERATURE REVIEW

The utilization of Artificial Intelligence (AI) and Machine Learning (ML) in wearable devices within the realm of education has garnered increasing interest and study. AI's significance lies in its transformative potential across industries, promising to reshape conventional practices and boost efficiency (Cardona et al., 2023). With specialized use cases accessible for all major enterprises, artificial intelligence (AI) is a valuable ally in this era of digital acceleration. Artificial intelligence (AI) is rapidly being used in every industry in the economy, including supply chain management, transportation, public safety, retail, HR services, financial services, financial inclusion, robotics, healthcare, and governance. AI's revolutionary influence is changing conventional paradigms in these sectors (Shaikh, et al, 2022, 2024, Sheela, P et al, 2024, Selwin et al, 2024, Gupta et al, 2024, Kusuma, K., et al,2024, Panigrahi, Rashmi Ranjan, et al,2023 and Sarea, A.et al, 2021). AI involves constructing computer systems capable of human-like tasks, such as learning, reasoning, problem-solving, perception, and language comprehension. Wearable biosensors have notably influenced education, providing educators with valuable resources and students with beneficial tools. Wearable Biosensor Technology (WBT) has emerged as a game-changer, offering new ways to enhance the learning experience (Hernández-Mustieles et al., 2024). The integration of AI and data in education, as highlighted by the Directorate-General for Education (2022), aims to help educators harness AI's benefits while being mindful of potential risks, encouraging positive, critical, and ethical engagement with AI systems. Emphasizing a human-centered approach, AI is seen as a collaborator that enhances human capabilities rather than replacing them. It is crucial to underscore the pivotal role of individuals in education, ensuring that AI systems complement rather than supplant teachers (Ruiz and Fusco, 2024). AI offers opportunities to address educational priorities more effectively, at scale, and with reduced costs, particularly crucial in overcoming learning disparities exacerbated by the pandemic. Leveraging AI can enhance the adaptability of learning resources to meet individual student needs (Cardona et al., 2023). Wearable devices, diverse in form and tailored for specific applications, are compact, designed for continuous monitoring, and expected to collect and process data to improve quality of life (Nahavandi et al., 2022). The integration of machine learning and AI technologies is pivotal in smart wearables due to their embedded architecture (Seng et al., 2023). Digital automated strategies, coupled with virtual reality, enhance safety management in educational contexts by streamlining education, planning, and inspection processes (Regona et al., 2022). Machine learning, a subset of AI, is increasingly influencing various fields, including social sciences like education. A systematic review of literature on machine learning and AI techniques in education is essential for understanding their applications (Forero-Corba & Bennasar, 2024). AI is revolutionizing wearable sensor accuracy and efficiency by identifying and rectifying data collection errors (Shajari et al., 2023). Intelligent tutoring systems emphasize personalized learning experiences tailored to individual students through adaptive interfaces. Data-driven AI systems excel in processing complex data streams in real-time, emphasizing the importance of user interfaces for next-generation ITS (Tuomi, 2018). The growing accessibility of AI through cost-effective smart devices is set to influence education by enhancing teaching methods, educator roles, and educational content (Doina et al., 2023). AI streamlines various teaching tasks traditionally performed by educators, such as grading assignments and recommending educational resources for professional development (Ahajjam et al., 2022). Wearable devices typically rely on AI to interpret biomarkers, with AI algorithms processing data collected by sensors (Gutierrez De Blume, 2022). Investigating teachers' perceptions of teaching AI is critical for successful AI implementation in educational settings (Trotsko et al., 2019). Technology's impact on teaching and learning processes underscores the importance of leveraging AI applications to optimize teaching and learning experiences (Tuomi, 2018). Student attitudes toward smart wearable devices in education are generally positive, indicating their popularity and perceived effectiveness in learning (Fu et al., 2021). Wearable devices hold the potential to usher in a new learning era (Bower & Sturman, 2015), with cost-effective wearables enhancing engagement and improving learning outcomes in education (Ferrier et al., 2022). The ALTAI checklist evaluates the trustworthiness of AI systems like AI4SPP, designed to predict student performance reliably (Fedele et al., 2024). Ensuring voluntary participation in wearing wearables and providing necessary support for digital health seminars are crucial for student engagement (Kinny et al., 2024). Physical and virtual sensors can be integrated into educational systems for data processing and analysis (González Crespo & Burgos, 2019). AI technology enables real-time data transmission, processing, and analysis via wearable sensors linked to web and database servers (Yang, 2024). AI facilitates personalized learning by recommending resources, identifying improvement areas, and adjusting task difficulty levels to support individual student needs (Harry & Sayudin, 2023). AI-assisted personalized learning systems positively impact student learning outcomes (Tiwari, 2023). Smart education promotes environmental awareness and sustainability through digital resources, aligning with sustainability goals by reducing paper usage and physical infrastructure (Aggarwal et al., 2023). Educational institutions play a crucial role in promoting sustainability through courses, sustainable practices, and green campus initiatives (Shishakly et al., 2024). Clear policies governing student data collection, usage, and protection are imperative for educational institutions (Al-rayes et al., 2024). AI feedback is valuable for assessing content knowledge in assignments, while application of knowledge and creative thinking should be evaluated by instructors (Naz & Robertson, 2024). Education and sustainable development are vital for Bahrain's economic and social growth, emphasizing the importance of education in fostering human resources, innovation, and competitiveness (Khalifa et al., 2023).

3. THE EMERGENCE OF WEARABLE TECHNOLOGY

3.1 Artificial Intelligence (AI)

AI, a computer science branch, uses hardware, algorithms, and data to create "intelligence" for tasks like decision-making, pattern recognition, and action execution. As a broad term, AI encompasses more specific concepts. AI systems are built primarily through human-provided rules (rule-based systems), or machine learning algorithms (Ruiz & Fusco, 2024). Many modern AI systems use machine learning (Cardona et al., 2023).

3.2 Machine Learning (ML)

Machine learning is a field that creates algorithms for AI systems. It's a subset of the broader AI concept. In ML, algorithms find patterns in data without humans spelling them out. They build decision-making models as they process information (Forero-Corba & Bennasar, 2024).

3.3 Wearable Technology: A Game-Changer for Education

Wearable technologies have emerged as transformative tools in the educational landscape, offering innovative ways to enhance learning experiences and improve academic outcomes. These devices, which can be worn as accessories, clothing, or even implants, have the potential to revolutionize how students engage with educational content and how educators deliver instruction (Bower & Sturman, 2015). Artificial Intelligence (AI) and Machine Learning (ML) are reshaping wearable devices, offering a myriad of transformative possibilities. By mimicking human tasks like learning and reasoning, these technologies enhance the user experience. Wearable biosensors, powered by AI, revolutionize educational processes, health monitoring, and more (Cardona et al., 2023). Through AI and ML, wearables provide accurate data insights, personalized experiences, and improved efficiency. This convergence opens avenues for enhanced teaching, learning outcomes, and sustainable practices. The fusion of AI, ML, and wearables signals a promising future, emphasizing the importance of responsible integration for a tech-enhanced lifestyle.

3.4 Types of Wearable Devices Used in Education

Several types of wearable devices have found applications in educational settings (Ruiz & Fusco, 2024, *Nahavandi et al., 2022) as shown inFigure 1.*

Figure 1. Classification of wearable devices

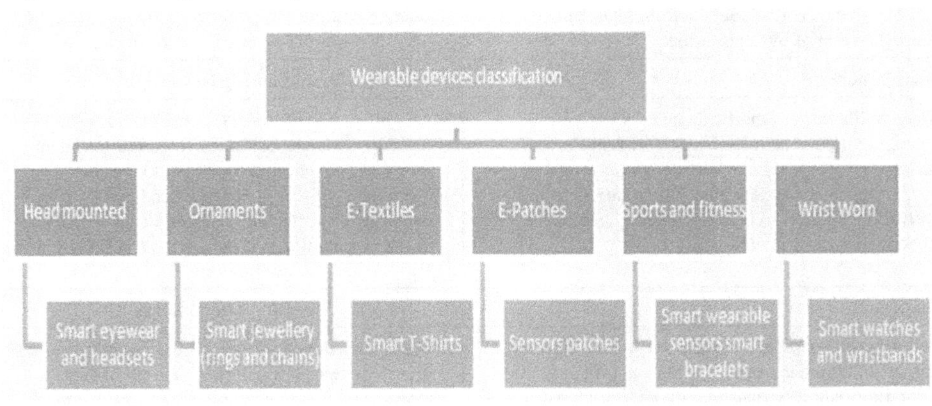

Source: Author

Figure 1 illustrates the diverse applications of wearable devices in education, showcasing how these technologies are revolutionizing learning experiences.

Wearable technologies are revolutionizing education by offering innovative tools for immersive learning experiences. Smart watches, with features like speech-to-text and voice recognition, facilitate real-time learning and feedback in various educational settings. Smart glasses, exemplified by Google Glass, introduce augmented reality for modeling and telemonitoring, providing learners with interactive guidance. Virtual Reality (VR) headsets transport users to virtual worlds, enabling simulations of historical events and high-stakes scenarios, enhancing experiential learning. Brain sensor headbands, equipped with EEG sensors, monitor cognitive performance, aiding in focus training Ruiz & Fusco, 2024, *Nahavandi et al., 2022*. Studies highlight smart bands as a prevalent wearable choice, underscoring their importance in educational applications (Seng et al., 2023; Abd-alrazaq et al., 2024).

With an anticipated annual growth rate (CAGR) of 18%, the wearables technology sector is projected to expand to a market value of USD 265.4 billion by 2026. Within this framework, we will explore the ways in which wearable technology can enrich education by facilitating virtual, interactive, and personalized learning experiences (Saiyed, 2024).

Table 1. Cases of wearable tech for learning

VR Headsets	**Smart Glasses**
Innovative Educational Experiences: Through virtual field trips, students explore historical sites and museums dynamically. Immersive simulations in VR replicate real-world scenarios for training in various fields. Interactive learning involves engaging with 3D models and virtual environments, enhancing comprehension of intricate subjects.	Enhanced Learning with AR: Smart glasses provides augmented reality, blending virtual information seamlessly with the real world for immersive experiences. Hands-free education enables students to receive visual and auditory guidance without physical interaction, simplifying learning in practical environments.
Smart Watches	Brain Sensing Headbands
Effective Planning: Smart watches aid students in organizing their schedules with reminders for tasks, classes, and study periods. Health Monitoring: Advanced smart watches track fitness metrics, motivating students to embrace healthy habits and enhance their overall well-being.	Cognitive Monitoring: Headbands track brain activity, assisting educators in gauging student attention and learning patterns more effectively. Stress Management: Scientific headbands monitor stress levels, offering feedback to aid students in stress reduction and enhancing their learning capabilities.
Wearable Cameras	Wearable Microphones
Capturing Experiments: Wearable cameras empower students to document experiments and projects, facilitating seamless sharing with peers and instructors. On-field Investigations: Wearable cameras support conducting research in diverse conditions, capturing real-world scenarios accurately regardless of weather or time constraints.	Language Acquisition: Wearable microphones can elevate language learning, enabling students to hone speaking and listening skills with feedback on pronunciation and fluency. Collaborative Dialogues: Wearable microphones enhance group discussions by ensuring effective communication, particularly in noisy settings.

Source: (Saiyed, 2024).

4. METHODOLOGY

The author has used descriptive research in this work. The study utilized secondary data from respected academic journals, industry reports, case studies, and relevant publications on AI and ML in wearables used in the educational sector. The criteria for selection included a publication date within the past 7 years. The selected literature was thematically analyzed to identify themes, trends, and patterns. This involved categorizing the data to identify similarities and differences. The comparison of studies revealed consistency, inconsistencies, and gaps in literature.

Additionally, secondary data from relevant sources will be utilized to provide a comprehensive overview of Bahrain's education sector and the existing projects related to the integration of AI and ML in wearable technologies.

5. RESULTS AND DISCUSSIONS

RQ1: Do Wearable Technology Enhances Learning Outcomes for Students, and Provide More Customized Learning Opportunities?

The integration of artificial intelligence and wearable technologies has ushered in a new era of personalized learning experiences (Seng et al., 2023). In education, AI and ML in wearable tech enhance learning through real-time tracking, personalized insights, and behavior analysis. Wearables boost student engagement with gamification and rewards. They ensure inclusivity by aiding diverse learners and bridging gaps. Predictive analytics enable early interventions, supporting students proactively (*How Wearable Technology Is Revolutionizing the Education Sector?* 2024) shown in figure 2.

Figure 2. Wearable technology in the education sector

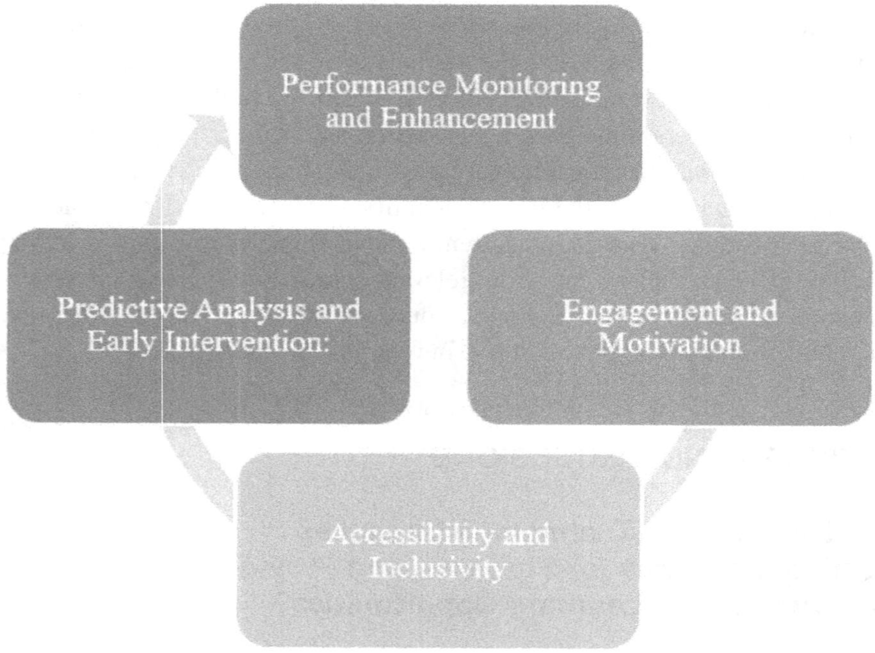

Source: Author.

In the field of education, wearable technologies, as illustrated in Figure 2, demonstrate four key interconnected activities within the education sector.

Table 2. 4 Squares- Wearable technologies in education sector

Square 1: Performance Monitoring and Enhancement	**Square 2: Engagement and Motivation**
• Real-time Performance Tracking • Student Performance Metrics • Data Analytics and Insights for Educators • Behavioral and Cognitive Pattern Analysis	• Gamification Elements in Wearable Devices • Social Learning Features • Reward Systems and Progress Tracking
Square 3: Accessibility and Inclusivity	Square 4: Analysis and Intervention
• Assistive Technologies for Special Needs Students • Bridging the Digital Divide	• Predictive Analytics for Early Intervention • Customized Content Delivery (tailored to intervene early)

Source: *How Wearable Technology Is Revolutionizing the Education Sector?* 2024),

Table 2 outlines key aspects of wearable technologies in education. Performance Monitoring and Enhancement involve real-time tracking, data analysis for insights, and behavioral pattern identification. Engagement and Motivation focus on gamification, social learning, and reward systems. Accessibility and Inclusivity enhance inclusiveness, while Predictive Analysis aids in early interventions and customized support.

Square 1: Performance Monitoring and Enhancement

Real-Time Performance Tracking

One of the key advantages of AI-powered wearables in education is their ability to provide continuous, real-time performance tracking. These devices can monitor various metrics, including attention span, comprehension levels, and even emotional states, to give educators a comprehensive understanding of each student's progress. Smartwatch apps are very much of help because they let students know right away when they've finished something or done well, which keeps them going by being happy about what they have been doing (Silva et al. 2023, and Abd-Rahman et al. 2024). In addition, it's amazing for teachers because they can see how everyone is doing in real-time, giving feedback straight away, which is vitally important if you desire to learn material well (Koutromanos & Kazakou, 2020). Then there's a particular one with intelligent and informed clothes that have these biometric sensors. They can sense through if someone's getting stressed out, so teachers can come in and give help before things get much — and there's also this gadget—an EEG headband—that can sense through when your brain is in the zone for learning by tracking how focused you are. Teachers can use that data to plan their lessons better or set up the best times to study so everyone takes the most advantage of it (Kinny et al. 2024). Using technology like this might really help school a lot by making fully sure everyone's getting help exactly when they need it, and keeping everyone excited to learn.

Student Performance Metrics

It may seem incongruous--but wrist devices and advanced glasses might actually change how teachers think through and watch over how well students are doing. These technology pieces are able to track phenomena happening today with a student's body, such as how tired they are from working too hard, or where their eyes are directed, and connect that data to how they're doing in educational organizations (Fu et al., 2021). For example, research has found that anxiety affects about 28 percent of students from kindergarten to 12th grade significantly, interfering with their ability to learn. By using these marvelous technology tools to watch for signs of stress or nerves during tests or class activities, teachers can spot students who might be struggling with these feelings. (Gutierrez de Blume, 2024).

Data Analytics and Insights for Educators

Educationists can now obtain highly useful data to enhance learning because wearable devices, such as smartwatches, are being utilized more frequently in educational organizations. By looking into the rich data from these gadgets, we can understand better regarding how students do in class, what they do, and how they think (Seo et al. 2021); these new technology tools are helping us explore analyzing data and finding amazing ideas in education.

Behavioral and Cognitive Pattern Analysis

To think through which sections of a lesson grasp students' focus the best, eye-tracking technology can be built into intelligent and informed glasses (Shishakly et al. 2024). Using this data, educators can improve their teaching approaches, making learning more interesting. By examining data from wearables, machine learning algorithms reveal how students react to different teaching styles and interact with study materials. Such deep dives can shed light on student behavior and thinking trends (Shajari et al. 2023).

Square 2: Engagement and Motivation

Gamification Elements in Wearable Devices

Wearable devices use elements from games to make learning activities fun. They bring in points, badges, leaderboards, competition with rules, and the excitement of trying to finish quickly to turn regular exercises into fun challenges or searches for hidden prizes for rewards, or into competitions with friends (Kinny et al. 2024). This idea, called gamification, is noticeably focused on putting parts of game design into things that aren't games -- and it really fits well with wearables for education (Deterding et al. 2011). Studies show that these game-like features can boost motivation and make physical activity significantly more engaging. The integration of gamification in wearables provides immediate positive reinforcement, helping students overcome initial hurdles in engaging with educational content and sustaining higher levels of activity in the long term (Nahavandi et al., 2022).

Social Learning Features

The social aspect of learning plays a crucial role in student engagement, and wearable technologies have been designed to leverage this natural human inclination towards social interaction. By incorporating features that facilitate peer-to-peer communication and collaboration, these devices create a sense of community and shared learning experiences (Deterding et al., 2011). One effective way wearables promote social learning is through leaderboards. These competitive elements not only introduce a sense of friendly rivalry but also increase individual accountability. When students join a leaderboard, they reveal their previously private levels of activity to others, allowing peers to hold them accountable and encourage improvement (Kinny et al., 2024).

Reward Systems and Progress Tracking

Wearable technologies excel in implementing robust reward systems and progress tracking features, which are essential for maintaining student motivation (*How Wearable Technology Is Revolutionizing the Education Sector?* 2024). These devices provide continuous, real-time performance tracking, monitoring various metrics such as attention span, comprehension levels, and even emotional states (Kinny et al., 2024). This comprehensive data collection gives educators a deeper understanding of each student's progress and allows for timely interventions. The ability to visualize progress through charts, graphs, or virtual representations creates a tangible sense of accomplishment, encouraging students to set and achieve new goals (Mator Design, 2024). Furthermore, some wearable technologies incorporate economic incentives to increase user engagement. For example, offering discounts on educational materials or wellness products based on academic performance or consistent device usage has been shown to increase students' willingness to use wearables and engage with educational content (Saiyed, 2024). By combining gamification domains, social learning features, and reward systems, wearable technologies create a holistic approach to enhancing a student's engagement and motivation (Deterding et al., 2011).

Square 3: Accessibility and Inclusivity

Assistive Technologies for Special Needs Students

Wearable technology, such as intelligent and informed watches and augmented reality glasses, are changing the industry for students who need some more assistance (Shajari et al., 2023 and Silva et al., 2023). These gadgets aren't only wonderful; they're making it easier for students with disabilities to be independent and feel more part of what's happening around them (Yang, 2024).: take the Dot smartwatch, for example. It's designed for people who can't see very well, letting them check the time with Braille they can feel. In addition, it connects to their phones so they can get texts and app updates without needing to see the screen (Nahavandi et al., 2022). Then there are SignAloud gloves that some college students, Thomas Pryor and Navid Azodi, came up with. They're not only regular gloves; they can take ASL signs and turn them into spoken words and text by thinking through the complexities of how the hands are moving, all sent over Bluetooth. This particular one could change the trade for deaf students trying to converse in class or with their peers (Mator Design, 2024).

Bridging the Digital Divide

wearable technologies offer massive potential for developing accessibility and inclusion, it's crucial to address the digital divide that exists in many communities (Seo et al., 2021).

The COVID-19 pandemic really showed how large of a gap there is with people getting online, especially for students who can't afford good internet at home for material such as remote learning (Trotsko et al., 2019). To tackle the related issues, various initiatives have been launched . For instance, the Federal Communications Commission is leading efforts to close the connectivity gap by making broadband more affordable and accessible in lower-income populations, rural areas, and school systems. Additionally, organizations like Defeat the Digital Divide are working to connect school districts with resources and funding to provide reliable internet service to their students (*Bridging the Digital Divide in Education | LapSafe News & Articles*, 2024).

Square 4: Analysis and Intervention

Predictive Analytics for Early Intervention

Wearable AI looks extremely promising for helping in educational organizations, especially because it can catch when students are starting to struggle before things become awful. They do a spectacular thing where they look at material, such as how fast a student's heart is beating and other body signals, plus how they're doing in their classes, to think through which students might start having a tough time. This essentially gives teachers a heads-up to help out the students who really need it, before they're in deep trouble. AI gadgets are actually very good at seeing when students are stressed—hitting the nail on the head 75% of the time (Abd-alrazaq et al., 2024). That's not perfect—but it's surely a beginning. Even though they must become better at spotting stressed-out students, it's clear wearable AI could be revolutionary in making completely sure students don't fall through the cracks.

Customized Content Delivery

AI-powered wearables excel in delivering customized content that aligns with each student's interests, learning preferences, and goals. By analyzing data on a student's past performance, engagement levels, and personal interests, these devices can curate a unique learning experience tailored to the individual (Tiwari, 2023). Virtual Reality headsets are very impressive because they can make learning significantly more interesting for everyone, no matter how you like to learn. If you're someone who gets more out of seeing things, you could dive into some truly amazing 3D models and simulations--but then, if you're the type that prefers listening, there's content for you too—like getting to hear more explanations and being part of interactive discussions. This way, no one feels left out or bored because everything's made just for their own manner of learning. Now, specifically AI-powered wearables, they're essentially like the best helpers in the study world. They keep track of what each person likes, how they learn best, and even how well they've been doing up until now--but let it be known, it's not only just regarding keeping tabs on you (Abd-alrazaq et al., 2024).

RQ2: What are the Benefits, Challenges, and Impact of Incorporating AI And ML Into Wearable Technology in Bahrain's Educational System?

Integrating AI and ML into wearable technology within Bahrain's educational framework yields advantages such as tailored learning experiences, immediate feedback, and heightened engagement. Obstacles encompass data privacy issues and infrastructure preparedness. The result entails enhanced student performance, expanded accessibility, and the cultivation of a technology-driven educational milieu in Bahrain (Motti, 2019 and Amos, 2024).

Table 3. Benefits of using wearable devices in education

Benefits	Description
Engagement:	providing timely and personalized information, making learning more interactive and captivating for students.
Personalization	catering to individual learning needs and styles, enhancing the overall learning experience
Assistive Technologies	aiding students with visual impairments or intellectual disabilities, fostering inclusivity and accessibility in education.
Feedback and Communication	Augmented reality systems in wearables facilitate feedback mechanisms between students and instructors
User-Friendly Tools	Requiring less technical expertise to operate, making them attractive to educators and students alike
Improved Learning Environment	assisting in tasks, such as time management, reminders for exams, and in-class activities.
Safety and Accountability	contribute to safety, accountability, and efficiency both at home and in educational settings

Source: (Motti, 2019 and Amos, 2024).

The integration of wearable devices in education faces various challenges and limitations as some of them mentioned in the below given Table 4.

Table 4. Challenges and limitations of using wearable devices

Challenges/ limitation	Description
Privacy and Security	The collection of personal and biometric data raises important questions about data security and student privacy
Cost Implications	Acquiring and maintaining advanced wearable devices can be a barrier, especially for educational institutions with limited budgets. So, the expense of implementing wearable technology at a scale can be prohibitive for many educational institutions
Technical Complexity	Require specialized technical knowledge and support, which could pose challenges for educators lacking expertise in this area. Some issues such as battery life, connectivity, and device reliability can impact the effectiveness of wearable technologies in educational settings.
Pedagogical Integration	integrated into the curriculum and pedagogical practices can be a significant challenge, as educators must align their use with learning objectives and instructional strategies
User Acceptance	Adoption of wearable technologies by both students and educators may vary, impacting their effectiveness and utility in educational settings.
Maintenance and Support	Resource-intensive and require ongoing investment.
Distraction	Multifunctional nature of some wearable devices may potentially distract students from their learning tasks
Equity	Equal access to wearable technologies across diverse student populations

Source: (Motti, 2019 and Amos, 2024).

IMPACTS

As Bahrain advances in the adoption of technological innovations in education, a thorough exploration is necessary to understand the lasting impacts of wearable devices on student learning and academic progression in the Bahraini context. The seamless integration of wearable technology into Bahrain's educational framework is pivotal for maximizing its benefits and ensuring its enduring implementation to support the Sustainable Development Goals (SDGs). (Government of Bahrain, 2023). Moving from traditional computers to agile wearable technologies represents a transition towards personalized learning tools that expand individual capabilities. Wearable devices, such as those discussed by Bower & Sturman (2015) and Motti (2019), can facilitate collaborative learning, offer personalized support, and improve accessibility, benefiting students with special needs. These technologies empower both students and educators through various applications, including data aggregation, tailored notifications, and feedback, contributing to immersive learning experiences and heightened student engagement levels.The incorporation of smart devices into educational organization mounts unique learning opportunities by providing hands-free access, individualized learning pathways, and immediate contextual information. The adoption of advanced wearable devices in education can enhance student performance, increase motivation, and foster positive attitudes towards learning (Koutromanos & Kazakou (2020). By understanding and leveraging wearable technology effectively within Bahrain's educational landscape, the country can advance its educational goals, enhance student engagement, and contribute to the broader global objective of providing inclusive and equitable quality education for all, aligning with Sustainable Development Goal 4 - Quality Education. This integration will not only optimize educational outcomes but also support Bahrain's economic vision and commitment to sustainable development through education.

RQ3: How Does the Incorporation of AI and ML in Wearables Align With Ethical Considerations Within Regulatory Framework Towards SDGs?

The Rise of AI and ML in Education Sector

Bahrain is establishing itself as a notable center for AI innovation within the Middle East, placing increasing emphasis on incorporating artificial intelligence and machine learning across diverse sectors, notably in education. The nation's dedication to adopting these advancements is reshaping its educational sector, setting the stage for substantial advancements in teaching techniques and learning approaches.

Below given table 5 exhibits AI initiatives in key sectors reflect a regional commitment to technological advancement and economic growth (Middle East Briefing, 2024).

Table 5. AI initiatives in MENA region

Country	Key Initiatives and Investments	Impact and Future Outlook
Qatar	Qatar is leading AI development in the Gulf region with a substantial investment of US$2.4 billion to boost AI capabilities and attract global tech talent. The country aims to establish a smart economy, projecting digital investments to reach US$5.7 billion by 2026, a significant increase from US$1.65 billion in 2022.	Qatar's AI investments signify a strategic move towards digital transformation and economic diversification. Projects like the Fanar AI initiative are enhancing linguistic programs and supporting the nation's digital goals. The increased funding is expected to bolster innovation, efficiency, and productivity across various sectors, positioning Qatar as a key player in AI technology within the MENA region.
Saudi Arabia	Saudi Arabia is progressing with its National Strategy for Data and AI, targeting US$20 billion in investments by 2030. The plan includes training 20,000 AI specialists, creating over 30,000 AI-related jobs, and aligning with Saudi Vision 2030 objectives. In industries such as energy, healthcare, and government, they are starting to use AI a lot. For example, Saudi Aramco has this amazing computer program project that makes energy use and drilling significantly better.	Saudi Arabia is really focusing on making their economy better and creating more jobs with their AI strategy; they are trying to make things better and coming up with new ideas in important areas such as energy and healthcare; this is all part of their larger plan to develop the country. The country's ambitious plans to train AI specialists and create job opportunities demonstrate a commitment to technological advancement and positioning Saudi Arabia as a regional AI hub.
UAE	The UAE has established the Mohamed bin Zayed University of Artificial Intelligence, the first graduate-level AI university globally, underlining its dedication to becoming a prominent AI research and development hub. The transportation and logistics sectors are utilizing AI for traffic management and autonomous vehicles, while healthcare applications focus on diagnostics, patient care, and administrative efficiencies.	The UAE's emphasis on AI education and research is propelling the country towards global leadership in artificial intelligence. By leveraging AI in transportation, logistics, and healthcare, the UAE is enhancing operational efficiency and service delivery. The country's strategic investments in AI infrastructure and talent development are paving the way for continued innovation and technological advancements, positioning the UAE as a key player in shaping the future of AI technologies.

Source: (Middle East Briefing, 2024).

Current State of AI/ML Adoption

The adoption of AI and ML in Bahrain's education sector is gaining momentum, driven by the increasing global demand for these technologies growth has a significant influence on Bahrain's educational institutions, which are increasingly. The global market for machine learning is expected to reach USD 119.50 billion by 2024 (Gutierrez De Blume, 2024). This incorporates AI and ML technologies to enhance learning experiences and improve educational outcomes. Bahrain's strategic position as a gateway to the Gulf Cooperation Council (GCC) region has attracted global IT companies to establish a presence in the country. The availability of talent, cost competitiveness, and market access to the overall region have made Bahrain an attractive hub for AI and ML initiatives (Harry & Sayudin, 2023). This influx of expertise and resources has a positive impact on the education sector, providing opportunities for collaboration and knowledge transfer.

Government Initiatives Supporting AI/ML in Education

The government of Bahrain has demonstrated a strong commitment to promoting the integration of AI and ML in the education sector through various initiatives.

Some notable examples involve e.g. In collaboration with Tamkeen and Microsoft, Bahrain Polytechnic has inaugurated the Artificial Intelligence (AI) Academy, aimed at empowering students interested in pursuing the rapidly advancing field of technology as a career choice. (Faisal, 2020). Another a recent collaboration between the Labor Fund (Tamkeen), Amazon Web Services (AWS), and the Bahrain Institute of Banking and Finance (BIBF) to train 60 Bahraini individuals in Machine Learning and Artificial Intelligence technologies (Tamkeen to Empower 60 Bahraini Nationals with Machine Learning and Artificial Intelligence Training in Collaboration with AWS and BIBF, 2023).

These strategic initiatives aim to address the needs of the evolving digital economy and equip the local workforce with high-demand skills. Since 2022, Tamkeen has facilitated over 1,500 training opportunities in diverse domains such as information technology, software engineering, data science, and game development. These programs have yielded successful outcomes, with 65% of graduates going on to be employed, gain promotions, or start their own businesses (AI in Startups: Nurturing Innovation in Bahrain's Tech Ecosystem, 2024). This commitment to ICT skills development positions Bahrain as a hub for technological innovation in the region.

Privacy and Security Considerations

The integration of wearable technologies in education brings forth significant privacy and security concerns that must be addressed to ensure the ethical use of student information and compliance with regulations (*Data Protection in Schools*, 2024). As these devices collect vast amounts of personal data, including heart rate variability, electrodermal activity, and eye movements, it is crucial to implement robust data protection measures (Shishakly et al., 2024).

Data Protection Measures

To safeguard sensitive student information collected through wearable technologies, educational institutions must implement comprehensive security protocols. Educational organizations places have to make sure students' data, gathered by intelligent and informed watches and gadgets, is safe; they need to keep login details safe, tell when someone's tinkering with accounts they shouldn't, and take steps to stop anyone from breaking in and stealing data. With medical information getting strikingly large in the healthcare world, it's vitally important to make things even more secure to deal with all that data safely.

Educational organizations have to check their security measures regularly to find and fix any weak spots that could let someone look at or snatch student data. Using codes nobody can completely resolve and putting data in a secure online space can keep student details out of the wrong hands. Also, educational organizations must make rules regarding how long to keep students' information and when to remove it, making fully sure it's only around as long as it needs to be; the truth of this can't be ignored, because keeping student data safe is a major issue.

Ethical Use of Student Information

Educational organizations really need to be direct regarding using the data they get from items such as smartwatches or any other technology gear that students wear. They must make sure everyone's quite accepting of it—the students wearing the gear and the adults around them—before they start collecting any personal data or handing it over to others, such as the individuals or people who run exams (Tiwari, 2023). They can't simply be vague regarding it—they need to lay out what data they're after, what they're going to do with it, and how long they'll keep it around for. Being open regarding what's going on helps students and their families trust the school more. That means giving them the complete picture in a manner that's very easy to understand. The data shared should cover the details on how the school collects, uses, and protects student data. It must cover how students and their families can begin to understand the data, fix mistakes, or even ask for it to be destroyed (Koutromanos & Kazakou, 2020). Sometimes, yet rarely so, using this data can end up being a path that leads to trouble. Such as using someone's biometric data to decide grades, which can make things complicated: stressing out students or trampling over their privacy (Fedele et al., 2024).

Compliance with Regulations

Educational organization are starting to use more technology gear that collects data, which means they have to really be careful regarding following the rules for keeping that data safe (Sharma, 2024). They must look at what rules they need to follow; which, for example, could mean thinking scrupulously regarding things such as the GDPR if they're in Europe, or FERPA if they're in the U.S. And it's not only just about knowing the rules. These organizations have to do a variety of tasks to ensure they are properly handling your data, including: employing individuals specifically responsible for overseeing data; consistently monitoring how data is managed; and being prepared to address any issues, such as a data leak (Fedele et al. 2024). Also, if someone asks to see their data, the educational institutes needs to be ready to show them. In addition, if any of that data gets out when it shouldn't, the educational institutes need to tell the right people fast. To not completely fail with this, educational institutions have to come up with a clear plan of action that spells out who is responsible for what, how to deal with data properly, and what to do if things go very wrong. It's a lot to manage--but it's critical now with all the technology content becoming a larger part of learning every day. Regular staff training on data protection best practices and the specific requirements of applicable regulations is essential to maintain a culture of privacy and security within the institution. By addressing these privacy and security considerations, educational institutions can harness the benefits of wearable technologies while protecting student data and maintaining trust in the educational ecosystem. Bahrain has established a robust framework of regulations to safeguard the personal data of users, ensuring privacy and security in the digital realm. (*Bahrain Business Laws | Personal Data Protection Law*, n.d.)

Use of AI Supported Wearable Devices in Education and Sustainable Development Goals

The integration of wearable devices within the education sector creates a distinct opportunity to harmonize educational approaches with the Sustainable Development Goals (SDGs). Through the incorporation of these cutting-edge technologies, Bahrain has the potential to drive educational progress that significantly supports the SDG 4 objective of ensuring inclusive and high-quality education for all. Here's how wearable devices can facilitate this connection: (*Kingdom of Bahrain Portal for Sustainable Development Goals*, n.d.) *as shown inFigure 3.*

Figure 3. Wearable devices and SDGs

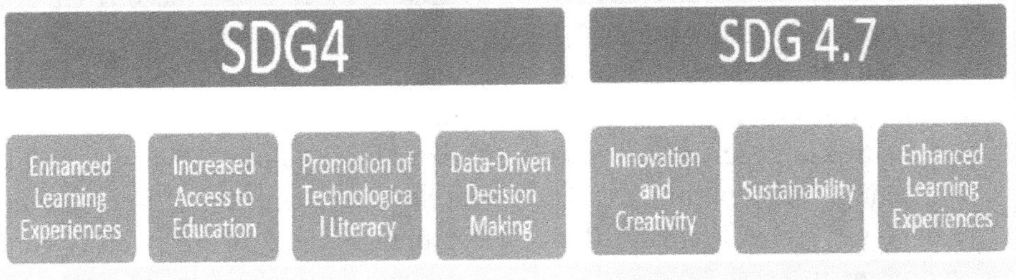

Source: Author

The above figure 3 exhibits the Wearable devices in education transform learning by personalizing experiences, enhancing access, promoting tech literacy, fostering innovation, enabling data-driven decisions, and supporting sustainability. Bahrain's adoption of wearables aligns with SDG 4 goals, ensuring quality education, inclusivity, and environmental responsibility, driving progress towards holistic and sustainable educational development.

6. FINDINGS AND CONCLUSION

Transformation of Academic Performance: The integration of AI-powered wearables in Bahrain's education sector can fundamentally altering academic performance by providing personalized learning experiences and real-time feedback to students and educators. This transformative approach can be useful in enhancing the quality of education delivery.

Enhanced Student Engagement: Wearable technologies, ranging from smartwatches to augmented reality glasses, are fostering increased student engagement through interactive learning experiences and access to real-time analytics. These tools are reshaping how students interact with educational content.

Inclusive Learning Environments: Wearable technology is helping make educational bodies more welcoming for everyone because it can change to fit how different people learn, and gives each student what they specifically need to learn better; this fits in with the goal of making fully sure education is fair and available to everyone.

Challenges and Considerations: For wearable technologies to really work in educational organization, it's not simply a matter of starting to use them. We must think regarding items such as keeping student information safe and making fully sure every student can start to use these tools. Solving problems related to privacy and access is vitally important to make these technology tools a good fit for learning.

Nurturing Innovation: Collaborative efforts among government bodies, educational institutions, and private sector entities are crucial in fostering an ecosystem that nurtures innovation in educational technology. This collaboration is essential for preparing students for the evolving demands of the digital age.

Balancing Technology and Privacy: Keeping student data private is vitally important if we want to keep making spectacular technological content for educational organizations without making everyone uncomfortable. Bahrain must handle using new technology to make learning better--and making completely sure they're not oversharing our private details. People will trust them more and we can keep improving with technology in educational organizations if they master this.

Prospects: The amalgamation of imminent wearable devices, such as neural interface wearables and smart clothing embedded with sensors, holds massive potential to further revolutionize the educational landscape. These developments offer prospects for even more immersive and personalized learning experiences, shaping the future of education globally.

In conclusion, the integration of AI-powered wearables in Bahrain's education sector represents a significant step towards enhancing the quality, inclusivity, and innovation in education. By taking on challenges, teaming up, and exploring new places that have special electronic tools, Bahrain's aiming to be a top spot for learning, not only in the Middle East but everywhere.

REFERENCES

Abd-alrazaq, A., Alajlani, M., Ahmad, R., AlSaad, R., Aziz, S., Ahmed, A., Alsahli, M., Damseh, R., & Sheikh, J. (2024). The Performance of Wearable AI in Detecting Stress Among Students: Systematic Review and Meta-Analysis. *Journal of Medical Internet Research*, 26, e52622. DOI: 10.2196/52622 PMID: 38294846

Aggarwal, D., Sharma, D., & Saxena, A. (2023). Adoption of Artificial Intelligence (AI) For Development of Smart Education as the Future of a Sustainable Education System Journal of Artificial Intelligence, Machine Learning and Neural Network Adoption of Artificial Intelligence (AI) For Development of Smart Education as the Future of a Sustainable Education System. Journal of Artificial Intelligence. *Machine Learning and Neural Network*, 03(06), 23–28. Advance online publication. DOI: 10.55529/jaimlnn.36.23.28

Ahajjam, T., Moutaib, M., Aissa, H., Azrour, M., Farhaoui, Y., & Fattah, M. (2022). Predicting Students' Final Performance Using Artificial Neural Networks. *Big Data Mining and Analytics*, 5(4), 294–301. DOI: 10.26599/BDMA.2021.9020030

. Ahmed, A., Aziz, S., Qidwai, U., Farooq, F., Shan, J., Subramanian, M., Chouchane, L., EINatour, R., Abd-Alrazaq, A., Pandas, S., & Sheikh, J. (2023). Wearable Artificial Intelligence for Assessing Physical Activity in High School Children. Sustainability (Switzerland), 15 (1), 1- DOI: 10.3390/su15010638

Alrayes, A., Henari, T. F., & Ahmed, D. A. (2024). ChatGPT in Education – Understanding the Bahraini Academics Perspective. *Electronic Journal of e-Learning*, 22(2), 112–134. DOI: 10.34190/ejel.22.2.3250

Amos, Z. (2024, October 25). The Futuristic Role of Wearables in Education. IoT for All. https://www.iotforall.com/the-futuristic-role-of-wearables-in-education

Bahrain Business Laws | Personal Data Protection Law. (n.d.). Bahrainbusinesslaws.com. https://bahrainbusinesslaws.com/laws/Personal-Data-Protection-Law

Bahrain Pioneers AI Regulation with New Standalone Law - HyScaler. (2023). HyScaler. https://hyscaler.com/insights/bahrain-pioneers-ai-regulation

Bahrain.bh. (2023). Main Template. Bahrain.bh. https://bahrain.bh/wps/portal/en/BNP/HomeNationalPortal/ContentDetailsPage/

Bower, M., & Sturman, D. (2015). What are the educational affordances of wearable technologies? *Computers & Education*, 88, 343–353. DOI: 10.1016/j.compedu.2015.07.013

Bridging the Digital Divide in Education | LapSafe News & Articles. (2024, February 22). Www.lapsafe.com. https://www.lapsafe.com/blog/bridging-the-digital-divide-in-education

Cardona, M. A., Rodríguez, R. J., & Ishmael, K. (2023). Artificial Intelligence and the Future of Teaching and Learning Insights and Recommendations. In Office of Educational Technology (pp. 18–34). U.S. Department of Education, Office of Educational Technology, https://tech.ed.gov/

Data protection in schools. (2024, August 7). Unicef.org. https://www.unicef.org/eca/reports/data-protection-schools

. Deterding, S., O'Hara, K., Sicart, M., & Dixon, D. (2011). Gamification: Using Game Design Elements in Non-Gaming Contexts. Gamification Summit and the Gamification Day at the 2011 GDC Serious Games Summit.

Faisal, S. (2020). Tamkeen, Bahrain Polytechnic & Microsoft collaborate to launch the 1st batch of the Artificial Intelligence Academy – StartUp Bahrain. Startupbahrain.com. https://startupbahrain.com/news/tamkeen-collaborates-with-bahrain-polytechnic-microsoft-to-kickstart-the-1st-batch-of-the-artificial-intelligence-academy/#:~:text=Bahrain%20Polytechnic%20has%20officially%20launched%20a

Fedele, A., Punzi, C., & Tramacere, S. (2024). The ALTAI checklist as a tool to assess ethical and legal implications for a trustworthy AI development in education. *Computer Law & Security Report*, 53, 105986. DOI: 10.1016/j.clsr.2024.105986

Ferrier, B., Lee, J., Mbuli, A., & James, D. A. (2022). Translational Applications of Wearable Sensors in Education: Implementation and Efficacy. [CrossRef] [PubMed]. *Sensors (Basel)*, 22(4), 1675. DOI: 10.3390/s22041675 PMID: 35214578

Forero-Corba, W. 1, & Bennasar, F. N. 1 1 U. de les I. B. (. (2024). Techniques and applications of Machine Learning and Artificial Intelligence in education: A systematic review. *ProQuest*, 27(1), 209–238. DOI: 10.5944/ried.27.1.37491

Fu, D., Chen, L., & Cheng, Z. (2021). Integration of Wearable Smart Devices and Internet of Things Technology into Public Physical Education. DOI: 10.1155/2021/6740987

González Crespo, R., & Burgos, D. (2019). Advanced Sensors Technology in Education. *Sensors (Basel)*, 19(19), 4155. DOI: 10.3390/s19194155 PMID: 31557927

Government of Bahrain. (2023). Home. Bahrain.bh. https://www.bahrain.bh/wps/portal/en/

Gupta, S., Srivastava, R., Shaikh, Z. H., & Irfan, M. (2024). Revolutionizing Rural Finance: Exploring the Impact of FinTech on Financial Inclusion in India. In Irfan, M., Muhammad, K., Naifar, N., & Khan, M. A. (Eds.), *Applications of Block Chain technology and Artificial Intelligence. Financial Mathematics and Fintech*. Springer., DOI: 10.1007/978-3-031-47324-1_3

Harry, A., & Sayudin. (2023). Role of AI in Education. Injuruty: Interdiciplinary Journal and Humanity. https://injurity.pusatpublikasi.id/index.php/in

Hernández-Mustieles, M. A., Lima-Carmona, Y. E., Pacheco-Ramírez, M. A., Mendoza-Armenta, A. A., Romero-Gómez, J. E., Cruz-Gómez, C. F., Rodríguez-Alvarado, D. C., Cruz-Garza, J. G., Arceo, A., Ramírez-Moreno, M. A., & de Lozoya-Santos, J. (2024). *Wearable Biosensor Technology in Education: A Systematic Review*. https://doi.org/DOI: 10.20944/preprints202403.0831.v1

How Wearable Technology is Revolutionizing the Education Sector? (2024). VLink. https://www.vlinkinfo.com/blog/how-wearable-technology-is-revolutionizing-the-education-sector/

Kabudi, T., Pappas, I., & Olsen, D. H. (2021). AI-enabled adaptive learning systems: A systematic mapping of the literature. *Computers and Education: Artificial Intelligence*, 2, 100017. DOI: 10.1016/j.caeai.2021.100017

Khalifa, M., Alrasheed, R., Khalifa, M., Alrasheed, R., & Aldada, A. (2023). Education in Light of the Sustainable Development Goals in the Kingdom of Bahrain "An Analytical Study" Education in Light of the Sustainable Development Goals in the Kingdom of Bahrain "An Analytical Study.". *Journal of Statistics Applications & Probability*, 12(S1), 1419–1423. DOI: 10.18576/jsap/12S103

Kingdom of Bahrain Portal for Sustainable Development Goals. (n.d.). Sdgs.gov.bh. https://sdgs.gov.bh

Kinny, F., Schlottau, S., Ali Sherazi, B., Obarcanin, E., & Läer, S. (2024). Digital health in pharmacy education: Elective practical course integrating wearable devices and their generated health data. *Exploratory Research in Clinical and Social Pharmacy*, 15, 100465. DOI: 10.1016/j.rcsop.2024.100465 PMID: 38983639

Koutromanos, G., & Kazakou, G. (2020). The use of smart wearables in primary and secondary education: A systematic review. *Themes in ELearning*, 13, 33–53.

Kusuma, K., Sheela, P., Panigrahi, R. R., & Shaikh, Z. H. (2024). Blockchain Technology and Islamic Finance: A Decade of Growth Trends. In Irfan, M., Kadry, S., Sharif, M., & Khan, H. (Eds.), *Fintech Applications in Islamic Finance: AI, Machine Learning, and Blockchain Techniques* (pp. 59–79). IGI Global., DOI: 10.4018/979-8-3693-1038-0.ch004

Mator Design. (2024, March 5). Thomas & Navid invented gloves that can translate sign language into text or speech. Medium. https://matorconsultancy.medium.com/thomas-navid-invented-gloves-that-can-translate-sign-language-into-text-or-speech-601baca9ae91

Middle East Briefing. (2024, August 7). GCC Countries Prioritize Investing in AI Capabilities. Middle East Briefing. https://www.middleeastbriefing.com/news/gcc-countries-invest-in-ai-capabilities/

Motti, V. G. (2019). Wearable Technologies in Education: A Design Space. *Lecture Notes in Computer Science*, 11591, 55–67. DOI: 10.1007/978-3-030-21817-1_5

Nahavandi, D., Alizadehsani, R., Khosravi, A., & Acharya, U. R. (2022). Application of artificial intelligence in wearable devices: Opportunities and challenges. *Computer Methods and Programs in Biomedicine*, 213(1), 106541. DOI: 10.1016/j.cmpb.2021.106541 PMID: 34837860

Naz, I., & Robertson, R. (2024). Exploring the Feasibility and Efficacy of hatGPT3 for Personalized Feedback in Teaching. *Electronic Journal of e-Learning*, 22(2), 98–111. DOI: 10.34190/ejel.22.2.3345

Negoi ă, D. O., & Popescu, M. A. M. (2023, December). The use of artificial intelligence in education. In *International Conference of Management and Industrial Engineering* (Vol. 11, pp. 208-214).

Panigrahi, R. R., Dash, M., Shaikh, Z. H., & Irfan, M. (2023). Review of Machine Learning Techniques in the Supply Chain Management of Indian Industry: A Future Research Agenda. In Irfan, M., Elhoseny, M., Kassim, S., & Metawa, N. (Eds.), *Advanced Machine Learning Algorithms for Complex Financial Applications* (pp. 199–219). IGI Global., DOI: 10.4018/978-1-6684-4483-2.ch013

Regona, M., Yigitcanlar, T., Xia, B., & Li, R. Y. M. (2022). Opportunities and Adoption Challenges of AI in the Construction Industry: A PRISMA Review. *Journal of Open Innovation*, 8(1), 45. DOI: 10.3390/joitmc8010045

Ruiz, P., & Fusco, J. (2024, March 31). Glossary of Artificial Intelligence Terms for Educators – CIRCLS. Circls.org. https://circls.org/educatorcircls/ai-glossary

Sarea, A., Subramanian, S., Alareeni, B., Shaikh, Z. H., Hawaldar, I. T., & Elshaker, A. H. (2021). Financial Disclosures by Using Machine Learning Analysis. In Alareeni, B., Hamdan, A., & Elgedawy, I. (Eds.), *The Importance of New Technologies and Entrepreneurship in Business Development: In The Context of Economic Diversity in Developing Countries. ICBT 2020. Lecture Notes in Networks and Systems* (Vol. 194). Springer., DOI: 10.1007/978-3-030-69221-6_5

Selwin, S., Saseendran, H., Vemavarapu, K. R., Shaikh, I., Shaikh, Z. H., & Harris, S. B. (2024). AI-Driven Employee Engagement: Transforming Banking Workforce Dynamics Using the BEULA Model. In Awwad, B. (Ed.), *The AI Revolution: Driving Business Innovation and Research. Studies in Systems, Decision and Control* (Vol. 525). Springer., DOI: 10.1007/978-3-031-54383-8_32

Seng, K. P., Ang, L.-M., Peter, E., & Mmonyi, A. (2023). Machine Learning and AI Technologies for Smart Wearables. *Electronics (Basel)*, 12(7), 1509. DOI: 10.3390/electronics12071509

Shaikh, Z., Irfan, M., Sarea, A., & Panigrahi, R. R. (2024). *The Emergence of Islamic Fintech and Bahrain: Prospect for Global Financial Sectors* (Vol. 503). Springer., DOI: 10.1007/978-3-031-43490-7_52

Shaikh, Z. H., Irfan, M., Nomran, N. M., & Lawin, M. L. (2024). The Effect of Strategic Partnership on Innovation and Business: Performance of the Fintech Industry in Bahrain. In Irfan, M., Kadry, S., Sharif, M., & Khan, H. (Eds.), *Fintech Applications in Islamic Finance: AI, Machine Learning, and Blockchain Tech-niques* (pp. 236–254). IGI Global., DOI: 10.4018/979-8-3693-1038-0.ch016

Shaikh, Z. H., Irfan, M., Nomran, N. M., & Ratnakaram, S. P. (2024). Unlocking the Po-tential of AI for Efficient Governance: Innovative Approaches of Bahrain. In Irfan, M., Elmogy, M., Gupta, S., Khalifa, F., & Dias, R. (Eds.), *AI-Driven Decen-tralized Finance and the Future of Finance* (pp. 115–142). IGI Global., DOI: 10.4018/979-8-3693-6321-8.ch005

Shaikh, Z. H., Sarea, A., & Irfan, M. (2022). *Islamic Banking Strategies in the World of Fintech: Success Story of Bahrain* (Vol. 423). Springer., DOI: 10.1007/978-3-030-93464-4_10

Shajari, S., Kuruvinashetti, K., Komeili, A., & Sundararaj, U. (2023). The Emergenceof AI-Based Wearable Sensors for Digital Health Technology: A Review. *Sensors (Basel)*, 23(23), 9498. DOI: 10.3390/s23239498 PMID: 38067871

Sharma, R. (2024, October 30). Future of Education: AI Compliance with FERPA and GDPR - Hurix Digital. Digital Engineering & Technology | Elearning Solutions | Digital Content Solutions. https://www.hurix.com/data-privacy-in-education-through-ferpa-and-gdpr-adherence/

Sheela, P., Kusuma, K., Panigrahi, R. R., & Shaikh, Z. H. (2024). Fintech and Islamic Banking: A Systematic View and Future Research Agenda. In Irfan, M., Kadry, S., Sharif, M., & Khan, H. (Eds.), *Fintech Applications in Islamic Finance: AI, Machine Learning, and Blockchain Techniques* (pp. 1–23). IGI Global., DOI: 10.4018/979-8-3693-1038-0.ch001

Shishakly, R., Almaiah, M. A., Lutfi, A., & Alrawad, M. (2024). The influence of using smart technologies for sustainable development in higher education institutions. *International Journal of Data and Network Science*, 8(1), 77–90. DOI: 10.5267/j.ijdns.2023.10.015

Silva, L. M., Cibrian, F. L., Monteiro, E. M., Bhattacharya, A., & Beltran, J. A. (2023). *Clarisse Bonang, Epstein, D. A., Schuck, S., Lakes, K. D., & Hayes, G. R.* Unpacking the Lived Experiences of Smartwatch Mediated Self and Co-Regulation with ADHD Children., DOI: 10.1145/3544548.3581316

Tamkeen to empower 60 Bahraini nationals with Machine Learning and Artificial Intelligence training in Collaboration with AWS and BIBF. (2023, December 5). https://www.tamkeen.bh/en/tamkeen-to-empower-60-bahraini-nationals-with-machine-learning-and-artificial-intelligence-training-in-collaboration-with-aws-and-bibf/#:~:text=The%20program,%20in%20collaboration%20with%20Amazon

Tiwari, R. (2023). The integration of AI and machine learning in education and its potential to personalize and improve student learning experiences. *INTERANTIONAL JOURNAL of SCIENTIFIC RESEARCH in ENGINEERING and MANAGEMENT*, 07(02). Advance online publication. DOI: 10.55041/IJSREM17645

Trotsko, A., Kharkiv, H., Olena, B., Kirilenko, G., Halyna, B., & Trush, O. (2019). TEACHERS' PROFESSIONAL SELF- IMPROVEMENT IN THE CONDITIONS OF DISTANCE LEARNING IMPLEMENTATION IN HIGHER EDUCATION INSTITUTIONS Liudmyla S. Rybalko. *Information Technologies and Learning Tools*, 72(4). Advance online publication. DOI: 10.33407/itlt.v72i4.3088

Tuomi, I. (2019). *The Impact of Artificial Intelligence on Learning, Teaching, and Education: Policies for the Future. JRC Science for Policy Report.* European Commission.

Yang, Y. (2024). Application of wearable devices based on artificial intelligence sensors in sports human health monitoring. *Measurement. Sensors*, 33, 101086. DOI: 10.1016/j.measen.2024.101086

Chapter 6
Engaging Minds:
Gamification and Interactive Learning With Wearable Devices

Habeeb Ur Rahiman
https://orcid.org/0000-0002-8042-7375
Kingdom University, Bahrain

Rashmi Kodikal
https://orcid.org/0000-0001-7026-6654
Graphic Era University, India

Fathima Mehnaz
https://orcid.org/0009-0006-1066-3686
Sahyadri College of Engineering and Management, India

Gaman R. G.
http://orcid.org/0009-0005-9580-4136
NMAM Institute of Technology, Nitte University, India

ABSTRACT

The research paper examines the importance of gamified education system and relationship between elements, features and experiences of wearable devices in gamified education system though user motivation and engagement on learning outcomes and experience. The study administered structural equation model to identify the relationship between these variables. To reach the objective the of the research, the study applied various statistical techniques such as factor analysis, Kaiser–Meyer–Olkin (KMO), Bartlett's test, correlation and regression analysis, descriptive statistics. The outcome of the result indicated significant relation between these variables on study outcomes and student engagement. The study highlighted the importance of gamification in the higher education by embedding in to learning

DOI: 10.4018/979-8-3693-7817-5.ch006

management system. The technicality of the study notified the role of the gamified element, and its features significantly contributed to the student's engagement and learning and assessment.

1. INTRODUCTION

In a recent academic environment, incorporation of technology into educational delivery has changed a higher educational landscape. Gamification and wearable devices are one such revolutionary attempts in the post covid era that changed the perception of higher education amongst educators and students. Gamification involves incorporating game elements, such as point systems, leaderboards, and badges, into traditional learning activities to boost engagement and motivation (Lin et al., 2018). Gamified educational approach has successfully contributed to gaining attention from various outcome-based education. This approach has gained popularity because it leverages people's natural competitiveness and desire for recognition. In education, gamification can make learning more engaging, encouraging students to participate more frequently and for longer periods, ultimately improving outcomes (Annansingh, 2017; Figg & Jaipal-Jamani, 2018a; Stansbury & Earnest, 2017). The world has witnessed tough time during the pandemic and lockdown in most part of the world disturbed on campus class room to curb the spread of covid. Although educational institutional adapted online mode of delivery as part of alternative, many institutions identified online have a future opportunity with the help of embedded technology and gamified tools. For instance, in a Physics course, an online discussion forum might use a badge system where students earn badges like "Ptolemy" after 10 posts, "Galileo" after 20, "Kepler" after 30, and "Einstein" after 40. These badges are visible to all, fostering a sense of camaraderie or competition among students. On the other hand, game-based learning involves creating learning activities where the structure and principles of games are embedded within the activity itself(Lambert & Watkins, 2013; Tsay & Kofinas, 2017). Studies recognised that, in hybrid class approach, incorporating game like elements are proved to be important elements and strategy for increasing engagement and motivations. For example, in an Economics course, students might take part in a virtual stock-trading competition, while in a Political Science course, they could engage in mock negotiations by role-playing a labor dispute.

Previous studies identified that minimal interaction in the classroom possibly hinder students' concentration and motivation, leading to increased anxiety and stress (Pereira et al., 2016; Villagrasa et al., 2014). In a higher educational environment, students' participation and involvement in all means is highly essential where opportunities to practice and receive feedback are crucial. Therefore, it is essential for educators to find ways to engage students and reduce anxiety. Traditional learning methods alone are insufficient for addressing these challenges, to enhance engagement, motivation, and academic performance while reducing anxiety. Considering the current educational system and tech-based generation which sophisticated knowledge over various gadgets and platforms, it is essential to adapt game-based learning or gamification in routine educational practices is a vital recommended approach to establish innovative learning practice (Tan, 2013; Villagrasa & Duran, 2013).

Teaching through gamified system is an innovative and required educational approach that combines game elements with instructional content to boost students' enthusiasm, participation, and learning effectiveness (Caponetto et al., 2014; Ortiz-Rojas et al., 2017). As information technology rapidly advances and educational concepts evolve, gamification is becoming a crucial aspect of educational reform. Unlike traditional teaching methods, where students often passively receive knowledge with limited motivation, gamified teaching introduces game design, competitive elements, and incentives to ignite students' interest and enhance their learning outcomes.

The concept of educational games emerged in the 1980s from the fusion of video games and instructional design (Goshevski et al., 2017; Mora et al., 2018). Current international research focuses on both the theory and practice of educational games. Notably, Kristian Kiili from Finland developed the experiential game model, while Alan Amory and Robert Seagram introduced the game achievement model. The success of educational games relies on finding a balance between game theory and educational theory, ensuring that both elements work together effectively (Hung, 2017; Titus & Ng'ambi, 2014). Therefore, the current study is intended to explore the impact of gamification and interactive learning with wearable devices towards students' cognitive engagement, learning outcome and user engagement in higher education system.

Theoretical Framework

Multiple psychological and educational theories are rooted in gamification in higher education that explain its effectiveness in enhancing learning and motivation. One of the fundamental theories the study adopted is Self-determination theory developed by Deci & Ryan (1985), posits that individuals are motivated when their basic psychological needs are fulfilled. This theory provides a robust explanation of human motivation, suggesting that individuals are highly motivated when *autonomy, competence, and relatedness,* i.e. the three basic psychological needs are fulfilled. The current research identifies how gamification and interactive learning with wearable devices contributes to the enhancement and productivity of higher education. The SDT has been considered after ensuring its alignment that gamified environments not only engage students but also address their core psychological needs, as proposed by SDT. Studies have shown that such alignment significantly boosts motivation and learning outcomes (Erenli, 2013).

The other theory considered for this study was considered Experiential Learning Theory (ELT) proposed by Kolb et al. (2014) which highlights the centrality of hands-on experiences and reflective processes in effective learning. The ELT theory focuses on how knowledge is structured through a systematic cycle of concrete experiences, reflective and profound observation, abstract conceptualization, and functional experimentation. This cycle facilitates deeper understanding, skill retention, and application of knowledge in real-world contexts.

The theory of ELT advocates and emphasises the application of wearable devices, such as augmented reality (AR) glasses, virtual reality (VR) headsets, and haptic feedback tools, are most effective tools incorporating gamified learning environment in classrooms and these instruments create real time experience and opportunities for learners in improving their knowledge and transformative skills. These technologies immerse learners in interactive scenarios that mimic real-world situations or simulated environments, making learning more dynamic and impactful.

Kolb et al. Experiential Learning Theory, when combined with wearable technologies and gamification, provides a robust framework for interactive learning. This approach helps the learners in understanding the real-world or simulated environments, wearable devices foster hands-on experiences. The outcome of this approach promotes critical thinking, skill development, and knowledge retention, ensuring that learning is not only effective but also engaging and applicable to practical challenges.

2. LITERATURE REVIEW

2.1. Gamification in Education

There are remarkable studies identified in gamification in education. Many studies in this domain focused on the role of gamification in higher education and its impact on classroom engagement. The literature also focused on how gamification can enhance students' classroom participation and attention in delivery content. Although previous studies focused on various aspects, the current study identified multiple gaps in existing research. One of the major gaps the study observed is how gamification and wearable devices induce behavioral changes in students.

The gamified educational model is an innovative and creative teaching approach that seamlessly integrates game design elements with educational content (De Freitas et al., 2017). By incorporating features like level progression, reward systems, and role-playing, this method aims to spark students' interest, boost their motivation, and enhance learning outcomes. In and hybrid leaning environment, many institutions adopted tech-embedded learning management systems to make continues engagement of students and their participation. Multiple studies noted that this transform learning into a more engaging and enjoyable experience, allowing students to acquire knowledge and develop skills in a relaxed and pleasant environment (Figg & Jaipal-Jamani, 2018b; Lamrani et al., 2018). Additionally, gamification encourages the development of students' independent learning abilities and problem-solving skills, fostering active participation in the learning process(Cliffe Schreuders & Butterfield, 2016; Loos & Crosby, 2017).

Duggal & Gupta (2020) noticed a novel gamification-based approach to enhance classroom engagement. In this model, educational content is designed with game-like characteristics, such as levels, rewards, and role-playing, making the learning process more engaging and appealing. These elements help ignite students' enthusiasm, increase their initiative, and make learning more conducive to memory retention and understanding (Boskic & Hu, 2015; Daud et al., 2017). In an extended study, researcher identified that the gamification approach can be tailored to different levels of difficulty, allowing students to improve their abilities through continuous challenges and attempts (Costello, 2017). Through LMS, such system and approach also provide immediate feedback and rewards, enhancing students' motivation and efficiency.

Another vital aspect of the research in the past is the role of gamified or wearable devices in supplying feedback. As students complete various tasks and challenges, they experience a sense of achievement and receive rewards, further fueling their motivation and enthusiasm (Mora et al., 2016; Schoenenberger et al., 2016). The gamified teaching process can incorporate competition and teamwork to encourage interaction among students and foster a dynamic learning environment. Competition stimulates students' drive and initiative, cultivates their problem-solving skills, and adds interest and challenge to learning(Bellotti et al., 2013; Clements et al., 2017). Meanwhile, cooperative learning promotes communication and mutual support, improves efficiency, and nurtures teamwork, social skills, and leadership abilities (Rahiman et al., 2023).

The combination of competition and cooperation in teaching is not mutually exclusive; rather, it corresponds to competent stimulating students' excitement for studying and promoting their overall development (Tsay et al., 2018). The gamification model also emphasizes the importance of timely feedback, which helps students to regulate their learning approaches and reinforces the learning effect. The study also noticed that timely feedback plays a critical role in teaching, positively influencing students' progress by providing them with clear insights into their performance and areas for improvement (Sulphey, 2017). This feedback enhances motivation and encourages students to stay engaged in the learning process. With the engaging and interactive nature of games, the gamified education model enables students to achieve their educational goals in a relaxed and enjoyable atmosphere. It also offers students greater autonomy in selecting their learning content and path, enhancing the adaptability and personalization of their educational experience.

2.2. Gamification and Students' Engagement

Engagement, while not as recent a concept as gamification, remains a topic of significant debate. The study of student engagement is extensive, yet the challenge lies in creating a clear and concise framework due to the abundance of theoretical constructs (Herbert et al., 2011). As various interpretations of student engagement have emerged, they have become crucial to modern measures of success in higher education. Landers suggested that gamification influences behaviors and attitudes that, in turn, affect learning outcomes (Hew et al., 2016; Signori et al., 2018). To effectively use gamification for influencing these behaviors and attitudes, it's essential to understand them fully. Focusing on student engagement, Landers described it as a target attitude (often linked with fun) that can be modified through gamification (Lockley & Boyle, 2014; Noran, 2016). This modification has the potential to mediate the relationship between instructional content and learning outcomes. We conducted a literature review to identify research discussing student engagement as it relates to psychological influences on learning, which aligns with Landers' work.

The most straightforward framework available for describing student engagement is provided by (Gordon et al., 2013). Kahu identifies various instances where engagement is often conflated with its causes or effects, such as motivation or learning. By examining different approaches to engagement, Kahu resolves these conflations and highlights how the remaining aspects fit together. Two key approaches are the psychological and behavioral perspectives, both of which align with Landers' theory (Vanduhe et al., 2020).

In the context of higher education, student engagement is often measured by changes in student behavior prompted by effective teaching practices (Altmann et al., 2018; Glez Almaguer et al., 2021). This behavioral perspective is the first element of Kahu's framework to be considered. There are parallels between Landers' idea of student engagement as a modifiable attitude through gamification and the behavioral perspective, which views engagement as influenced by teaching practices (dos Reis Lívero et al., 2021; Siklander & Harmoinen, 2021). However, this behavioral definition of engagement is limited, as it focuses solely on observable behaviors, overlooking important factors like affect. Additionally, it fails to consider the dynamic and situational nature of student engagement, focusing only on what higher education institutions can control. The complexities of engagement and its influences are also challenging to disentangle from this perspective, which is a limitation shared with Landers' theory (Al-Azawi et al., 2019; Saputra et al., 2021).

2.3. Gamification and Students' Performance

Gamification is a key aspect to enhance students' performance in routine class assessment and their participation in classroom. Plethora of studies notified various gamified tools encouraged students to actively participate in various activities in class more effectively in compared to traditional approach (Järveläinen, 2021; Venter, 2020). Recent research in education highlights the growing importance of acquiring practical skills and engaging in active learning. Previous studies emphasize the value of innovative educational methods, such as collaborative learning, service learning, simulation-based experiential learning, and movie analysis (Huang & Hew, 2018). In management education, numerous studies stress the importance of providing students with experiential learning opportunities, immersing them in real-world business decision-making. For instance, one study on simulation learning through a stock market analysis project shows that this hands-on approach enhances student satisfaction (Rincon-Flores & Santos-Guevara, 2021). This method not only discourages surface learning but also fosters deeper learning strategies. Active teaching strategies that give students firsthand experience and a central role in their learning process have become crucial for improving academic performance in today's educational environment.

3. RESEARCH METHODS

3.1. Research Design

The current study relied on survey and observation-based approach by collecting samples from selected states from India and Kingdom of Bahrain. The study was conducted in the graduation and master's degree program classroom with various gamified tools namely Kahoot and Manti-meter. The students were asked to participate in classroom activities through this tool and identify their performance, engagement and level of satisfaction in teaching and learning tasks.

3.2. Data Collection and Sampling

The data was collected in various classes considering its nature (theoretical and practical) in both graduation and master's degree programs. A total of 270 samples were collected and classified (159 from India) and (111 from Bahrain). The data was obtained from students who participated in a gamified learning environment using platforms such as Moodle and Brightspace, which serve as the learning management systems (LMS) for these universities.

Apart from the sample survey, the researcher also identified the students' participation in the classroom activities. All the activities were noted, and the level of participation was monitored. Based on the tools applied in the classroom, all the activities and student's performance were recorded to identify the level of performance and engagement in the classroom. This allowed for a detailed analysis of the students' overall engagement and performance, helping to determine the effectiveness of the gamified learning approach.

4. RESULTS

General Characteristics of the Samples

The samples of the respondents collected from India and Bahrain. From India 350 respondents including 240 from graduation program and 110 from Master program. Similarly, from Bahrain, 170 respondents listed and 117 from Graduation level and 53 from Master program.

The Table 1 indicates Kaiser–Meyer–Olkin (KMO) Measure the adequacy of the sampling for the dataset found to be 0.823, point at a high level of adequacy for factor analysis. Furthermore, Bartlett's Test of Sphericity reached statistical significance (Chi-Square 2816.437, df =176, Sig.=0.001), indicating that the correlation matrix is not an identity matrix, confirming the suitability of the data for factor analysis.

Table 1. KMO and Bartlett's test

Kaiser–Meyer–Olkin Measure of Sampling Adequacy	**0.823**
Bartlett's Test of Sphericity	2816.437
df	176
sig	0.001

Sources: Data Analysis

Table 2. Factor Matrix

Variables	Items	1	2	3	4	5	6
Gamification Elements	GL1	0.768					
	GL2	0.897					
	GL3	0.689					
Interactivity Features	IF1				0.890		
	IF2				0.765		
Experience with Wearable Devices	EWD1			0.786			
	EWD2			0.750			
	EWD3			0.786			
User Motivation and Cognitive Engagement	LO1		0.890				
	LO1		0.756				
	LO1		0.690				
Learning Outcomes	SE1					0.897	
	SE2					0.765	
	SE3					0.679	
User Experience	UE1						0.785
	UE2						0.780
	UE3						0.765

Source: Data analysis

The researcher has administered rotated factor matrix through Maximum Likelihood extraction (MLE) and Varimax rotation with Kaiser Normalization. The results indicate different factors for different variables. Significant loadings are identified for variables related to gamification elements, demonstrating a cohesive factor associated with effective systematic gamification elements in the process. Similarly, the second factor captures variables related to Interactivity Features, with strong loadings on Interactivity Features1 and Interactivity Features2. The outcome further identified factors represents Experience with Wearable Devices, as evidenced by substantial loadings on EWD1 and EWD1. Further, prominent loadings on User Motivation and Cognitive Engagement and in extended the associated with variables related to Learning Outcomes, as indicated by loadings on SE1, SE2 and SE3. Finally factor six represents User Experience. The rotation method converged after six repetitions, solidifying the stability of the factor solution. The results of rotated factor matrix indicate an effective interpretation of the underlying factors influencing the observed variables, enabling a more nuanced understanding of the associations within the dataset (Table 2).

To test the relationship between the variables, a study used structural equation modeling (SEM) and CFA. We performed SEM and CFA using SPSS and AMOS to perform statistical analysis and structural equation models. In summary, Structural Equation Modeling (SEM) is a group of multivariate statistical techniques used to analyze complex relationships between measured variables and latent constructs. Confirmatory Factor Analysis (CFA) is employed to validate the factor structure of a set of observed variables. The results from SEM offer valuable insights into the relationships between key constructs, such as learning outcomes and user experience. The path coefficients, standard errors, critical ratios and p-values offer a comprehensive understanding of the associations between independent and dependent variables.

The result explores CFA found to be a positive correlation between gamification elements and learning outcomes and user experience (0.54), interactivity features (0.48), experience with wearable devices (0.38), user motivation and cognitive engagement (0.43). These results clearly indicate that students engaged in effective gamified systems are more engaged in class tend to have better academic performance. Similarly, user motivation and cognitive engagement are the most significant factors influencing performance.

The results of standardized factor loadings for items within six constructs i.e. gamification elements, interactivity features, experience with wearable devices, user motivation and cognitive engagement, learning outcomes and user experience. The results shows that gamification elements construct exhibits effective standardized factor loadings ranging from 0.693 to 0.796, supporting a high Cronbach's Alpha of 0.873, demonstrating good internal reliability. The Interactivity Features construct exhibits loadings between 0.765 and 0.890, resulting in a Cronbach's Alpha of 0.789. Experience with Wearable Devices, with loadings from 0.726 to 0.796, shows a high Cronbach's Alpha of 0.789. User Motivation and Cognitive Engagement, indicated by loadings between 0.790 and 0.856, has a Cronbach's Alpha of 0.897. Learning Outcomes, featured loading from 0.765 to 0.897, has a Cronbach's Alpha of 0.745. Finaly, User Experience, featuring loadings from 0.665 to 0.885, displays a Cronbach's Alpha of 0.870.

The results show that Composite Reliability (CR) values are consistently high across all constructs, ranging from 0.845 to 0.897. The Average Variance Extracted (AVE) for each construct is also satisfactory, ranging from 0.231 to 0.423, confirming convergent validity. The model fit indices, including a chi-square of 423.201 with 245 degrees of freedom, a chi-square/df ratio of 1.344, an RMSEA of 0.092, and a CFI of 0.784, suggest that the model fits the data reasonably well. Discriminant validity was assessed using the (Fornell & Larcker, 1981) criteria, where the square root of AVE must be greater than the inter-variable correlations. As this condition is met, we can conclude that the variables demonstrate good discriminant validity (Table 3).

Testing of Hypothesis

The pathway results indicate that there is a significant contribution between key variables (Figure1). The outcome of the SEM indicates that Learning Outcomes is positively influenced by Gamification Elements where (β =0.450, $p < 0.001$), interactivity features (β=0.633, $p < 0.001$), experience with wearable devices (β =0.350, $p < 0.001$), and user motivation and cognitive engagement (β=0.533, $p < 0.001$) signifying that effective gamified system with interactive features and motivation has an effective influence towards learning outcome. The further reveals that user motivation and cognitive engagement and learning outcome is positively associated with user experience (β = 0.350, $p < 0.001$), indicating that more motivated and engaged students tend to have better academic performance with user experience.

The result of the hypothesis indicates level of acceptance of each hypothesis proposed by the study. The Hypothesis 1, asserting an influence of Gamification Elements through User motivation and engagement enhances the learning outcome is supported, indicating that effective gamified tools tend to exhibit higher levels of engagement and results in better learning outcome. H2, which posits an influence of Interactivity Features on User motivation and engagement enhances the learning outcome is also supported, suggesting that effective Interactivity Features in given gamified tools contribute positively to student engagement and learning outcome. Similarly, H3, proposed influence of wearable devices experience through User motivation and engagement enhances the learning outcome is supported. This result indicates that there is a significant impact of experience with wearable devices on student engagement and learning outcome in the context of gamification. Finally, H4, the results identified that the influence of user motivation and cognitive engagement in gamification on student learning and user experience is supported. This outcome indicates that higher levels of use motivation and cognitive engagement are associated with better learning outcome and user experience in the gamified learning environment. These results specify constructive insights into the factors influencing student engagement and performance in a gamified educational environment.

Table 3. Reliability and convergent validity

Variables	Items	Factor Loading	Cronbach's alpha	Composite reliability	Average variance extracted
Gamification Elements	GL1	0.768	0.873	0.879	0.356
	GL2	0.797			
	GL3	0.693			

continued on following page

Table 3. Continued

Variables	Items	Factor Loading	Cronbach's alpha	Composite reliability	Average variance extracted
Interactivity Features	IF1	0.890	0.789	0.865	0.423
	IF2	0.765			
Experience with Wearable Devices	EWD1	0.796	0.789	0.897	0.378
	EWD2	0.750			
	EWD3	0.726			
User Motivation and Cognitive Engagement	LO1	0.790	0.897	0.886	0.234
	LO1	0.856			
	LO1	0.790			
Learning Outcomes	SE1	0.897	0.745	0.845	0.324
	SE2	0.765			
	SE3	0.779			
User Experience	UE1	0.885	0.870	0.876	0.231
	UE2	0.780			
	UE3	0.665			

Sources: Data Analysis

Figure 1. Structural Equation model

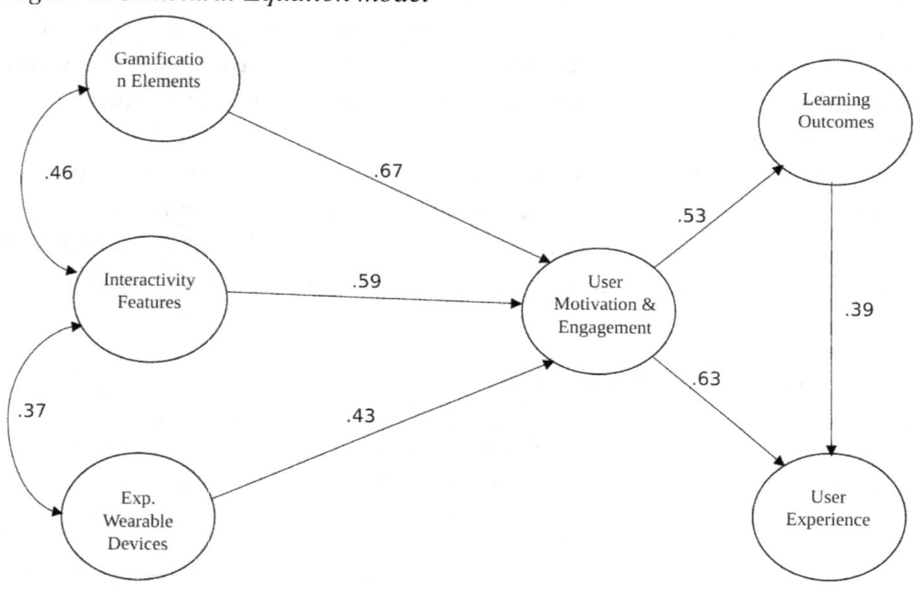

Source: Data analysis

5. DISCUSSION

The study results indicate highly relevance of gamified systems in classroom environment and its impact on students' engagement in class and their learning outcome as well as experience. The study has proposed three hypotheses in association with independent and dependent variables and identifies supportive results.

The element of gamified system is essential in the context of gamified classroom. The results indicated effective gamification elements enhanced the user motivation and cognitive engagement in classrooms. The results have been supported by existing literature like (Kumar & Al-Besher, 2022; Rivera & Garden, 2021) notified that generating systematic gamified framework with elements of gamified teaching and learning methods enhances the student's engagement. Similarly, in an extended study, (Sridhar et al., 2024) argued that the gamification application with metaverse technology enhances the work performance.

The study further identified the association between influence of interactivity features on user motivation and cognitive engagement. The results indicate positive outcome having efficient interactive features in a gamified teaching tool. The results also support the previous research outcome indicated that institutions with proper interactive features that embedded with the learning management system of the teaching process tend to have more outcome oriented than institution with relying upon normal traditional teaching methods (Hou et al., 2025; Lee et al., 2019).

The confirmed relationships between the influence of experience with wearable devices through user motivation and engagement enhances the learning outcome is also supported. The results indicate that virtual reality, argument reality, smartwatches, smart clothing and accessories and other online platforms are essential to enhance the user engagement and learning outcome. The results also indicate that utilizing such tools is the part of the essential learning scenario and required to enhance the user enhancement. In the same line of the observation have been made by (Bower & Sturman, 2015; Ghawail & Yahia, 2022).

User motivation and cognitive engagement in gamification on student learning outcome and user experience are essential aspects of the gamified learning environment. The results indicate that user motivation by implementing tools like Kahoot and Minti meter in classrooms play an effective role in enhancing student engagement to improve their transformation skills. Then user experience on the other hand in the learning process enhanced the implication of effective engagement strategy in study environment. The results backed by other notable research outcomes identified the role of these gamified tools and engagement strategies (Ghawail & Yahia, 2022; Kohnke & Moorhouse, 2021; Wirani et al., 2022).

These findings provide important insights into the growing body of research on gamification in education. The outcome of the research advocates the effectiveness of study environment by applying gamified study and learning management in higher education institutions. The educators and instructional designers can use this information to guide the creation and application of gamified learning strategies, highlighting the key roles of these listed variables to improve the individual student engagement in achieving successful educational outcomes.

6. CONCLUSION

The results of this study have identified how the gamification and interactive learning with wearable devices in the learning environment enhanced the user motivation and cognitive engagement and thereby improved the learning outcome and user experience. The result of the study indicates that gamification elements, interactivity features associated with gamified learning system, experience with wearable devices in the learning system are the key parameters that enhance the user engagement and learning outcomes. The results contribute significantly to proving the essentiality of the gamified system in the entire classroom performances of the students. The results investigated the role of technology and interaction in gamified environments and explored other factors like engagement and learning experience that improved the effectiveness of gamification in educational settings.

7. THEORETICAL IMPLICATIONS

Gamification is a transformative educational strategy that blends play with learning to engage today's learners. It draws on notable psychological theories like self-determination, flow, and narrative psychology (Rahiman et al., 2023). Gamification boosts cognitive engagement and learning experience of the learning by promoting deeper understanding, critical thinking, and problem-solving. It also enhances intrinsic motivation, leading to better academic performance through increased focus, engagement, and resilience (Rahiman & Kodikal, 2024). Additionally, gamification in a learning environment encourages social interaction, supporting both collaboration and healthy competition among students and boosts their transformative skills. By fostering intrinsic motivation, it can inspire a lifelong passion for learning, extending beyond short-term academic goals. However, its effectiveness depends on the context, and further research is needed to adapt it to changing learning preferences. The key outcome of the study is it contribute the essential of gamified education to educators, policymakers, and researchers to optimize teaching, improve student outcomes, and keep education relevant in an evolving world.

8. PRACTICAL IMPLICATIONS

The study provides a reliable framework for measuring student engagement and performance in educational environments. The outcome of the study explores the possible area in the learning management system to be upgraded to enhance the teaching process more effectively with gamified standard. The study highlights the need for innovative design to fully leverage gamification's potential. Future research can possibly explore the optimal conditions for implementing gamification, considering design, user characteristics, and the educational context. These findings can help make informed decisions about using gamification in education, promoting intrinsic motivation and aligning with learning objectives. The outcome of the study also emphasises the implication of gamified education system through the teaching and learning policy of the institution and thereby educating the faculty members on possible usage and its application in their routine curriculum. Since the finding explores the positive outcome for user experience and learning outcome, it is suggested to embed the gamified system in education, and classroom assessment.

Key findings suggest that wearable devices can enhance engagement by providing real-time performance feedback and promoting active learning through gamified tasks. The results highlighted in the results notify the role of gamification in adapting various successive environments in an organisation. Furthermore, students demonstrate higher motivation levels and improved learning retention when gamification elements are integrated into the curriculum. The research also highlights the potential challenges related to privacy, data security, and accessibility that need to be addressed for broader implementation.

The Outcome of the results identified the how the gamification and interactive learning with wearable devices in the learning environment enhanced the user motivation and cognitive engagement and thereby improved the learning outcome and user experience. This result indicated that effective implementation of the gamified elements through systematic interactive features in curriculum delivery would generate productive learning outcome. The results also emphasized that users' experience with wearable devices in the learning system are the key parameters that enhance the user engagement and learning outcomes. The results further elaborate the role of people dealing with wearable devices in implementing gamified systems significantly contributes to the outcome of learning engagement and performance. This result provides a significant contribution to various theories in developing academic contribution. The results contribute significantly to proving the essentiality of the gamified system in the entire classroom performances of the students. The results investigated the role of technology and interaction in gamified environments and explored other factors like engagement and learning experience that improved the effectiveness of gamification in educational settings.

The theoretical approach contributes to the advocate of the system embedded gamified system that boosts the cognitive engagement and learning experience of the of the learning by promoting deeper understanding, critical thinking, and problem-solving. It also enhances intrinsic motivation, leading to better academic performance through increased focus, engagement, and resilience. Additionally, gamification in a learning environment encourages social interaction, supporting both collaboration and healthy competition among students and boosts their transformative skills. By fostering intrinsic motivation, it can inspire a lifelong passion for learning, extending beyond short-term academic goals. However, its effectiveness depends on the context, and further research is needed to adapt it to changing learning preferences. The key outcome of the study is that it contributes the essential of gamified education to educators, policymakers, and researchers to optimize teaching, improve student outcomes, and keep education relevant in an evolving world.

The practical implications of the research are essential to understanding student psychology in their learning approach. This study significantly contributes to the academic world in considering gamified system to enhance learning experience, program development and certifying gamified atmosphere in learning more effectively and professionally. The results also stress the importance of a fun-learning approach which is successful in many aspects of academic settings. The study further contributes to the policy makers and strategic administrators of the institute considering this system as one of the study requirements is their strategic planning and part of the key performance index. The application of gamified system in higher education is most relevant in this prospective world.

REFERENCES

Al-Azawi, R., Joe, S. A., Al-Obaidy, M., & Westlake, J. (2019). The Use of Gamification Technique in Agile Development Methodology. In Sanchez, G., Rodríguez-González, S., Liberona, D., & Uden, L. (Eds.), *Communications in Computer and Information Science* (Vol. 1011, pp. 3–13). Springer Verlag., DOI: 10.1007/978-3-030-20798-4_1

Al Ghawail, E. A., & Ben Yahia, S. (2022). Using the E-Learning Gamification Tool Kahoot! to Learn Chemistry Principles in the Classroom. *Procedia Computer Science*, 207, 2667–2676. https://doi.org/https://doi.org/10.1016/j.procs.2022.09.325. DOI: 10.1016/j.procs.2022.09.325

Altmann, A., Ebersberger, B., Mössenlechner, C., & Wieser, D. (2018). The Disruptive Power of Online Education: Challenges, Opportunities, Responses. In *The Disruptive Power of Online Education: Challenges, Opportunities, Responses*. Emerald Group Publishing Ltd. https://www.scopus.com/inward/record.uri?eid=2-s2.0-85115278629&partnerID=40&md5=26565d3224daa5c211b62e480f0ad9ca

Annansingh, F. (2017). An investigation into the gamification of e-learning in higher education. In *Disability and Equity in Higher Education Accessibility* (pp. 268–284). IGI Global., DOI: 10.4018/978-1-5225-2665-0.ch012

Bellotti, F., Berta, R., De Gloria, A., Lavagnino, E., Antonaci, A., Dagnino, F. M., & Ott, M. (2013). A gamified short course for promoting entrepreneurship among ICT engineering students. *Proceedings - 2013 IEEE 13th International Conference on Advanced Learning Technologies, ICALT 2013*, 31–32. DOI: 10.1109/ICALT.2013.14

Boskic, N., & Hu, S. (2015). Gamification in higher education: How we changed roles. In L. Kolas & R. Munkvold (Eds.), *Proceedings of the European Conference on Games-based Learning* (Vols. 2015-January, pp. 741–748). Dechema e.V. https://www.scopus.com/inward/record.uri?eid=2-s2.0-84955154619&partnerID=40&md5=b0fcf9bcb3a235f668a0fff8aa1a224a

Bower, M., & Sturman, D. (2015). What are the educational affordances of wearable technologies? *Computers & Education*, 88, 343–353. https://doi.org/https://doi.org/10.1016/j.compedu.2015.07.013. DOI: 10.1016/j.compedu.2015.07.013

Caponetto, I., Earp, J., & Ott, M. (2014). Gamification and education: A literature review. In C. Busch (Ed.), *Proceedings of the European Conference on Games-based Learning* (Vol. 1, pp. 50–57). Dechema e.V. https://www.scopus.com/inward/record.uri?eid=2-s2.0-84923559781&partnerID=40&md5=84bb4279ab429280fa63ad588e2e6552

Clements, A. J., Ahmed, S., & Henderson, B. (2017). Student experience of gamified learning: A qualitative approach. In Pivec, M., & Grundler, J. (Eds.), *Proceedings of the 11th European Conference on Games Based Learning, ECGBL 2017* (pp. 88–94). Academic Conferences and Publishing International Limited., https://www.scopus.com/inward/record.uri?eid=2-s2.0-85036452359&partnerID=40&md5=9f24fc6f67145e6969bd3c6974a90ea1

Cliffe Schreuders, Z., & Butterfield, E. (2016). Gamification for teaching and learning computer security in higher education. *2016 USENIX Workshop on Advances in Security Education, ASE 2016, Co-Located with the 25th USENIX Security Symposium*. https://www.scopus.com/inward/record.uri?eid=2-s2.0-85081683632&partnerID=40&md5=3d08087cb4e82536a51f8c9c113e1963

Costello, R. (2017). Gaming innovations in higher education: Emerging research and opportunities. In *Gaming Innovations in Higher Education: Emerging Research and Opportunities*. IGI Global., DOI: 10.4018/978-1-5225-2981-1

Daud, R., Salam, S., Mohamad, S. N. M., & Yusoff, A. M. (2017). Modeling a mobile gamification model to increase student engagement: An analysis using analytic hierarchy process. *Advanced Science Letters*, 23(9), 8707–8712. DOI: 10.1166/asl.2017.9955

De Freitas, S., Irving, L., Verbert, K., Gibson, D., Star, K., Alvarez, V., & Charleer, S. (2017). How to use gamified dashboards and learning analytics for providing immediate student feedback and performance tracking in higher education. *26th International World Wide Web Conference 2017, WWW 2017 Companion*, 429–434. DOI: 10.1145/3041021.3054175

Deci, E. L., & Ryan, R. M. (1985). The general causality orientations scale: Self-determination in personality. *Journal of Research in Personality*, 19(2), 109–134. DOI: 10.1016/0092-6566(85)90023-6

dos Reis Lívero, F. A., da Silva, G. R., Amaral, E. C., de Souza, A. N. V., Baretta, I. P., Diegues, M. E. M., Arpini, E., & Lovato, E. C. W. (2021). Playfulness in the classroom: Gamification favor the learning of pharmacology. *Education and Information Technologies*, 26(2), 2125–2141. DOI: 10.1007/s10639-020-10350-w

Duggal, K., & Gupta, L. R. (2020). Hope Enabler: A Novel Gamification-Based Approach to Enhance Classroom Engagement. In *Lecture Notes in Networks and Systems* (Vol. 121, pp. 501–519). Springer. DOI: 10.1007/978-981-15-3369-3_38

Erenli, K. (2013). The impact of gamification-recommending education scenarios. [IJET]. *International Journal of Emerging Technologies in Learning*, 8(S1), 15. Advance online publication. DOI: 10.3991/ijet.v8iS1.2320

Figg, C., & Jaipal-Jamani, K. (2018a). Developing teacher knowledge about gamification as an instructional strategy. In *Ophthalmology: Breakthroughs in Research and Practice* (pp. 227–255). IGI Global. DOI: 10.4018/978-1-5225-5198-0.ch013

Figg, C., & Jaipal-Jamani, K. (2018b). Developing teacher knowledge about gamification as an instructional strategy. In *Teacher Training and Professional Development: Concepts, Methodologies, Tools, and Applications* (Vol. 3, pp. 1215–1243). IGI Global. DOI: 10.4018/978-1-5225-5631-2.ch056

Fornell, C., & Larcker, D. F. (1981). Evaluating structural equation models with unobservable variables and measurement error. *JMR, Journal of Marketing Research*, 18(1), 39–50. DOI: 10.1177/002224378101800104

Glez Almaguer, C. A., Aguirre Acosta, A. C., & Roman Jimenez, O. R. (2021). "Wow" Experience with Immersive Reality: Gamification in the Tec Virtual Campus. *Future of Educational Innovation Workshop Series - Machine Learning-Driven Digital Technologies for Educational Innovation Workshop 2021*. DOI: 10.1109/IEEECONF53024.2021.9733757

Gordon, N., Brayshaw, M., & Grey, S. (2013). Maximising gain for minimal pain: Utilising natural game mechanics. *ITALICS Innovations in Teaching and Learning in Information and Computer Sciences*, 12(1), 27–38. DOI: 10.11120/ital.2013.00004

Goshevski, D., Veljanoska, J., & Hatziapostolou, H. (2017). A review of gamification platforms for higher education. *ACM International Conference Proceeding Series, Part F130953*. DOI: 10.1145/3136273.3136299

Herbert, B. J. P., Charles, D., McNeill, M., Moore, A., & Charles, M. (2011). Dynamic virtual learning landscapes to enhance student reflective processes. In M. Meimaris & D. Gouscos (Eds.), *Proceedings of the European Conference on Games-based Learning* (Vols. 2011-January, pp. 691–702). Dechema e.V. https://www.scopus.com/inward/record.uri?eid=2-s2.0-84938566496&partnerID=40&md5=d0854015789791774ab5505c0f680960

Hew, K. F., Huang, B., Chu, K. W. S., & Chiu, D. K. W. (2016). Engaging Asian students through game mechanics: Findings from two experiment studies. *Computers & Education*, 92–93, 221–236. DOI: 10.1016/j.compedu.2015.10.010

Hou, Y., Li, Z., & Li, H. (2025). Sensor based interactive digital entertainment and gamified training to alleviate basketball player fatigue. *Entertainment Computing*, 52, 100838. https://doi.org/https://doi.org/10.1016/j.entcom.2024.100838. DOI: 10.1016/j.entcom.2024.100838

Huang, B., & Hew, K. F. (2018). Implementing a theory-driven gamification model in higher education flipped courses: Effects on out-of-class activity completion and quality of artifacts. *Computers & Education*, 125, 254–272. DOI: 10.1016/j.compedu.2018.06.018

Hung, A. C. Y. (2017). A critique and defense of gamification. *Journal of Interactive Online Learning*, 15(1), 57–72. https://www.scopus.com/inward/record.uri?eid=2-s2.0-85020689516&partnerID=40&md5=8556a4eb9a9170e270389dc58fbe48b8

Järveläinen, J. (2021). Collective Mindful and Mindless Behavior in Game-Based Learning – Comparing Physical and Digital Educational Escape Rooms. *42nd International Conference on Information Systems, ICIS 2021 TREOs: "Building Sustainability and Resilience with IS: A Call for Action."* https://www.scopus.com/inward/record.uri?eid=2-s2.0-85159101926&partnerID=40&md5=999bdac99f198f4b41a344f39d831652

Kohnke, L., & Moorhouse, B. L. (2021). Using Kahoot! to Gamify Learning in the Language Classroom. *RELC Journal*, 53(3), 769–775. DOI: 10.1177/00336882211040270

Kolb, D. A., Boyatzis, R. E., & Mainemelis, C. (2014). Experiential learning theory: Previous research and new directions. In *Perspectives on thinking, learning, and cognitive styles* (pp. 227–247). Routledge. DOI: 10.4324/9781410605986-9

Kumar, K., & Al-Besher, A. (2022). IoT enabled e-learning system for higher education. *Measurement. Sensors*, 24, 100480. https://doi.org/https://doi.org/10.1016/j.measen.2022.100480. DOI: 10.1016/j.measen.2022.100480

Lambert, N., & Watkins, L. (2013). Meet Mohammed: Using simulation and technology to support learning. *The Journal of Mental Health Training, Education and Practice*, 8(2), 66–75. DOI: 10.1108/JMHTEP-03-2012-0001

Lamrani, R., Abdelwahed, E. H., Chraibi, S., Qassimi, S., & Hafidi, M. (2018). Gamification and serious games based learning for early childhood in rural areas. In Golfarelli, M., Bellatreche, L., Jean, S., Nakamatsu, K., Ordonez, C., Mery, D., Benslimane, D., & Abdelwahed, E. H. (Eds.), *Communications in Computer and Information Science* (Vol. 929, pp. 79–90). Springer Verlag., DOI: 10.1007/978-3-030-02852-7_7

Lee, U., Han, K., Cho, H., Chung, K.-M., Hong, H., Lee, S.-J., Noh, Y., Park, S., & Carroll, J. M. (2019). Intelligent positive computing with mobile, wearable, and IoT devices: Literature review and research directions. *Ad Hoc Networks*, 83, 8–24. https://doi.org/https://doi.org/10.1016/j.adhoc.2018.08.021. DOI: 10.1016/j.adhoc.2018.08.021

Lin, D. T. A., Ganapathy, M., & Kaur, M. (2018). Kahoot! It: Gamification in higher education. *Pertanika Journal of Social Science & Humanities*, 26(1), 565–582. https://www.scopus.com/inward/record.uri?eid=2-s2.0-85045525024&partnerID=40&md5=2f289beb827fad108109309ed6388c45

Lockley, A., & Boyle, A. (2014). Towards a game-based learning ecosystem: An institutional strategy. In C. Busch (Ed.), *Proceedings of the European Conference on Games-based Learning* (Vol. 1, pp. 312–320). Dechema e.V. https://www.scopus.com/inward/record.uri?eid=2-s2.0-84923587008&partnerID=40&md5=bbaaab7906d3490e3c2c20d4f35389bf

Loos, L. A., & Crosby, M. E. (2017). Gamification methods in higher education. In P. Zaphiris & A. Ioannou (Eds.), *Lecture Notes in Computer Science (including subseries Lecture Notes in Artificial Intelligence and Lecture Notes in Bioinformatics): Vol. 10295 LNCS* (pp. 474–486). Springer Verlag. DOI: 10.1007/978-3-319-58509-3_37

Mora, A., Planas, E., & Arnedo-Moreno, J. (2016). Designing game-like activities to engage adult learners in higher education. In F. J. Garcia-Penalvo & R. I. for E. S. (IUCE) University of Salamanca Department of Computer Science and Automatics Plaza de los Caidos s/n Salamanca (Eds.), *ACM International Conference Proceeding Series* (Vols. 02-04-November-2016, pp. 755–762). Association for Computing Machinery. DOI: 10.1145/3012430.3012603

Mora, A., Tondello, G. F., Nacke, L. E., & Arnedo-Moreno, J. (2018). Effect of personalized gameful design on student engagement. *IEEE Global Engineering Education Conference, EDUCON, 2018-April*, 1925–1933. DOI: 10.1109/EDUCON.2018.8363471

Noran, O. (2016). On gamification in action learning. *ACM International Conference Proceeding Series, 01-05-February-2016*. DOI: 10.1145/2843043.2843344

Ortiz-Rojas, M., Chiluiza, K., & Valcke, M. (2017). Gamification in computer programming: Effects on learning, engagement, self-efficacy and intrinsic motivation. In Pivec, M., & Grundler, J. (Eds.), *Proceedings of the 11th European Conference on Games Based Learning, ECGBL 2017* (pp. 507–514). Academic Conferences and Publishing International Limited., https://www.scopus.com/inward/record.uri?eid=2-s2.0-85036465514&partnerID=40&md5=a3e843cf56ebf4ea0f51d4f209fb385d

Pereira, A. S., Moreira, A. A., Chaló, P., Sancho, L., Varela, A., & Oliveira, C. (2016). Development challenges of a full integrated app in higher education. In *Gaming and Technology Addiction: Breakthroughs in Research and Practice* (Vol. 1, pp. 117–139). IGI Global. DOI: 10.4018/978-1-5225-0256-2.ch001

Rahiman, H. U., & Kodikal, R. (2024). Revolutionizing education: Artificial intelligence empowered learning in higher education. *Cogent Education*, 11(1), 2293431. Advance online publication. DOI: 10.1080/2331186X.2023.2293431

Rahiman, H. U., Kodikal, R., & Suresh, S. (2023). Game on: Can gamification enhance productivity? *F1000 Research*, 12, 818. DOI: 10.12688/f1000research.131579.2 PMID: 38434668

Rincon-Flores, E. G., & Santos-Guevara, B. N. (2021). Gamification during Covid-19: Promoting active learning and motivation in higher education. *Australasian Journal of Educational Technology*, 37(5), 43–60. DOI: 10.14742/ajet.7157

Rivera, E. S., & Garden, C. L. P. (2021). Gamification for student engagement: A framework. *Journal of Further and Higher Education*, 45(7), 999–1012. DOI: 10.1080/0309877X.2021.1875201

Saputra, N., Satispi, E., & Herlina, M. G. (2021). Growing Learning Agility by Combining Learning Culture with Knowledge Management. *2021 International Conference on Advanced Computer Science and Information Systems, ICACSIS 2021*. DOI: 10.1109/ICACSIS53237.2021.9631346

Schoenenberger, C.-A., Korkut, S., Jaeger, J., & Dornberger, R. (2016). BioTourney: Gamifying a biology class by applying a content-independent learning game framework. In L. Boyle, L. Boyle, T. M. Connolly, & T. M. Connolly (Eds.), *Proceedings of the European Conference on Games-based Learning* (Vols. 2016-January, pp. 575–583). Dechema e.V. https://www.scopus.com/inward/record.uri?eid=2-s2.0-84997018044&partnerID=40&md5=6ab4bce210b027b2dc922d71f7df4702

Signori, G. G., De Guimarães, J. C. F., Severo, E. A., & Rotta, C. (2018). Gamification as an innovative method in the processes of learning in higher education institutions. *International Journal of Innovation and Learning*, 24(2), 115–137. DOI: 10.1504/IJIL.2018.094066

Siklander, P., & Harmoinen, S. (2021). Ice Age is Approaching: Triggering University Students' Interest and Engagement in Gamified Outdoor Playful Learning Activities. In *Science and Drama: Contemporary and Creative Approaches to Teaching and Learning* (pp. 125–143). Springer International Publishing. DOI: 10.1007/978-3-030-84401-1_8

Sridhar, S., Saraswathi, S., Bharathi, S. V. S., & Devipriya, A. (2024). Gamification applications based on metaverse for healthcare industry. In Dhanaraj, R. K., Vashishtha, S., Sathyamoorthy, M., Balusamy, B., & Cengiz, K. (Eds.), *Metaverse Technologies in Healthcare* (pp. 233–244). Academic Press., https://doi.org/https://doi.org/10.1016/B978-0-443-13565-1.00011-7 DOI: 10.1016/B978-0-443-13565-1.00011-7

Stansbury, J. A., & Earnest, D. R. (2017). Meaningful Gamification in an Industrial/Organizational Psychology Course. *Teaching of Psychology*, 44(1), 38–45. DOI: 10.1177/0098628316677645

Sulphey, M. M. (2017). Game based learning as an aid for extenuating higher education sector issues-the case of Saudi Arabia. *International Journal of Simulation: Systems, Science and Technology, 18*(1), 6.1-6.10. DOI: 10.5013/IJSSST.a.18.01.06

Tan, C. T. (2013). Towards a MOOC game. *ACM International Conference Proceeding Series*. DOI: 10.1145/2513002.2513040

Titus, S., & Ng'ambi, D. (2014). Exploring the use of digital gaming to improve student engagement at a resource poor institution in South Africa. In C. Busch (Ed.), *Proceedings of the European Conference on Games-based Learning* (Vol. 2, pp. 742–748). Dechema e.V. https://www.scopus.com/inward/record.uri?eid=2-s2.0-84923531536&partnerID=40&md5=790e0c18815d4118d44671efc7c203e4

Tsay, C. H.-H., & Kofinas, A. (2017). Making sense of student engagement in a technology-mediated gamified course. In Pivec, M., & Grundler, J. (Eds.), *Proceedings of the 11th European Conference on Games Based Learning, ECGBL 2017* (pp. 696–704). Academic Conferences and Publishing International Limited., https://www.scopus.com/inward/record.uri?eid=2-s2.0-85036451601&partnerID=40&md5=eeffe085ecf0e7f14567561530cc31f4

Tsay, C. H.-H., Kofinas, A., & Luo, J. (2018). Enhancing student learning experience with technology-mediated gamification: An empirical study. *Computers & Education*, 121, 1–17. DOI: 10.1016/j.compedu.2018.01.009

Vanduhe, V. Z., Nat, M., & Hasan, H. F. (2020). Continuance Intentions to Use Gamification for Training in Higher Education: Integrating the Technology Acceptance Model (TAM), Social Motivation, and Task Technology Fit (TTF). *IEEE Access : Practical Innovations, Open Solutions*, 8, 21473–21484. DOI: 10.1109/ACCESS.2020.2966179

Venter, M. (2020). Gamification in STEM programming courses: State of the art. In A. Cardoso, G. R. Alves, & T. Restivo (Eds.), *IEEE Global Engineering Education Conference, EDUCON* (Vols. 2020-April, pp. 859–866). IEEE Computer Society. DOI: 10.1109/EDUCON45650.2020.9125395

Villagrasa, S., & Duran, J. (2013). Gamification for Learning 3D Computer Graphics Arts. *ACM International Conference Proceeding Series*, 429–433. DOI: 10.1145/2536536.2536602

Villagrasa, S., Fonseca, D., Redondo, E., & Duran, J. (2014). Teaching case of gamification and visual technologies for education. *Journal of Cases on Information Technology*, 16(4), 38–57. DOI: 10.4018/jcit.2014100104

Wirani, Y., Nabarian, T., & Romadhon, M. S. (2022). Evaluation of continued use on Kahoot! as a gamification-based learning platform from the perspective of Indonesia students. *Procedia Computer Science*, 197, 545–556. https://doi.org/https://doi.org/10.1016/j.procs.2021.12.172. DOI: 10.1016/j.procs.2021.12.172

Chapter 7

Investigation of Smart Wearable Technology for Healthcare Educational Analytics and Applications

Nilamadhab Mishra
https://orcid.org/0000-0002-1330-4869
VIT Bhopal University, India

Anand Motwani
https://orcid.org/0000-0003-0823-0292
VIT Bhopal University, India

Saroja Kumar Rout
https://orcid.org/0000-0001-9007-3665
Vardhaman College of Engineering, Hyderabad, India

Arul Kumar Natarajan
https://orcid.org/0000-0002-9728-477X
Samarkand International University of Technology, Uzbekistan

ABSTRACT

As technology advances, so do its diverse and creative inventions. One of those innovations includes smart wearable technology that goes by different names, such as AI-based smart wearable devices or smart wearables. This paper delves into the multifaceted landscape of smart wearable technology, focusing on its integration and susceptibility in lesser-explored domains such as education, healthcare, and health

DOI: 10.4018/979-8-3693-7817-5.ch007

industrial applications. The paper explores smart wearables' evolution, highlighting their transformative impact on various educational analytics and application cases. It examines the benefits and risks associated with smart wearable technology in healthcare educational settings, shedding light on its widespread adoption's potential opportunities and challenges. By analyzing these aspects, this paper aims to provide insights into the modern-age susceptibility to compact smart wearables and propose measures to address security concerns while harnessing the potential of smart wearable technology for healthcare educational analytics and applications.

INTRODUCTION

Smart wearable technology or smart wearable technology can be described as a category of electronic device that sends and receives data while worn as an accessory, embedded in clothing, or implanted in the user's body. Smart wearables often provide real-time data that enhances accuracy in healthcare, fitness, etc. Their seamless interface establishes efficient interaction between humans and machines. These interactions range from user-desktop interaction to advanced visualization experiences such as Virtual reality (Smart wearable Sensors for Learning Enhancement in Higher Education). Smart wearables have evolved significantly within IoT (Internet of Things) devices. They are critical for life-changing applications (Zhou, Z. X., et al. (2019, November)).

Smart wearable devices, once only possible in science fiction, have become integrated into almost every aspect of our lives, growing microscopic to the eyes with each new amendment or advancement of technology. From fitness trackers as wristbands to augmented reality glasses that transform how we perceive our surroundings, smart wearable devices have incorporated themselves into various aspects of our tech-driven lives. Since the onset of the popularity of these devices, much attention has been directed towards their remarkable contribution to healthcare. This technology has proven to be effective in providing clinically sensitive data for more informed patient assessment (From A to Z: Smart wearable technology explained), early detection, and management of medical conditions (A novel framework for quality care in assisting chronically impaired patients with ubiquitous computing and ambient intelligence technologies), etc. (Wu, M., & Luo, J. (2019); Thakkar, H. K., et al. (2022)). However, the lesser-known applications of smart wearable technology within health education, education, and training exist. They showcase these smart wearables' versatility and potential beyond the healthcare field. This paper sheds light on the overlooked potential of smart wearable technology in health education, education, and industrialization. By exploring these lesser mainstream applications and analyzing their unique challenges and opportunities, we seek to highlight the broader impact of smart wearables on society and pave the way for future innovations in this rapidly evolving field. Furthermore, we analyze the susceptibility of these devices to modern-age threats.

EVOLUTION OF SMART WEARABLE TECHNOLOGY

Before delving into the nuances of various applications, it is essential to understand the evolution of smart wearables and how they have shaped our current landscape. The idea of smart wearable technology can be traced back centuries, beginning with abacus rings worn by the ancient Chinese and moving on to the pocket watches of the 17th century. Until the latter half of the 20th century, smart wearables started incorporating technical aspects fueled by the advancements in electronics, computing, and miniaturization, i.e., a Survey on Smart wearable Technology: History, State-of-the-Art, and Current Challenges. The advent of microprocessors and integrated circuits in the 1970s paved the way for the first generation of smart wearable devices, albeit in rudimentary forms. Early examples included calculator watches and digital hearing aids, marking the initial foray into smart wearable computing. Throughout the ensuing decades, technological progress accelerated at an unprecedented pace, leading to the development of increasingly sophisticated smart wearable devices (Almusawi H. A. et al. (2021); Chau et al., (2019); Ozkan-Yildirim et al., T. (2021); Sultan, N. (2015)).

The turn of the 21st century resulted in a proliferation of smart wearable technology driven by advancements in sensors, wireless connectivity, and data processing capabilities. Smartphones greatly contributed to the smart wearable revolution by providing a platform for seamless integration with smart wearable devices, establishing possibilities for connectivity and interaction. With the onset of smartwatches, notifications were delivered on the go; smart wearable cameras enabled users to capture moments from a first-person point of view. Furthermore, smart wearables have evolved from monitoring basic physical parameters of the users to physiological metrics such as tracking heart rate, sleep patterns, and even stress levels. Today, smart wearables have led to the transformation of wearing devices to wearing technology embedded within clothing. In the healthcare domain, smart wearables have proven to be significant in early disease detection by monitoring vital signs that are not physically visible and transmitting the data to healthcare professionals, thus enabling remote patient monitoring and quicker intervention (Xie, Y. et al. (2021); Sazonov, E. (Ed.). (2020)).

Undergoing recent developments, smart wearable technology encompasses diverse applications beyond the traditional realms of healthcare and fitness. This can be demonstrated by their capabilities in the field of virtual reality fashion-forward smart wearables that possess the formula to integrate technology into everyday attire. However, despite the widespread influence of smart wearables in these diverse applications, they often remain overshadowed by their more prominent counterparts (Huifeng, W., Kadry, S. N., & Raj, E. D. (2020)).

LITERATURE REVIEW

As specified earlier, smart wearable technology has emerged as a significant innovation with a wide- range of applications across various domains. These smart wearable devices are an important division of IoT (Internet of Things) that makes life easier and more secure, especially in a tech-dependent world. The role of sensors in enabling all these varied functionalities in smart wearable technology is very critical. They can detect and respond to different changes in their environment. The input source can range from motion, temperature, humidity, light, sound, distance, pressure, etc. These sensors come in different shapes, types, and sizes and have a vast number of applications such as smart jewelers and fabrics, body implantable, smart skin tattoos, Google Glass, smart TVs, VR (virtual reality) headsets, Fitbit fitness bands, and smartwatches, My Halo™(Smart wearable medical alert monitor), air pollution detectors, and smart security devices, etc. (Iqbal, M. H., et al. (2016); Borthwick et al., et al (2015)).

The contribution of this technology in the modern world is quite immense as it has enabled not just smart individuals to use smart wearables but also smart industries and smart enterprises. Allowing as far as space, this branch of applied science has encapsulated most operations in the world, from as basic as GPS tracking to complex applications such as that in optical computing. This section overviews existing research and literature on smart wearables, focusing on their applications and implications (Lindberg, R., Seo, J., & Laine, T. H. (2016)).

Various research studies have explored the use of smart wearables for health monitoring, disease management, and remote patient care. The development of smart wearable devices has lately been focused on detecting and managing adverse health conditions in aging populations. For instance, some research has focused on the potential of smart wearables to collect and analyze the manner of walking that could lead to fall prevention. The work provides a comprehensive overview of smart wearable technology in medicine. Their review highlights smart wearables and their applications in various sub-domains of medicine in the category of body sensors. This ranges from vital signs monitoring in the chest, shoulder, wrist, and abdomen to rehabilitation assessing devices. Later, some research conducted reviews of different types of non-invasive smart wearables and the involvement of smart wearables in drug delivery systems. The authors highlight the efficiency in the disease monitoring aspect of smart healthcare wearable devices and their application for diagnostic and treatment purposes (Holland, J. (Ed.). (2016); Greiwe, J., & Nyenhuis, S. M. (2020)).

Smart wearable technology promotes fitness and wellness through activity tracking, sleep monitoring, and nutrition management. The work evaluated the validity of commercially available smart wearable devices to monitor jump load with elite volleyball players. Considering sports experience design, the research used the design framework to assess the impact of smart wearable fitness technology (WFT) on the management and perceptions of service experiences. Their findings indicated that integrating WFT enhances the service experience via increased social interaction, gamification, and accountability. The work (Khosravi, S., Bailey, S. G., Parvizi, H., & Ghannam, R. (2021)) on the Effectiveness of Smart wearable Trackers on Physical Activity in Healthy Adults demonstrated the effectiveness of smart wearable fitness trackers in motivating individuals to adopt healthier lifestyles. The authors emphasized the importance of personalized feedback and goal setting in enhancing user engagement and adherence to fitness routines.

The fashion and design industry adapted smart wearables as accessories, with the onset of smart jewelers and fitness trackers embedded into clothing. The works (Burnham, J. P., et al. (2018); Sandall, B. K. (2016); Motti, V. G. (2019)) explored the currently available smart wearable technologies in the fashion industry, serving various purposes, from aesthetics to the comfort of the wearer. They also highlight the future of smart wearable technology in the fashion industry. The existing literature review provides valuable insights into the versatility of smart wearables and their applications across various fields. Further in this study, we will delve into the details of lesser-known smart wearable technology applications (Yacef, K., Caillaud, C., & Galy, O. (2018)).

SMART WEARABLE TECHNOLOGY IN HEALTHCARE EDUCATIONAL APPLICATIONS

Smart wearables have proven their immense value in lesser-known fields of application. Their hands-free nature has become one of the greatest creations of the 21st century by providing many benefits to mankind. Smart wearable technologies work in a web-based form. It tracks different tasks and enhances the user experience based on user needs and wishes. Customizations are based on the data that it collects, processes, and analyses, mostly in real time. Sensors controlled by microcontrollers are embedded in smart wearable devices to facilitate wireless data transfer between hardware and software components (Brown, W. J., & Reid, C. (2022)). This section delves into two lesser-known fields, exploring smart wearables' role in education and health education.

Smart Wearable Devices in Education

Smart wearable devices' hands-free and portable nature makes them ideal for facilitating quality classroom learning. These devices can exchange data through WIFI and communicate information over the internet, thus enabling a smooth connection in environmental settings such as a classroom within the institution. Smart wearable technologies are a major factor in creating a connected and comprehensive learning experience. As a leading area of innovation in the field of eLearning and smart wearable app development trends, smart wearables are all set to offer significant benefits to their usage in educational sectors (Godfrey, A., Hetherington, V., Shum, H., Bonato, P., Lovell, N. H., & Stuart, S. (2018)). One of the more popular implementations of smart wearables in educational settings is smartwatches. With the onset of companies like Google and Apple releasing smartwatches with features such as 'schooltime', students find it extremely beneficial to integrate smartwatches during their lectures. A smartwatch could be used to record lectures for educators, organize tests, send notes, send voice messages, etc. Some research delved into the psyche behind students' use of smartwatches to aid with their studies. They further elaborated on the various ways students found incorporating smartwatches in their studies useful and explored the limitations they offered. Virtual reality headsets are another impressive smart wearable technology that substantially impacts a classroom. They help learners have a hands-on learning experience without being present in the classroom, university, or any institution. They could also help communicate lectures to students if the professor cannot be physically present for the class. In the field of medical science, students can practice medical procedures in a virtual environment. The smart glass technology, though not implemented commercially, is a most successful attempt at catalyzing the possibility of augmented reality. Google Glass, which has been closer to successfully implementing this possibility, enabled users to project information through their glasses in their current environment. Learners, including the instructors, could also use these smart glasses to observe situations from multiple perspectives (Zhang, M., Luo, M., Nie, R., & Zhang, Y. (2017)).

A noteworthy utilization of smart wearable technology in eLearning involves the utilization of brain-sensing headbands. These headbands, like Muse, employ EEG sensors to monitor brain activity. While this technology is sophisticated and not typically applicable to routine learning scenarios, it provides valuable insights into brain function and responses to various stimuli. These headbands can be particularly effective for focus training, offering a means to address mental health issues. This entails detecting periods of stress, fatigue, or anxiety in the brain and employing guided meditations to develop strategies for managing these conditions. The emergence of smart wearable health education represents a significant fusion of health education experiences with smart wearable devices, including smartwatches, augmented reality (AR) glasses, fitness trackers, and virtual reality (VR) headsets. This integration facilitates a hands-free, immersive health education encounter wherein users can interact seamlessly with digital content while fully engaging with their surroundings. Unlike conventional health education setups that rely on controllers or keyboards, smart wearable health education brings the health education interface directly onto the user's body, enabling intuitive and instinctive interactions (Pope, Z. C., Barr-Anderson, D. J., Lewis, B. A., Pereira, M. A., & Gao, Z. (2019)). As smart wearable technology continues its rapid advancement, it leaves an indelible mark on various industries, with the health education sector being a prime beneficiary. In 2023, the incorporation of smart wearable technology into game development has emerged as a notable trend. Smart wearable devices, ranging from smartwatches to VR headsets, have become indispensable components of our daily lives, offering convenience and unlocking new health education potentials. The evolving landscape of smart wearable health education technology presents both opportunities and challenges for developers striving to deliver immersive health education experiences. With the proliferation of AR, VR, and fitness-oriented games, smart wearables are revolutionizing how we engage with health education content. The trajectory of smart wearable technology promises further innovation, shaping the future landscape of health education experiences (Loncar-Turukalo, et al., (2019)).

The health education industry has experienced a profound metamorphosis driven by IoT technology and the advent of motion-based interfaces like gesture and voice control. This technological revolution aims to elevate user experiences in health education and beyond, particularly through integrating health education smart wearable technology. Within this revolution, the Internet of Things (IoT) is pivotal, empowering gamers, developers, designers, and manufacturers alike. The utilization of IoT in health education leverages advanced software and hardware found in smartphones, leading to an abundance of IoT health education projects. These projects cater to a broader audience and elevate the health education experience to new heights. It created design themes and health education smart wearable concepts that further led to actionable design implications that can be adopted by designers and researchers for designing health education smart wearables. A major area is the development of gesture- and voice-controlled smart wearable technology, which represents a natural progression in human-computer interaction. These innovations liberate users from conventional input methods like mice and controllers, offering more natural interfaces aligned with how humans naturally interact with a focus on Smart wearable health education technology; as these technologies continue to evolve, they hold the potential to reshape the health education landscape and further blur the lines between physical and digital worlds. Smart wearable health education, i.e., using smart wearable technology, provides several advantages for an immersive and convenient user experience. It is portable, which allows movement, while health education further enhances the connection between the player and the character that they are playing in role-playing games (RPG). Smart wearables in health education also promote physical activity, thus involving fitness and exercise in the health education experience (Kirk, M. A., Amiri, M., Pirbaglou, M., & Ritvo, P. (2019)).

The potential integration of smart wearable health education into healthcare and patient rehabilitation is worth noting. With approximately 1.2 billion individuals worldwide engaging in health education, the popularity of games has surged significantly. Virtual reality (VR) has emerged as a valuable tool in rehabilitation, particularly for patients recovering from stroke and experiencing upper limb motor disturbances. VR-based rehabilitation systems offer a framework comprising several games designed to engage patients actively in motor skill practice. These systems encourage participation and compliance with therapeutic exercises by immersing patients in virtual environments. Patients interact with virtual objects in real time, engaging in activities tailored to exercise specific motor skills. This approach provides a novel and engaging rehabilitation experience and a means to track progress and customize treatment plans for individual patients on Smart Wearable Devices for Motivating Patients with Upper Limb Disability Via Health Education and Home Rehabilitation (Hernández-Mustieles et al. (2024)).

VULNERABILITIES OF SMART WEARABLES: EDUCATION AND HEALTH EDUCATION

A safe and secure environment is crucial for the welfare of any society as it affects all aspects of an individual, business, and community. Smart wearable technology has provided many solutions to ensure secure surroundings that certainly help improve quality of life. Their application in the branch of privacy/security has proved to be highly essential by either warning or stopping potential problems before they become bigger. No matter the advancement in the smart wearable technology field, associated vulnerabilities pose a non-dismissible threat to individual and communal security. As smart wearable technology completely depends on sensors and the data it handles, their stability and accurate analytical procession are critical. Most of the defects that are presently hindering the smart wearables' total efficiency are their short/bad battery life as well as slow recharging, handling, and safeguarding of the big data generated, poor communication and workflow, slow data transfer rates and data mining, material as well as associated requirements, and the size and rigidity of sensors. This section explores the susceptibility of smart wearables to threats in the modern age. We will delve into these vulnerabilities in education and health education. The findings will demonstrate the possible limitations in implementing this technology in the mentioned domain and also discuss adaptability to these concerns (Araújo, A. A. et al. (2024)).

Smart Wearables in Education: Security and Privacy Implications

Benefits and Risks of Smart Wearables in Education

Incorporating smart wearable technology into educational settings offers a range of potential benefits, from enhancing student engagement and personalized learning experiences to providing real-time performance tracking. However, alongside these advantages, inherent risks and challenges are associated with adopting smart wearables in education. In this section, we explore both the benefits and risks of smart wearables in education, shedding light on their implications for students, educators, and educational institutions. Smart wearable technology helps students to have an engaging and immersive experience as these devices act as communicating agents between students and instructors and enable active participation from both ends. This effectively impacts the students psychologically and might boost their motivation for new learning experiences. There is growing potential for implementing smart wearable devices in educational applications, enhancing students' learning and schooling skills. The progression of learning environments is enhanced with smart wearable technology, facilitating improved student learning and performance. Educators can monitor students' progress and well-being, including stress levels, during high-pressure situations such as exams. The gathered data is then analyzed to provide tailored learning experiences, fostering greater educational growth and achievement opportunities. In addition to monitoring students' academic performance, Smart jewellers or watches can send notifications to both students and teachers should conditions within a laboratory become hazardous. Additionally, a bracelet equipped with smart wearable technology can track various metrics such as heart rate, hydration levels, step count, and even breathing rate. Teachers can oversee this data through a tablet, ensuring students receive adequate physical activity while maintaining safety (Evans, S., & Willis, C. (2024)).

While smart wearable devices offer promising opportunities to enhance the educational experience, their integration into learning environments also poses certain risks and challenges. One significant concern is the potential invasion of students' privacy and security. As smart wearable devices collect a wealth of personal data, including biometric information and activity patterns, unauthorized parties risk using or accessing this data. Additionally, the constant monitoring enabled by smart wearables raises questions about the erosion of student autonomy and the blurring of boundaries between educational institutions and students' personal lives. This loss of privacy and autonomy may lead to feelings of surveillance and discomfort among students, undermining the trust between educators and learners. In addition to privacy concerns, another risk associated with smart wearable devices in education is the exacerbation of inequality and discrimination. While some students may have access to the latest smart wearable technology, others may not have the financial means to afford such devices. This access discrepancy can widen educational outcomes disparities, creating a digital divide between affluent and marginalized students. Moreover, the reliance on smart wearable devices for learning activities may inadvertently disadvantage students who cannot use or afford them, further perpetuating inequities in educational opportunities. Addressing these risks requires careful consideration of accessibility and equity in technology implementation and robust safeguards to protect students' privacy and mitigate potential discrimination.

Adoption and Vulnerabilities in Educational Settings

Despite the numerous benefits of smart wearables in educational settings, certain vulnerabilities remain associated. As smart wearables deal with access to data and information, they are susceptible to data breaches, information leaks, privacy concerns, etc. Additionally, there are adoption criteria for incorporating these feasible devices in day-to-day education. This section explores the above-mentioned issues. Owing to its initial stage of development and the limited market for smart wearable education, the devices tend to be expensive. Moreover, integrating smart wearable technology at a large scale for students or organizations can lead to added costs, thus leading to investors' reluctance. Adopting smart wearable technology in a classroom also requires the dependency of both students and instructors on the device. The usage of smart wearable technology in everyday learning can obstruct independent thinking. This overreliance on technology could lead to an irreversible psychological impact on students. For instance, students could lack motivation to participate in their respective classrooms without tech involvement. They might reject the idea of traditional learning altogether. Users might resort to unethical examination methods due to their dependence on smart wearables like smartwatches (Rich E. et al., (2024)).

Now that we have explored the complexities of involving smart wearables in education let us explore some vulnerabilities. Google Glass, one of the closest successful attempts at AR in an educational environment, displayed vulnerability regarding security and privacy. At the time of its implementation, Google Glass did not have a secure PIN or authentication system. It also possessed some issues regarding user privacy, as videos and pictures could be recorded without the user's consent. In addition to all these concerns, there have been multiple security issues, such as scanning malicious QR codes and WIFI Hijacking. Connectivity is one of the major components of a practical classroom setting. A network is established between the institution or instructor and the students using smart wearables such as VR headsets or smartwatches. However, this network can use a shared internet space or a Bluetooth connection for a smooth eLearning experience. Bluetooth connections in smart wearables are known to have risk factors; it explains how Bluetooth connection with a Fitbit and personal device possesses various vulnerabilities, including minimum key size for encryption, lack of user authentication, lack of end-to-end security for links, maximum visibility, and connectivity of the personal device.

Smart Wearables in Health Education: Security Challenges and Privacy Concerns

Integration and Risks in the Health Education Industry

Integrating smart wearable technology into the health education industry introduces exciting opportunities for immersive health education experiences and poses several risks and challenges. Smart wearable health education devices, such as augmented reality (AR) glasses and motion-tracking sensors, offer new ways for players to interact with digital content and experience games in their physical surroundings. This integration can revolutionize the health education industry by blurring the lines between virtual and real-world experiences, enhancing player engagement, and fostering novel forms of gameplay. With motion technology and gesture control, players can physically interact with the games. Integration of smart wearable technology enables players to use these hands-free features to enhance their gameplay. Smart wearables not only provide enjoyment factors to games but are also valuable for monitoring vital signs in players, i.e. heart rate and body rate. This data could help adjust the gameplay challenges and regulate the immersion factor in gamers. Next Industries has introduced a device called Tactigon that incorporates AI and IoT, taking smart wearable health education to the next level. This multifunctional smart wearable device has AI technology that seamlessly integrates motion and voice control features. This wireless device, designed with ergonomic considerations, fits comfortably in the hand and provides swift and accurate AI-driven recognition of both voice commands and gestures, irrespective of the surrounding environment or circumstances (Jose, B., et al., (2024)).

Smart wearable technology monitors and assesses player performance, providing valuable information. Ranging from heart rate to stress levels, these devices vividly illustrate an athlete's physical and mental condition during intense matches. Consequently, coaches and players can adjust training routines, reduce fatigue, and enhance performance, resulting in a more competitive and thrilling eSports environment. However, along with these opportunities come significant risks. One major concern is the potential for augmented reality (AR) and virtual reality (VR) health education experiences to blur the boundaries between the virtual and physical worlds, leading to safety hazards and real-world consequences. Players immersed in AR or VR environments may become disoriented or unaware of their surroundings, increasing the risk of accidents or injuries. Moreover, collecting sensitive biometric data by smart wearable health education devices raises privacy and security concerns, as this data could be vulnerable to breaches or misuse. Additionally, the reliance on smart wearable devices for health education experiences may exacerbate issues of digital addiction and excessive screen time, particularly among younger players. Addressing these risks requires careful consideration of safety measures, privacy protections, and responsible health education practices to ensure that smart wearable technology enhances, rather than compromises, the health education experience.

Accessibility remains one of the important challenges to smart wearable health education. Higher quality smart wearable health education devices are expensive and are expected to have a high return value. But this limits accessibility to a huge audience. Furthermore, health education smart wearables are rarely standalone devices, requiring a powerful system to render high-quality health education experiences. Pokemon Go introduced health education bracelets that could be accessible but could not be incorporated into a wide variety of games. Social factors also affect the realm of smart wearable health education. With an immersive health education experience, the devices might isolate users. The development of health education smart wearables has always focused on performance functions and efficiency without considering individual gamers' social acceptability or socialization. This risks the unavailability of social stimulation in users susceptible to seclusion. Wearing these devices in a social setting could lead to a feeling of embarrassment (Szeto, K., Arnold, J., & Maher, C. (2024)). Multi-player video games are prone to cyberattacks. If smart wearables are involved in sharing data with health education systems, these attacks could be very personal as the user is already immersed in the world of health education through virtual reality. The connectivity of these headsets to their PCs could also lead to data leaks and malware installation.

Strategies for Mitigating Security Risks in Smart Wearable Health Education

Ensuring the security of smart wearable health education devices is paramount to safeguarding both user data and health education experiences. This section explores various strategies for mitigating security risks in smart wearable health education to protect users from potential threats and vulnerabilities. By implementing robust security measures, developers and users can enjoy the benefits of smart wearable health education while minimizing the risks associated with data breaches and cyberattacks.

A) Cybersecurity attacks could be mitigated by organizing routine risk assessments. In the case of preventing cyber-attacks, risk assessments, followed by an incident response plan, have proven beneficial (Implement strategies to mitigate technological security risks).
B) To promote social acceptability in wearing devices during health education, several conventions shall be arranged to promote communal growth. This will encourage like-minded users to come together and engage in enjoyable aspects of smart wearable health education.
C) The researchers point out various general security concerns in smart wearable devices. The approach to identifying vulnerabilities and mitigating suggestions could be considered for identifying and exemplifying attack vectors that affect commercial smart wearable products.
D) Accessibility should be promoted through PR programs or Esports contests that could promote smart wearable health education to a larger audience. Game developers could also gather feedback through performance and psychological monitoring, which could also help design concepts that contribute to the development of health education smart wearables.

DISCUSSIONS

Investigating smart wearable technology for healthcare involves exploring how these devices can be used in educational analytics and applications. This field encompasses developing, implementing, and assessing wearable devices designed to monitor various health metrics and support educational initiatives. The potential trends include the following three prospects.

✓ Advancements in Technology: Emerging technologies and their potential impact on healthcare education.
✓ Integration with Other Systems: Combining wearable technology with other healthcare and educational systems for comprehensive solutions.
✓ Policy and Regulation: Anticipating policy and regulation changes that may affect wearable technology use in healthcare education.

Critical Research Discussions

1. Effectiveness of Wearable Technology in Healthcare Education

    Current Evidence and Findings: Studies show that wearable technology can significantly enhance healthcare education by providing real-time data and immersive learning experiences. For instance, smartwatches and fitness trackers have been used to teach medical students about patient monitoring and chronic disease management. Additionally, VR and AR devices offer interactive and detailed simulations for surgical training and anatomy lessons.

    Challenges and Limitations: However, the effectiveness of these technologies is often limited by the accuracy and reliability of the data they collect. Wearable devices, especially consumer-grade ones, may not always provide clinically accurate data. This discrepancy can lead to incorrect assumptions and learning outcomes, potentially affecting patient care when these students enter the workforce.

    Future Directions: To address these challenges, further research is needed to improve the accuracy of wearable devices. Collaboration between tech companies and healthcare professionals can help develop more precise sensors and algorithms. Additionally, integrating wearables with other health information systems can enhance data reliability and provide a more comprehensive learning experience.

2. Privacy and Ethical Considerations

    Data Privacy Concerns: The widespread use of wearable technology in healthcare education raises significant data privacy and security concerns. Wearables collect vast amounts of personal health data, which, if mishandled, can lead to privacy breaches and unauthorized access to sensitive information.

    Ethical Implications: Ethically, there is a fine line between leveraging data for educational purposes and infringing on individual privacy. The potential for continuous monitoring through wearables also raises concerns about surveillance and autonomy, as students and patients might feel constantly watched.

    Proposed Solutions: To mitigate these issues, robust data protection measures and clear ethical guidelines must be established. Educators and developers should prioritize user consent data collection and usage transparency. Implementing strong encryption, anonymizing data, and adhering to regulations are crucial to protecting user privacy.

3. Cost and Accessibility

Economic Barriers: While wearable technology has the potential to democratize healthcare education, cost remains a significant barrier. High-quality wearables and VR/AR devices are often expensive, making them inaccessible to many institutions and students, particularly in low-resource settings.

Impact on Education Equity: This cost barrier can exacerbate existing disparities in healthcare education, where well-funded institutions have access to cutting-edge technology while others do not. Such inequities can lead to a skills gap in the healthcare workforce, with some professionals better prepared than others.

Strategies for Inclusion: To address these disparities, strategies such as subsidized pricing for educational institutions, grants, and public-private partnerships can be explored. Developing cost-effective alternatives and open-source software for wearable technology can also help broaden access. Additionally, mobile-based solutions that leverage existing smartphones can offer many students a more affordable entry point.

4. Integration with Existing Educational Frameworks

Current Integration Efforts: Integrating wearable technology into existing healthcare educational frameworks has shown promise in enhancing learning outcomes. However, many programs struggle with effectively incorporating these technologies into their curricula due to a lack of standardized guidelines and training for educators.

Challenges in Implementation: Resistance to change and the steep learning curve of new technologies can hinder adoption. Educators need to be trained in using the devices to interpret the data and integrate it into their teaching methods.

Recommendations for Effective Integration: To facilitate effective integration, comprehensive training programs for educators and clear guidelines for incorporating wearables into the curriculum are essential. Developing interdisciplinary courses that combine technology with traditional healthcare education can also help bridge the gap. Creating a feedback loop where educators and students can share their experiences and suggestions can help refine and improve the integration process.

5. Long-Term Impact and Sustainability

Potential for Long-Term Benefits: In the long term, wearable technology promises to transform healthcare education by providing continuous, personalized learning experiences. These technologies can help create a more adaptive and responsive educational environment where students receive instant feedback and can track their progress in real time.

Sustainability Concerns: However, sustainability is a critical concern. The rapid pace of technological advancement can lead to obsolescence, where devices quickly become outdated. This raises questions about the long-term viability of investments in wearable technology for education. **Sustainable Practices:** Institutions should focus on scalable and upgradable technologies to ensure sustainability. Investing in modular devices that can be updated with new software or components can extend their lifespan. Additionally, fostering a culture of continuous learning among educators and students can help them stay abreast of technological advancements and incorporate them effectively into their practices.

Investigating smart wearable technology in healthcare educational analytics and applications presents a landscape of significant potential interwoven with substantial challenges. Addressing data accuracy, privacy, cost, integration, and sustainability issues is crucial for harnessing these technologies' full benefits. Future research and collaborative efforts will play a pivotal role in overcoming these barriers and ensuring that wearable technology contributes positively to the future of healthcare education.

CONCLUSION

In conclusion, integrating smart wearable technology into various sectors offers immense potential for enhancing efficiency, productivity, and user experiences. However, it also brings forth significant challenges related to privacy, security, and accessibility. By understanding the benefits and risks associated with smart wearables in education, health education, and industrial applications, stakeholders can make informed decisions to maximize the positive impacts while mitigating potential drawbacks. Meeting both efficiency and security becomes difficult with limiting budgets while dealing with intricate technologies. Capital requirement increases while sustainable resources are preferred. The task of meeting the expectation seems open to constant research and evolution today, but reaching the goal of a perfect smart wearable at hand does not seem that far. After all, smart wearables enhance the quality of life and have the potential to contribute to the welfare of society. Moreover, implementing robust security measures and adopting responsible practices are crucial to ensuring smart wearable technology's safe and effective integration into our daily lives. As we continue to innovate and evolve in the smart wearable technology landscape, addressing these challenges will be essential to harnessing the full potential of smart wearables in shaping the future of human-computer interaction. This structured approach covers the critical aspects of smart wearable technology in healthcare education and analytics, providing a comprehensive understanding of its potential, challenges, and future directions.

REFERENCES

Almusawi, H. A., Durugbo, C. M., & Bugawa, A. M. (2021). Wearable technology in education: A systematic review. *IEEE Transactions on Learning Technologies*, 14(4), 540–554. DOI: 10.1109/TLT.2021.3107459

Araújo, A. A., Gardim, L., Salma, J., Stephen, T., dos Santos, S. S., Silva, Í. R., de Godoy, S., & Mendes, I. A. C. (2024). Advancing Nursing Education through Wearable Electronic Devices: A Scoping. *Nurse Education in Practice*, 79, 104032. DOI: 10.1016/j.nepr.2024.104032 PMID: 38964082

Borthwick, A. C., Anderson, C. L., Finsness, E. S., & Foulger, T. S. (2015). Special article personal wearable technologies in education: Value or villain? *Journal of Digital Learning in Teacher Education*, 31(3), 85–92. DOI: 10.1080/21532974.2015.1021982

Brown, W. J., & Reid, C. (2022). Implementing a cost effective and configurable hybrid simulation platform in healthcare education, using wearable and web-based technologies. *Smart Learning Environments*, 9(1), 20. DOI: 10.1186/s40561-022-00201-1

Burnham, J. P., Lu, C., Yaeger, L. H., Bailey, T. C., & Kollef, M. H. (2018). Using wearable technology to predict health outcomes: A literature review. *Journal of the American Medical Informatics Association : JAMIA*, 25(9), 1221–1227. DOI: 10.1093/jamia/ocy082 PMID: 29982520

Chau, K. Y., Lam, M. H. S., Cheung, M. L., Tso, E. K. H., Flint, S. W., Broom, D. R., Tse, G., & Lee, K. Y. (2019). Smart technology for healthcare: Exploring the antecedents of adoption intention of healthcare wearable technology. *Health Psychology Research*, 7(1). Advance online publication. DOI: 10.4081/hpr.2019.8099 PMID: 31583292

Evans, S., & Willis, C. (2024). The use of digital technologies in the sport and physical education lesson: Fostering need-supportive behaviours in physical education teachers. *Wearable Technologies*, 5(1), 2527. DOI: 10.54517/wt.v5i1.2527

Godfrey, A., Hetherington, V., Shum, H., Bonato, P., Lovell, N. H., & Stuart, S. (2018). From A to Z: Wearable technology explained. *Maturitas*, 113, 40–47. DOI: 10.1016/j.maturitas.2018.04.012 PMID: 29903647

Greiwe, J., & Nyenhuis, S. M. (2020). Wearable technology and how this can be implemented into clinical practice. *Current Allergy and Asthma Reports*, 20(8), 1–10. DOI: 10.1007/s11882-020-00927-3 PMID: 32506184

Hernández-Mustieles, M. A., Lima-Carmona, Y. E., Pacheco-Ramírez, M. A., Mendoza-Armenta, A. A., Romero-Gómez, J. E., Cruz-Gómez, C. F., Rodríguez-Alvarado, D. C., Arceo, A., Cruz-Garza, J. G., Ramírez-Moreno, M. A., & Lozoya-Santos, J. D. J. (2024). Wearable Biosensor Technology in Education: A Systematic Review. *Sensors (Basel)*, 24(8), 2437. DOI: 10.3390/s24082437 PMID: 38676053

Holland, J. (Ed.). (2016). *Wearable technology and mobile innovations for next-generation education*. IGI Global. DOI: 10.4018/978-1-5225-0069-8

Huifeng, W., Kadry, S. N., & Raj, E. D. (2020). Continuous health monitoring of sportsperson using IoT devices based wearable technology. *Computer Communications*, 160, 588–595. DOI: 10.1016/j.comcom.2020.04.025

Iqbal, M. H., Aydin, A., Brunckhorst, O., Dasgupta, P., & Ahmed, K. (2016). A review of wearable technology in medicine. *Journal of the Royal Society of Medicine*, 109(10), 372–380. DOI: 10.1177/0141076816663560 PMID: 27729595

Jose, B., Strawderman, L., Burch, R. F., Saucier, D., & Kikuta, R. (2024). Passive vs. active wearable technology monitoring trunk flexion in elementary teachers. *International Journal of Human Factors and Ergonomics*, 11(2), 172–187. DOI: 10.1504/IJHFE.2024.139190

Khosravi, S., Bailey, S. G., Parvizi, H., & Ghannam, R. (2021). Learning enhancement in higher education with wearable technology. arXiv preprint arXiv:2111.07365.

Kirk, M. A., Amiri, M., Pirbaglou, M., & Ritvo, P. (2019). Wearable technology and physical activity behavior change in adults with chronic cardiometabolic disease: A systematic review and meta-analysis. *American Journal of Health Promotion*, 33(5), 778–791. DOI: 10.1177/0890117118816278 PMID: 30586996

Lindberg, R., Seo, J., & Laine, T. H. (2016). Enhancing physical education with exergames and wearable technology. *IEEE Transactions on Learning Technologies*, 9(4), 328–341. DOI: 10.1109/TLT.2016.2556671

Loncar-Turukalo, T., Zdravevski, E., da Silva, J. M., Chouvarda, I., & Trajkovik, V. (2019). Literature on wearable technology for connected health: Scoping review of research trends, advances, and barriers. *Journal of Medical Internet Research*, 21(9), e14017. DOI: 10.2196/14017 PMID: 31489843

Motti, V. G. (2019). Wearable technologies in education: a design space. In Learning and Collaboration Technologies. Ubiquitous and Virtual Environments for Learning and Collaboration: 6th International Conference, LCT 2019, Held as Part of the 21st HCI International Conference, HCII 2019, Orlando, FL, USA, July 26–31, 2019 [Springer International Publishing.]. *Proceedings*, 21(Part II), 55–67.

Ozkan-Yildirim, S., & Pancar, T. (2021). Smart wearable technology for health tracking: What are the factors that affect their use?. IoT in healthcare and ambient assisted living, 165-199.

Pope, Z. C., Barr-Anderson, D. J., Lewis, B. A., Pereira, M. A., & Gao, Z. (2019). Use of wearable technology and social media to improve physical activity and dietary behaviors among college students: A 12-week randomized pilot study. *International Journal of Environmental Research and Public Health*, 16(19), 3579. DOI: 10.3390/ijerph16193579 PMID: 31557812

Rich, E., Gray, S., Camacho Miñano, M. J., & MacIsaac, S. (2024). Digital health education: Critical perspectives. *Health Education Journal*, •••, 00178969241255438.

Sandall, B. K. (2016). Wearable technology and schools: Where are we and where do we go from here? Journal of Curriculum, Teaching. *Learning and Leadership in Education*, 1(1), 9.

Sazonov, E. (Ed.). (2020). *Wearable Sensors: Fundamentals, implementation and applications*. Academic Press.

Sultan, N. (2015). Reflective thoughts on the potential and challenges of wearable technology for healthcare provision and medical education. *International Journal of Information Management*, 35(5), 521–526. DOI: 10.1016/j.ijinfomgt.2015.04.010

Szeto, K., Arnold, J., & Maher, C. (2024). The Wearable Activity Tracker Checklist for Healthcare (WATCH): A 12-point guide for the implementation of wearable activity trackers in healthcare. *The International Journal of Behavioral Nutrition and Physical Activity*, 21(1), 30. DOI: 10.1186/s12966-024-01567-w PMID: 38481238

Thakkar, H. K., Chowdhury, S. R., Bhoi, A. K., & Barsocchi, P. (2022). *Applications of wearable technologies in healthcare: An analytical study. In 5G IoT and Edge Computing for Smart Healthcare*. Academic Press.

Wu, M., & Luo, J. (2019). Wearable technology applications in healthcare: A literature review. *On-Line Journal of Nursing Informatics*, 23(3).

Xie, Y., Lu, L., Gao, F., He, S. J., Zhao, H. J., Fang, Y., Yang, J., An, Y., Ye, Z., & Dong, Z. (2021). Integration of artificial intelligence, blockchain, and wearable technology for chronic disease management: A new paradigm in smart healthcare. *Current Medical Science*, 41(6), 1123–1133. DOI: 10.1007/s11596-021-2485-0 PMID: 34950987

Yacef, K., Caillaud, C., & Galy, O. (2018). Supporting learning activities with wearable devices to develop life-long skills in a health education app. In Artificial Intelligence in Education: 19th International Conference, AIED 2018, London, UK, June 27–30, 2018 [Springer International Publishing.]. *Proceedings*, 19(Part II), 394–398.

Zhang, M., Luo, M., Nie, R., & Zhang, Y. (2017). Technical attributes, health attribute, consumer attributes and their roles in adoption intention of healthcare wearable technology. *International Journal of Medical Informatics*, 108, 97–109. DOI: 10.1016/j.ijmedinf.2017.09.016 PMID: 29132639

Zhou, Z. X., Tam, V., Lui, K. S., Lam, E. Y., Yuen, A., Hu, X., & Law, N. (2019, November). Applying deep learning and wearable devices for educational data analytics. In 2019 IEEE 31st International Conference on Tools with Artificial Intelligence (ICTAI) (pp. 871-878). IEEE.

Chapter 8
Exploring the Immense Role of Wearable Devices and Smart Technologies in Revolutionizing Healthcare and Medical Education

Mohd Shadab
https://orcid.org/0000-0002-8907-4596
Arabian Gulf University, Bahrain

Zainab Husain Salman
Arabian Gulf University, Bahrain

Ronni Mol Joji
https://orcid.org/0000-0002-1343-0945
Arabian Gulf University, Bahrain

Zainab Salman Saleh
Arabian Gulf University, Bahrain

Mohammad Shahid
https://orcid.org/0000-0002-4684-3607
Arabian Gulf University, Bahrain

ABSTRACT

This chapter discusses the immense role of wearable health devices and the smart

DOI: 10.4018/979-8-3693-7817-5.ch008

technology in healthcare and medical education. Wearable devices are reshaping the health management system specifically after the COVID-19. People are using these devices to track their vitals such as blood pressure, oxygen saturation, heart rate etc. These health gadgets with smart biosensors allow user to track his health record reducing the hospital visits. Wearable gadgets have come a long way historically from a pacemaker and glucose monitors to the highly sophisticated artificial intelligence (AI) based system which not only can help in wellness tracking, but also can help in chronic disease management and telemedicine. Chapter also delves into how these wearable devices and smart technologies are shaping the medical education by integrating the virtual and augmented realities. Despite of having some challenges which should be overcome in the future such as data privacy, accuracy, cost etc. These devices and technologies are paving the way for smart healthcare solutions and personalized treatment.

INTRODUCTION

Since the advention of artificial intelligence (AI), tech industries are moving toward incorporating AI and machine learning (ML) into their devices. Technology is merging with new large language models (LLMS), so is the healthcare industry. Wearable health devices played a great role in revolutionizing the current trend of healthcare industry. Wearable devices are the gadgets attached to part of our body or worn with clothing. These tech gadgets are integrated with different kind of sensors to monitor the health conditions like temperature, pulse, oxygen saturation, sugar level, heartbeat etc. These wearable devices keep recording the data throughout the day providing the healthcare personnel a better and comprehensive view of the patient's health status. Checking the vitals of a sick person is a crucial part of health monitoring. For this, patient has to visit the healthcare facility several times. Wearable health gadgets have reduced the patient's visits dramatically. He not only can check his vitals at home, but he can also share and consult it with his family physicians remotely. The important parameters for which these gadgets are mostly exploited are:

- Blood pressure
- Oxygen saturation level
- Heart rate
- Physical activity
- Blood sugar

Use of wearable health devices helps clinicians to give a personalized and more suitable treatment. Most of these devices are available easily to the consumers in the format of wristbands, watches, patches, headgears etc. But smart watches share the largest segment of these gadgets. Apple and Samsung are the main giants manufacturing these smart watches although cheaper versions are also available in the market for a layman.

Overall use of wearable devices surged after the COVID-19 pandemic. Healthcare facilities were chockablock those days which triggered the use of personal health monitoring devices. Since then, the demand of wearing these gadgets is souring among public. As per the survey of the Stallarix, the global market value of wearable technology is estimated at 96.5 billion U.S. dollar in 2022 which will expand with a 15.47% rate hitting a value of 198.09 billion U.S. dollar by 2027 (Challenges in Wearable Technology in Healthcare | Stellarix, n.d.).

Objective

The objective of this chapter is to give a detailed description of various types of wearable devices and technologies like AI, IoT, AR, VR etc. in healthcare and how they are used. We also delved into how these devices and technologies can be applied to healthcare education and medical training. The key points discussed in this chapter are as follows:

- Historical Background
- Wearable Devices in Healthcare
- Smart Technologies in Healthcare
- Integration and Implementation of Wearable Devices and Smart Technologies
- Industry-academia collaboration in technology-enhanced educational opportunities
- Challenges

HISTORICAL BACKGROUND

Wearable technologies have been evolving rapidly, starting from simple devices to more sophisticated and multipurpose wearables. The earliest steps were taken by Roger Bacon who invented spectacles in the 13th century to correct vision impairments (*A Brief History of Wearable Computing*, n.d.). The first wearable computer was the creation of Edward Thorp and Claude Shannon in the 1960s. Even though it was not intended for use in healthcare, this innovation has set the stage for future medical applications. Starting from 1980s, the technological advancements in computing and sensors has paved the way for the modern wearable health monitoring devices. In the 21st century, wearables have entered the healthcare field in a meaningful way. Bluetooth headsets, which were developed in the early 2000s, has allowed wireless communication in medical settings. Digital pacemakers, which were also introduced in in the 2000s, has greatly improved patient care as it enabled the constant monitoring of the heart's activity and offered real-time interventions. Fitbit released its first device, Fitbit Classic, in 2008. It was able to monitor and collect data related to health parameters such as calories burnt, physical activity, and sleep stages (Bhaltadak et al., 2024).

Diabetes management also has greatly benefited from wearable technologies. glucose monitoring can now be done without the need for finger pricking. the first prototype of continuous glucose monitoring (CGM) systems was introduced in the late 1990s. however, it was facing biocompatibility problems, which hindered its success. Products like Medtronic Guardian were among the first CGM devices to be made commercially accessible in the early 2000s. CGMs then continued to evolve into modern, more accurate systems characterized by accuracy and comfort and were able to be integrated with smartphones and insulin pump. In 2014, Abbott launched it FreeStyle® Libre flash glucose monitoring system. This device allows users to scan a sensor for real time glucose level without a finger prick (Cappon et al., 2017).

Another outstanding achievement in the early 21st century was the improvement in healthcare technology represented by the Apple Watch and Fitbit devices. Fitbit was founded by James Park and Eric Friedman in 2007 (*When Fitbit Was Young: The Early Years*, n.d.). Fitbit released it first device 'Fitbit Classic' in 2008 that was able to monitor and collect data related to health parameters such as calories burnt, steps & distance, and sleep stages (Evenson et al., 2015). Apple smartwatch was launched in 2015. Steve Jobs got the idea when he noticed the difficulties in monitoring outpatients while he was fighting pancreatic cancer (Steve Jobs Was Real Inspiration Behind Creation of Apple Watch To Improve Health Care, n.d.). This period also witnessed the introduction of smart eyewear like Google Glasses, which, although not entirely a healthcare device, it has set the stage for modern AR tools in medical practice (Bhaltadak et al., 2024).

2019 saw the introduction of smart clothes with embedded sensors, making it possible to continuously monitor vital signs and physical performance. Despite being created initially for sports training, these devices are being used more and more in healthcare settings to track patients' long-term health and wellbeing (Bhaltadak et al., 2024).

In the year of 2021, the wearables with AI technology are starting to infiltrate the healthcare industry. These devices use algorithms for health tracking, analyzing data, and even suggestion of personalized treatments according to user. Their use in chronic disease management, wellness, and telehealth is growing quickly (Bhaltadak et al., 2024).

WEARABLE DEVICES IN HEALTHCARE

A significant rise in the usage of wearable devices is noticed due to the shift towards preventative healthcare and active personal healthcare management. Those devices can measure different physiological parameters including blood pressure (BP), heart rate (HR), and respiratory rate (RR). These devices can also contribute to managing various chronic illnesses by the telemedicine functions they offer, by allowing continuous monitoring and feedback from healthcare providers.

Depending on their applications and functions, wearable devices can be categorized into two types: monitoring devices and therapeutic devices. A comprehensive flow chart of wearable health devices categories is given in figure 1.

Figure 1. A comprehensive categorization of wearable health devices

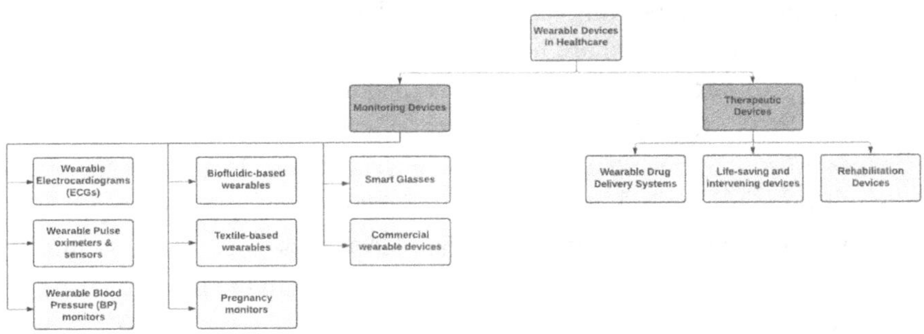

Monitoring Devices

Most applications of wearable devices in healthcare focus on monitoring functions. The devices are designed to be used within various environments, ranging from healthcare settings to home use.

Wearable ECGs

Multiple companies have designed wearable electrocardiograms (ECGs) that perform monitoring of the patients' heart well-being and detect arrhythmias (Al-Alusi et al., 2019). Since the introduction of the concept of ambulatory ECGs, wearable echocardiograms have undergone many advancements. A conventional form of wearable ECGs is the one designed by Norman Jefferis Holter in the 1950s. This ECG is designed to be worn as electrodes attached to the skin and connected to a monitor. The monitor can be worn as a belt, over the shoulder, and even incorporated into clothing. Holter's ECGs can perform continuous cardiac monitoring, up to 48 hours, making them helpful to monitor patients with unexplained syncope, detect arrhythmias, and monitor patients post-operatively, Miniaturization has transformed Holter's ECG from being a bulky equipment to a more wearable and user-friendly form. Recent models of Holter ECGs are also now integrated with smartphones to allow continuous monitoring and tracking by the healthcare professionals (Kennedy, 2013; *MyHeart: Your Guide to Heart Health*, n.d.).

A distinctive progress in the area of wearable ECGs is the textile-based ECG monitor designed by Yapici et al. Textile-based wearable devices incorporate sensors into clothing. In contrast to the conventional ECG monitors that use uncomfortable gel-based Ag/AgCl electrodes (wet electrodes), Yapici et al. opted for graphene-functionalized clothing embedded with electrodes that allow continuous cardiovascular monitoring and are more comfortable to the user.

This ECG monitor can be worn as a wristband and neckband. In the case of the wristband, the ECG measurement is done by wearing one on each wrist with each having a sensing electrode and one also having a reference electrode. The neckband is made up of three parts of conductive textile, with two acting as sensing electrodes and one acting as a reference electrode. Research indicates that graphene-functionalized textile electrodes have a correlation of up to 97% with the traditional Ag/AgCl electrodes (Iqbal et al., 2021; Yapici & Alkhidir, 2017).

Pulse Oximeters and Pulse Sensors

Traditionally, pulse oximeters, which are devices that measure the amount of oxygen in the blood, use finger clips connected to a monitor to measure the amount of oxygen in the blood. Pulse oximetry works by a process called photoplethysmography (PPG) that shines two LEDs (red and infrared) through the skin. Hemoglobin, the molecule that holds oxygen, absorbs this light, and the unabsorbed light is detected by photodiodes. The difference in the amount of light detected between the two wavelengths of the red and infrared LEDs is utilized to determine the amount of oxygen in the blood (Baker et al., 2017). Pulse oximeters are now incorporated into wrist worn devices like smartwatches **(Figure 2-A)**. Pulse oximeters now can also be worn as rings **(Figure 2-B)**, forehead sensors **(Figure 2-C)**, ear clip sensors **(Figure 2-D)**, and can also be applied to the toe or foot of infants and neonates **(Figure 2-E & F)** (Medtronic, n.d.; ECRI, n.d.; Wellue, n.d.; Rickards et al., 2021). Pulse rate and monitors use a similar principle to detect the pulse rate, and they are also now can be measured from wrist sensors. However, it is important to mention that these devices are not intended for medical use and should not be used to diagnose any illness (Baker et al., 2017).

Figure 2. Examples of some Pulse Oximeters & Pulse Sensors (A) Apple Smartwatch pulse oximeter (Apple Support (BH), n.d.), (B) Wellue ring O2 Ring Oximeter (Wellue, n.d.), (C) Nellcor™ SpO2 Forehead Sensor (Medtronic, n.d.), (D) Turner Medical Ear-clip pulse oximeter (Turner Medical, n.d.), (E&F) Covidien Specialty SpO2 Sensors for infants and neonates (Tiger Medical, n.d.).

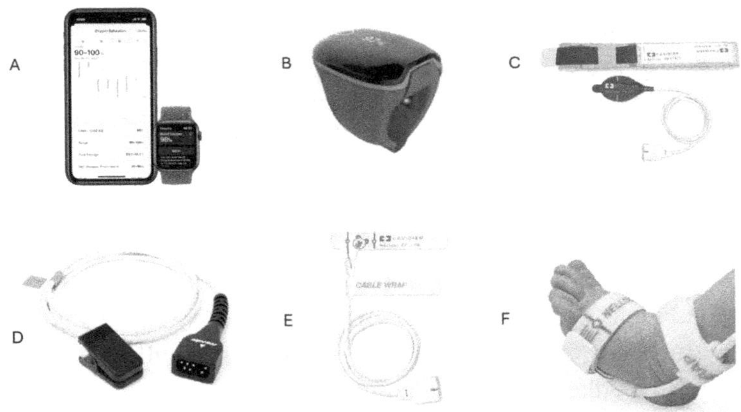

Wearable Blood Pressure Monitors

Cuff-less BP monitoring devices that go around the patients' wrist or fingers eased the process of having ongoing monitoring of their blood pressure. They gather information on the pulse timing and characteristics through different methods like photoplethysmography (PPG) and applanation tonometry, then calculate the blood pressure based on certain algorithms (Hu, Martin, et al., 2023). The Valencell Fingertip™ Blood Pressure Monitor is one of the best devices in this field for its features that include calibration free operation using PPG sensors and AI algorithms for accurate readings from the fingertip (Panula et al., 2023). Unlike the conventional sphygmomanometers, which depend on cuff inflation, cuff-less BP monitors are more comfortable and support long-term monitoring of BP.

However, cuffless BP monitoring devices have several limitations, including the inability to measure the BP directly since it is estimated as a function of derived variables related to the time and the characteristics of our pulse. These variables themselves rely on multiple sensors and algorithms and require calibration. These devices are also limited to the fact that there is no ideal referent standard for them to be validated. Moreover, cuff-less BP devices when calibrated while in a seated, resting position, may result in inaccurate reading during a dynamic state and will ultimately lead to estimation drifts over time. Also, cuff-less BP devices worn on the wrist can fail to account for variations in different body circumstances that could affect the accessibility of the radial artery, such as obesity, large wrist circumference, and peripheral artery diseases. Moreover, the fact that these devices depend on the heart rate to estimate BP initiates a problem as pulse-BP dissociation can occur in many medical conditions, a side effect of medications, and even in physiological states (Hu, Martin, et al., 2023). Despite these limitations, cuff-less BP monitors can benefit both the patients and healthcare providers as they offer continuous BP monitoring and enhance patient comfort (Hu, Park, et al., 2023).

Biofluidic Based Devices

The technology of microfluidics utilized in wearable devices is still not widely used but provides great opportunities for medical care. Wearable microfluidics devices are now utilizing microstructures are capable to handle and store small volumes of body fluids like sweat and tear for analysis.

Koh et. al has designed a microfluidic system for sweat analysis. This system can be used for the continuous measurement of critical biomarkers, including electrolytes like Chloride (Cl^-), which is important for monitoring cystic fibrosis. This device can also pH levels, as well as glucose and lactate levels, which are important for patients with diabetes. This device works by allowing the reaction of the biomarkers with specific chromogenic chemical reagents. A smartphone then can be used to take pictures of how the color of the chemical analytes changes in order to quantify the concentrations of lactate, glucose, and chloride. The concentrations of lactate, glucose, chloride and pH levels are quantified using the RGB method. In this method, the amount of red, green, or blue (RGB) in the image changes in relation to the concentration of each substance (Koh et al., 2016).

Tear-based wearable sensors are useful in several other aspects too, especially in the monitoring and control of diabetes. For instance, the smart lens that is developed in collaboration between Google and Novartis tracks glucose levels from tears using a lens (Hee Keum et al., 2020). This is considered a non-invasive method compared to testing blood with glucometers.

Lin et al., on the other hand, created a smart contact lens using phenylboronic acid (PBA) and hydroxyethyl methacrylate (HEMA) which swells and changes in thickness due to the presence of glucose. A light emitting diode is used to light the lens and the light is recorded by the smartphone. The smartphone then processes the images using embedded software and determines the value of glucose depending on how thick the lens is (Lin et al., 2018).

Textile-Based Wearable Devices

Textile-based wearable devices are the devices that incorporate sensors into clothing, allowing continuous monitoring of the various physiological data. An example is the Sensoria smart socks, which are especially help with diabetic patients. These sensors are incorporated with temperature sensors and can alert the patient and the healthcare provider when the temperature rises, which could potentially indicate inflammation and ulceration (*Siren Care*, n.d.).

Pregnancy Monitors

Wearable pregnancy monitors are helpful in monitoring and maintaining the well-being of both the mother and the newborn. They can provide continuous recording of the heartbeat and movements of a fetus which can help in preventing stillbirths and more complications (Alim & Imtiaz, 2023).

Smart Glasses

The eSight 4 is an example of smart glasses that can be used by persons with a high degree of central vision loss and even legal blindness, giving the subject an 20/20 vision. These smart glasses are designed using a camera that captures the visual scene in a fast way and with a high definition and sends the information the two OLED screens placed close to the eye at the same time. The device uses certain algorithm to process and improve the visual images, allowing the patient to see the scene clearly in front of them. The glasses also stimulate synaptic nerve activity in the user's remaining photoreceptor cells, allowing the user to accommodate to wide range of images and in the end increases their visual ability. This allows the users to maintain their peripheral vision and at the same time be able to perform daily routine tasks without assistance (*Functional Vision To Legally Blind Individuals - Interview With ESight's Director Of Product Marketing, Roland Mattern | Patreon*, n.d.; *How It Works | ESight Eyewear*, n.d.).

Commercially Available Wearable Devices

In the recent years, many commercial monitoring devices, like smartwatches (e.g., Apple Watch and Samsung Watch) and fitness trackers (e.g., Fitbit) have emerged in the market. They are characterized by being comfortable to the user, accurate in measurements, biocompatible, and with a long battery life (Haghi et al., 2017) .

These devices are mainly designed for keeping track of our wellbeing and monitoring activities like heart rate (HR), respiratory rate (RR), oxygen saturation, and calories burnt. Some devices can also monitor other parameters. For example, Fitbit Sense™ has a smart feature for stress monitoring and management. This is achieved by a special sensor called the electrodermal activity (EDA) scanner, that measures the changes in the conductivity of the skin due to sweat, which is an indicator of stress. And based on the data collected by this sensor, a stress management score is calculated, and breathing exercises and other techniques are customized to help the user calm down and reduce their stress (*Advanced Health Smartwatch | Fitbit Sense*, n.d.; Horvers et al., 2021). On the other hand, Apple Watch® has an ECG feature that can generate a single-lead ECG. It has been approved by the Food and Drug Administration (FDA) in 2020 (K201525-Luke Olson, n.d.). The single-lead ECG is generated has a strong correlation with the traditional 12-lead ECG (Saghir et al., 2020). Additionally, it also has the ability to detect falls and notify the emergency contacts and services about it, which can be useful for the elderly or patients with mobility issues (Healthcare - Apple Watch - Apple, n.d.).

Another example of commercially available wearable monitors is the continuous glucose monitors (CGMs). For insulin-dependent diabetic patients, it is challenging to maintain a normal level of blood sugar while minimizing the risk of hypoglycemia (Ikegami et al., 2022). However, due to advancements in continuous glucose monitoring (CGM) technologies and their integration with other devices, it is now easier to achieve control on diabetes with a minimal risk of hypoglycemic episodes. CGMs, like Dexcom G7 can monitor blood sugar levels continuously over the day and night and provides glucose readings frequently and sends data to its related application or smartphone via Bluetooth.

CGMs are typically worn as patches on the skin. To measure glucose levels, the sensors insert a painless microneedle beneath the skin. CGMs does not provide blood glucose readings. In contrast to a glucometer, CGMs measures the glucose levels in the interstitial fluid just below the skin. Studies have shown that CGM is more effective than self-monitoring of blood glucose levels (SMBG) in both type 1 and type 2 diabetes by leading to reductions in glycosylated hemoglobin (HbA1c) and hypoglycemic episodes (Dexcom Continuous Glucose Monitoring | Dexcom, n.d.; Manov et al., 2023; Mihai et al., 2021; Pratley et al., 2020).

Wearable Devices with a Therapeutic Function

Wearable devices with a therapeutic function can be classified into 3 main categories: drug delivery systems, devices for acute and chronic condition management, and devices that support the rehabilitation process. These devices often combine both therapeutic and monitoring functions, as they can monitor the various physiological and biochemical information and then adjust treatment accordingly.

Wearable Drug-Delivery Systems

An example of this system is the use of continuous glucose monitors (CGMs) with insulin pumps and other devices. Modern CGMs, such as Dexcom G7 is capable of providing accurate glucose readings and can alert for hypoglycemic and hyperglycemic states. CGMs can be integrated with multiple devices like connected insulin pens (CIPs) like Medtronic InPen that can provide features like calculators and reminders for injections. This can contribute to better adherence to treatment and help to prevent excessive insulin dosing and overall, improved glycemic control.

When combined with insulin pumps, this system can provide an automated insulin delivery with dose corrections based on the information gathered by the CGMs. Cloud-based technologies enable the integration of data from CGMs, insulin pumps, and CIPs which allows healthcare providers to analyze the glycemic trends and make immediate decisions regarding diabetes management. The future vision involves the development of automated insulin delivery (AID) systems that require minimal user input and contains AI-dependent tools to optimize insulin dosing and improve glycemic control and reduce the burden of diabetes self-management (Yoo & Kim, 2023).

Another example of a wearable drug delivery system is an ocular ring that delivers the drug bimatoprost continuously through the eye to manage glaucoma, a disease that damages the optic nerve due to increased intraocular pressure. This continuous infusion is particularly helpful for patient's non-adherent to bimatoprost eye drops (Brandt et al., 2017; Lee et al., n.d.).

Another application is the use of smart bandages to manage chronic wounds, like the ones caused by diabetic ulcers and pressure injuries. These smart bandages are able to monitor various wound conditions, such as temperature (increased temperature can be an indicator of infection), pH (alkaline conditions support bacterial growth), moisture and oxygen levels (optimal levels are crucial for wound healing).

It can also identify bacterial infections caused by bacteria like pseudomonas aeruginosa using techniques like impedance and immunoassays. Based on the wound conditions, these smart bandages can deliver drugs to the wound site. Even through smart bandages can help minimize complications related to wound management, like infections and necrosis, their use is still limited, and the work is in progress to make more biocompatible materials that do not disrupt the wound healing while improving their biosensing capabilities (Hossain et al., 2024).

Life-Saving and Intervening Devices

The Philips Respironics DreamStation™ Go is one of the respiratory support devices that helps manage obstructive sleep apnea (OSA), a disease in which the airway of the patient collapses during sleeping leading to a decrease in oxygen saturation or arousal from sleep. It delivers a continues positive pressure through the airway of the patient, preventing it from collapsing when asleep. It can also adjust the pressure according to the patient's respiratory patterns (DreamStation Go | Philips, n.d.; Yamauchi et al., 2024).

Likewise, defibrillators like LifeVest® can be useful for patients with high risk for sudden cardiac death. It can monitor the heart rhythm and then deliver an instant shock if a life-threatening arrhythmia was detected (Agarwal et al., 2018).

Rehabilitation Support Devices

Wearable devices are emerging in the field of rehabilitation. They can provide continuous support and monitoring for patients to assist their recovery. These devices can help many patients, ranging from those living with mobility deficits to athletes recovering injuries, or dealing with chronic pain (Rodgers et al., 2019).

These devices can be used to help in the recovery of various medical conditions, such as traumas and neurological disorders. Quell®, for example, is a wearable device that utilizes transcutaneous electrical nerve stimulation (TENS) method, a technique in which electrodes are used to deliver electrical waves across the skin. This helps to modulate the nerve activity and reduce the pain resulting from conditions like chronic low back and low extremity pain (Gibson et al., 2017; Jamison et al., 2019).

Integrating virtual reality with wearable devices is considered a step forward in rehabilitation. With the help of VR, these devices, for example, can create a stimulated environment for the daily activities of a patient recovering from a stroke, allowing them to practice these activities in a safe environment. Furthermore, using wearable devices for rehabilitation has been proven not only to help patients to improve their motor skills, but can also increase their commitment and motivation, which are both important to achieve a successful rehabilitation process (Stock et al., 2024).

The Real® System is a VR system that is used for the rehabilitation of patients recovering from strokes and patients with mobility issues. It consists of headsets that can be worn and sensors that can be attached to the body, thus having the ability to monitor the patients' moves and provide real-time feedback to direct the patients' movements and improve their motor skills. Based on the needs of every patient, this system creates a customized exercise program that helps patients improve their range of motion and muscle strength. There are multiple programs designed to help patients. The Real-y-series contains many activities to improve both the physical and cognitive skills of patients in rehabilitation. For example, the **Mindful Market™** activity is designed to help patients improve their cognitive abilities by stimulating daily-life activities by engaging in a virtual marketplace. On the other hand, **Sports Park™** Activities help patients to develop strength and endurance, by taking part in a range of interactive sports activities that target the upper and lower limbs (The REAL® System, Virtual Reality Rehabilitation Tool, n.d.).

However, it is important to mention that the role of VR in wearable rehabilitation devices is not limited to physical rehabilitation. It can also be used in the psychological and cognitive aspects, such as those patients suffering from anxiety and phobias. Through these devices, patients can take advantage from virtual reality to eventually deal with their fears in a safe, stimulated environment (Maples-Keller et al., 2017).

Smart Technologies in Healthcare

At present, artificial intelligence is considered as an emerging revolution in the field of medicine. It is present in several aspects of medicine including that of diagnosis and treatment. It is beneficial in personalized healthcare through assisting the physicians in drug adjustment and designing a targeted approach to each patient. AI also plays a role not only in the individual level but also at the population level, helping doctors in the anticipation of healthcare needs on a larger and a wider scale. It also contributes to population health management by identifying risks through predictive analytics.

AI in Diagnostics

Despite the advancements in the field of medicine, disease diagnosis remains challenging to physicians due to their complex nature overlapping of signs and symptoms. Diagnosis has always been dependent on the amount of knowledge a doctor they possess and their level of expertise. This can result in variations in diagnoses and possibilities of making errors and misinterpretations (Alowais et al., 2023).

However, the emergence of artificial intelligence (AI) and its subfields of machine learning (ML) and deep learning (DL), have made this process much easier.

ML is an AI that can learn from data by training algorithms to identify patterns and relationships. This allows the system to recognize patterns that may not be caught by human experts. This system is able to analyze and learn from data on its own and make judgements based on what it has learned. It can also learn from past data and improve over time, but usually requires human intervention to adapt to new datasets (Rana & Bhushan, 2023; Thomas, 2021).

On the other hand, DL is an advanced form of ML and is considered a "human-like" form of artificial intelligence. It resembles the human brain's structure and function, as it uses layered algorithms like artificial neural networks (ANNs). These algorithms can learn from data and has the ability to adapt and adjust their predictions independently without human input, unlike ML (Allegra et al., 2022).

Many medical specialties, especially pathology and radiology can greatly benefit from the use of AI, particularly the DL systems. Traditional pathology involves the manual examination of slides to identify any abnormalities. This leads to personal biases by clinicians and variations in diagnosis (Gurcan et al., 2009). However, deep learning is minimizing these inconsistencies by analyzing digital pathology images to detect abnormalities. Digital Pathology involves the conversion of histopathological glass slides into a digital form using whole-slide imaging (WSI) (Aeffner et al., 2017; Zarella et al., 2019). AI can also assist in quality control across the workflow in pathology lab, starting from the pre-analytic to the post-analytic stages. It can perform image quality control, assist with diagnosis, and suggest ordering of any further studies (Kim et al., 2022). AI also assists in integrating histopathological images with patients' data, which makes diagnosis and treatment predictions clearer (Tian, 2016). Studies have shown that pathologists who are assisted by AI outperformed the unassisted ones in the accuracy and efficiency of their work (Steiner et al., 2018). Despite its benefit, AI-guided pathology diagnosis faces multiple challenges and limitations. The results depend on the quality of the input, which is variable. They also require algorithm validation and regulatory considerations to ensure that they can be used safely in medical practice (Kim et al., 2022).

Deep learning models are also used in radiology to handle and evaluate the enormous amounts of imaging data, especially in the field of radiology. For example, AI is playing an essential role in oncological radiology by assisting doctors in (1) detecting abnormalities in imaging, (2) characterization, and (3) monitoring of diseases. AI technologies, especially deep learning has a better performance in detecting abnormalities over the traditional computer-aided detection (CAD) with a better sensitivity and accuracy. For characterization, AI can process a large amount of data to aid in diagnosis, staging, and quantifying radiological characteristics like the size, extent, and the internal texture of the lesion. This helps in analyzing radiological features that are challenging for radiologists to assess consistently. The traditional computer-aided Diagnosis (CADx) depends on predefined features to differentiate benign from malignant nodules. However, CADx integrated with DL can learn features automatically and have a better performance than the traditional CADx. Moreover, AI can analyze changes in imaging over time, which helps in monitoring disease and evaluating treatment responses. And beyond these three tasks, AI also helps in image reconstruction and registration, which improve the image quality and leads to a more accurate diagnosis (Hosny et al., 2018).

Aside from pathology and radiology, other domains of diagnostics such as genomic medicine and clinical laboratory sciences have also shown huge promise in the use of AI. For instance, AI tools are employed in genomic medicine as they help to analyze extensive genomic data searching for disease-associated variants such as autism spectrum disorder or cancer. In this way, AI is able to make prediction of phenotypes, disease susceptibility and even therapeutic response to different drugs, based on genetics which is a new shift in the era of personalized medicine. In addition, AI applications improve the speed and accuracy of various procedures in lab sciences including microbial identification and antibiotic susceptibility testing. Such AI technologies are capable of integrating genomic sequencing, microscopic imaging, and blood culture analysis to enable early diagnosis of pathogens and the initiation of proper treatment. Such progression makes it possible to speed up the process of obtaining laboratory results and improves the accuracy of the diagnosis of dangerous diseases such as sepsis and other infectious diseases (Alowais et al., 2023).

It is also important to embrace the role of AI is in various medical specialties like cardiology, orthopedics and many others. AI algorithms developed in cardiology are used for the automatic analysis of ECGs, which are effective in diagnosing arrhythmias of heart and other disorders at initial stages thereby enhancing treatment efficacy. It can also be employed in nuclear cardiology, through assisting in Single-photon emission computed tomography (SPECT) image processing for the diagnosis in coronary artery diseases (CAD) (Lopez-Jimenez et al., 2020). In orthopedics, AI applications has focused on the assessment of X-ray and MRI images to help with the detection of bone fractures, spine disorders, joint abnormalities, and malignancies in the bone and soft tissues (Teja Nallamothu & Bharadiya, 2023). In dermatology, cancer screening is aided by AI technology such as in melanoma detection (Patel et al., 2023).

The use of AI offers many advantages compared to the traditional diagnostic methods, including improved accuracy, cost and time efficiency, reduction of human errors, and providing assistance and support to the physicians during diagnosis (Kim et al., 2022).

AI in Treatment and Decision Support

Dose optimization refers to the determination of the most effective dose for each patient while minimizing the side effects (Korn et al., 2023). This process can be guided by AI algorithms, which can analyze various factors related to the drug pharmacokinetics and patients' biomarkers that helps to guide an individualized dosing for every patient. TDM, on the other hand, refers to the continuous monitoring of the drug concentrations in the patient's blood and adjusting the dose based on these levels to ensure that the drug remains within its therapeutic range with optimal tolerability (Alowais et al., 2023).

AI is highly applicable in TDM. One example is vancomycin, an antibiotic with a narrow therapeutic index, meaning it only has a small difference between the dose that offers a therapeutic effect and the dose that causes toxic effects (Tamargo et al., 2015). AI can analyze data from the patient, including the drug levels, renal functions, and other parameters and recommends adjustments to the dose or frequency of drug administration. This ensures drug efficiency while minimizing toxicity (Begg et al., 1999; Wang et al., 2022).

AI can provide clinicians with accurate information related to drugs which supports their decision making in treatment choice. It can also provide virtual assistance to doctors through AI-powered chatbots (GPT-4) to handle the medical queries of doctors regarding treatment choice and dose optimization. This helps reducing medical errors regarding wrong dosage and its side effects. These virtual assistants can also analyze patients' data, aid in diagnosis, and recommend an optimal treatment plan for each patient (Albogami et al., 2024; Li et al., 2021; Xu et al., 2020).

AI plays a major role in personalized treatment. It analyzes the data of patients including their genotype, lifestyle, environmental factors, and biomarkers. After that, it customizes treatment plans suited for the needs of each individual. For instance, using ML algorithms in analyzing the genomic information of patients can help in treatment selection as well as the prediction of patients' requirement of certain medications and their responses to treatment (Alowais et al., 2023).

In 2019, the department of health in Abu Dhabi, UAE, launched the "UAE's Genome Program". It is an example of personalized medicine as it creates a reference genome for UAE citizens to help doctors make personalized treatment plans based on the genetic profile of every individual. This project is using advanced genome sequencing technologies and AI to analyze the genome of Emirati population.

This kind of project will not only benefit patients by providing personalized healthcare but also it has a broader aim to predict and manage diseases in the whole population by identifying common genetic variants that predispose these people to certain diseases (The Emirati Genome Programme | The Official Portal of the UAE Government, n.d.).

AI in Predictive Analytics

Al could also play a role in population health management through predictive analysis and risk assessment. Through analyzing patients' data, it can contribute to lessening healthcare costs and the burden on healthcare system through identifying patients at risk of developing diseases and in need for hospital admissions with interventions to prevent their readmissions (Alowais et al., 2023).

Additionally, Al helps in improving the predictive models related to public health management. Traditional epidemiological methods such as susceptible-Infected-Recovered model used widely to understand disease spread are not capable of understanding the complexity of modern disease dynamics. However, tools like ML, DL, and NN that is monitored by Al for science (AI4S) is now merged with those traditional methods for better and more precise predictions; through integrating different data like epidemiological, environmental, social media, and mobility data and analyzing complex information, with an ability to adapt to new data to eventually give efficient prediction and sufficient preparation and management to infectious diseases (Scholastica Ijeh et al., 2024; Zhao et al., 2024).

AI in Drug Discovery

AI is changing the pattern of the traditional drug discovery pathway due to its ability to perform and accomplish tasks faster, more accurately and much efficiently than was possible before. For instance, machine learning techniques provides a better analysis of data and allow more accurate predictions of both efficacy and toxicities of drugs under development. Case studies demonstrate how AI has helped locate new antibiotics, cures for cancer, and sped up the search for COVID-19 treatments (Blanco-González et al., 2023).

AI in Clinical Trials

Artificial Intelligence (AI) is transforming clinical trials (CTs) by making the process more efficient and accelerated. Clinical trials require the use of multiple resources, and many processes involved in conducting those CTs can be lengthy, laborious, and time-consuming. In the preclinical and early phases of drug development, AI aids in selecting drug candidates and evaluating their safety and also limits the number of participants needed for the clinical trial and assists with matching qualified patients with adequate trials. After recruitment, AI also helps in the conduction of the clinical trial. Wearable devices aided by AI allow for constant tracking of patients and raise the safety levels as well as compliance with the required treatment in trial. Artificial Intelligence also makes it easier for experts to analyze trial data by reducing the number of manual tasks performed with increased precision (Askin et al., 2023).

Internet of Things in Healthcare

The term Internet of Things (IoT) refers to a network of connected devices that collect and exchange data together by communicating through the internet. In healthcare, however, this technology is known as the Internet of Medical Things (IoMT). It is a system consisting of multiple devices and applications that connect patients, doctors, and the healthcare system as a whole. In IoMT, every medical device is connected and monitored over the internet by healthcare professionals, which allows a more effective data collection, patient monitoring, and administration of health services. (Baker et al., 2017; Razdan & Sharma, 2022).

A model by Baker et. al has been proposed to shape IoMT. It consists of four parts: wearable sensors & central nodes, short-range communications, long-range communications, and secure cloud-storage architecture & machine learning. Wearable sensors measure the various physiological conditions like the pulse, RR, and body temperature, and these data are transmitted to a central node that process these data and can make preliminary decisions based on them and can forward it then to external sources. Smartphones are examples of a central node. However, having a dedicated central node is preferred on a smartphone as it will have an improved battery life due to reduced functionality, since it only focuses on IoT tasks.

The communication between the wearable sensors and the central nodes depends on a short-range communication method. This method must be safe to the human body, secure, and with a minimal latency. On the other hand, long-range communication is required to transmit data in the most secure and accurate way from central nodes to a database where it is accessible to doctors. Long-range communications must also be with a low latency and high availability to avoid delays in transmitting time-sensitive information and ensure data can be transmitted and received regardless of time and location. Lastly, cloud storage stores medical data securely for the use by healthcare professionals and supports machine learning by allowing the analysis of big datasets by identifying trends and recommending personalized treatment plan for every patient (Baker et al., 2017).

Current implementations of IoT in healthcare includes remote patient monitoring. Wearable devices like heart rate and glucose monitors use IoT to continuously send data to the doctors, allowing them to remotely monitor patients' health and intervene when it is needed. Also, IoMT serves as the technological foundation that allows the functionality of smart E-healthcare. At the same time, smart E-healthcare. Which includes smart hospitals as a key component, use IoMT and AI technologies (especially ML) to deliver patient care. In smart hospitals, data is collected into electronic health records (EHRs) and then analyzed to deliver personalized health regimens to patients (Razdan & Sharma, 2022). Also, smart hospitals with their advanced infrastructure and technologies allow for the implementation of telemedicine ("the use of technologies to remotely diagnose, monitor, and treat patients"), telehealth ("the application of technologies to help patients manage their own illnesses through improved self-care and access to education and support systems"), and remote robotic surgery (Kvedar et al., 2014).

Internet of Things devices, such as patient monitors, result in the formation of a large number of data related to the patients' vital signs, medication adherence, and physical activity. This large amount of data generated by the IoT devices result in the formation of what is known as "big data in healthcare", which is characterized by 3Vs: volume, velocity, and variety. This large amount of data is stored in cloud-based systems to be accessible to doctors and then depends on other technologies like machine learning to analyze these data and make predictions which helps doctors to make predictions which helps doctors to make informed decisions (Dash et al., 2019; Dimitrov, 2016).

Virtual Reality and Augmented Reality in Healthcare

Virtual reality (VR) and augmented reality (AR) allow the user to visualize, explore, and interact with digital information in a three-dimensional manner. VR makes the user completely separated from the real world in a 3D environment. However, AR allows the user to retain the connection with the real world by dealing with digital data super-imposed over the real world.

Even though VR and AR are expensive and currently have a limited use in hospitals and medical care, they hold a promising application in surgical planning, rehabilitation, pain management, and mental health management.

Augmented reality and virtual reality are used in surgical planning and training by displaying the patient's 3D data which gives the surgeon an accurate picture of the patient's anatomy before the procedure (Frajhof et al., 2018). Augmented reality can also help while performing procedures like vascular punctures and endotracheal intubation, by displaying the important information directly on the patient's body (Kanschik et al., 2023).

VR can be used for acute and chronic pain managements. It provides a virtual environment that can distract the patient's mind from the pain and anxiety. For instance, using VR platforms have been proved to decrease pain perception and increase pain tolerance in burn patients during wound care and dressing, patients during dental procedures, and patients undergoing physical therapy (Gupta et al., 2018). Beyond the management of acute pain, VR applications is expanding to help with chronic pain management and postoperative recovery (Ding et al., 2020; Wong et al., 2022).

VR-based rehabilitation (VRBR) is used to assist the rehabilitation of patients with motor dysfunctions resulting from a stroke or an injury. VRBR uses technologies that stimulate the daily life activities of patients and allows them to practice it in a controlled environment. This enhances patients' motivation and adherence to treatment and improve the motor skills of the patients such as balance, gait, and coordination. VRBR can be designed to each patient's needs, making it more adaptable than the conventional rehabilitation methods (Aderinto et al., 2023).

VR can also be used in reducing symptoms related to psychotic disorders, like paranoia, delusions, and hallucinations by exposure therapy and integrating cognitive behavioral techniques into the VR environment. Another application of VR here is to manage schizophrenia, by dealing with the cognitive deficits associated with it -like attention and memory-, through engaging and realistic stimulation (Lan et al., 2023).

INTEGRATION AND IMPLEMENTATION OF WEARABLE DEVICES AND SMART TECHNOLOGIES

Wearable technologies are described as sophisticated, lightweight, bendable gadgets that are equipped with a range of sensors to track bodily movement. While consumers do their everyday business, these gadgets can be continuously linked and tracked through applications. Because of these characteristics, using wearable technology in teaching-learning contexts provides many academic options, including administrative tasks, educational contexts, student engagement, capturing and collaborating, assessment and suggestions, and many more (Kuzu Demir & Demir, 2017). Wearable technological advancements have revolutionized education, enabling students to approach their studies in a more positive, engaged, and independent manner. With the help of wearable technologies, students can comprehend information more rapidly and with fewer mental processes or behaviours. It is crucial to remember that the use of wearable technology in the classroom is very different from the conventional teaching approaches, where pupils must attend classes at the designated time and place. Still, teachers must acquire the skills necessary to utilize wearables in a learning setting (M. Afshari" Cloud Tweaks –, n.d.). A few instances of wearable technology in higher education are the University of Western Australia's Self eHealth Challenge, which uses Fitbits, and the University of South Wales' usage of virtual reality (VR) head-mounted displays in engineering. It was verified that the use of this technology in educational institutions led to "deeper student knowledge and comprehension of scenario-based practices, and effortless integration into student learning processes" as well as "a greater connection with events and individuals" (Alvarez et al., 2016).

Wearable Technology in Healthcare Education

Education institutions were encouraged by COVID-19 to use cutting-edge edtech to implement entirely remote or hybrid learning. The emergence of digital technology, like as video conferencing tools and online learning management software that can be accessed on portable laptops, tablets, and smartphones, often takes years, but in the case of the pandemic and physical teaching constraints, it happened in a matter of months (Almusawi et al., 2021a). In the future, edtech will need to offer complete remote teaching and learning in the event of lockdowns similar to pandemics (Khosravi et al., 2022).

Imagination, inventiveness, inspiration, interaction, interconnectedness, and improvement are the new paradigms that are emerging from the educational pillars today. The digital age has given rise to several novel learning methods, including social, personalized, blended, interactive, immersive, adaptive, continuous, etc (authorCorporate:UNESCO, 2019). A new paradigm in international education is smart education. Enhancing learners' educational experiences and encouraging lifelong learning is the goal. It centers on personalized, contextual, and seamless education to encourage the development of students' intelligence. It makes possible the capacity to solve issues in intelligent settings. There are numerous obstacles facing smart education, including instructional technology, pedagogical, and teachers' education, educational institutions, and educational philosophy (Palanivel, 2020).

Based on the technologies available, smart education solutions are created to satisfy the needs of staff, parents, students, teachers, and other stakeholders. Additionally, it is built to scale, e-content with learning data capacity and a commercial or instructional role. The educational solution possesses a cloud platform for education that consists of smart devices, e-resources, smart campuses, social media and system integration services (Palanivel, 2020). The term "Internet of Things" (IoT) describes how real objects—such as cars, buildings, and other objects—are connected through network connectivity, software, and sensors that enable data collection and sharing. IoT can be used in education to build smart campuses and classrooms that provide improved learning opportunities via the use of networked devices, real-time data analytics, and customized content delivery (Adel, 2024). Advanced analytics is another category that includes methods like predictive and prescriptive analytics, which estimate future patterns and suggest the best course of action based on historical data. Advanced analytics can be used in the classroom to spot trends in students' behavior, performance, and engagement. This helps teachers make data-driven decisions that will help students succeed and enhance learning outcomes (Hooda et al., 2022). Even the study, development, and application of robotics technology results in the creation of autonomous or semi-autonomous robots that can help teach programming and concepts of engineering, provide practical training, and even support children with special needs in the classroom (Mubin et al., 2013).

The potential of wearable technology, or wearables, in education is gaining attention (Almusawi et al., 2021b). There are wearable devices that can be worn on the head, neck, chest, torso, waist, shoulders, arm, wrist, hand, finger, legs, and feet, among other body parts. Most of the devices in the market are wrist-worn and primarily intended for fitness applications. It's possible that wearer familiarity and comfort contributed to their success. Additionally, it has been demonstrated that wearable technology—such as smartwatches, smart glasses, and head-mounted displays—is advantageous for teaching. Figure 3 depicts some of the wearable gadgets that can be used in education. These tools were employed in a variety of educational settings, including real-world learning, medical education, and student participation (Khosravi et al., 2022).

Figure 3. Some head-worn wearables used for educational purposes modified from Khosravi et al.(Khosravi et al., 2022).

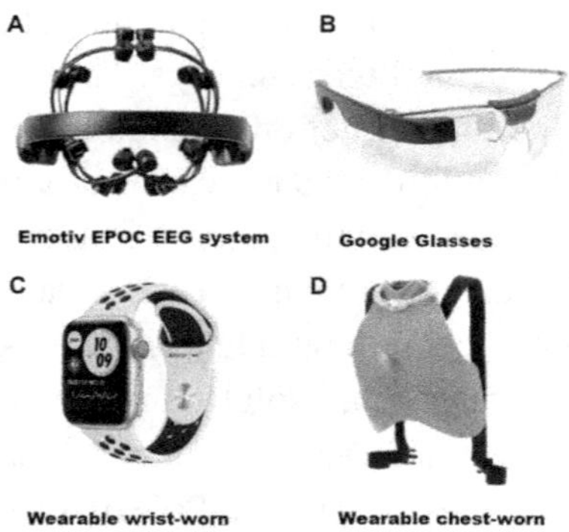

From on-site paramedic use to advanced care and consultations, Google Glass and other technologies play a ubiquitous role in the healthcare continuum and serve as a vital clinical tool (Nosta J, n.d.). Perhaps the first medical school to incorporate Google Glass into its curriculum is the University of California, Irvine School of Medicine. This medical school implemented wearable computer integration into its four-year curriculum, starting with clinical skills training and anatomy classes in the first and second years and moving on to hospital rotations in the third and fourth years (UC Irvine, n.d.). Teachers think that Glass's special features—such as its hands-free, smartphone-like information display, voice command capabilities for internet communication, and ability to broadcast and record patient care and student training activities using proprietary software that complies with the federal Health Insurance Portability & Accountability Act of 1996—will benefit students (Pelletier, 2014). Although Google Glass is the most well-known of these gadgets, many of the same functionalities are available in rival models such as the Optivent ORA and Epson Moverio. For teachers who wish to replicate a hands-on learning experience for pupils in the classroom, smart glasses are ideal since they record video and show it back on the lens' display. For instance, Lucien Engelen, the director of Radboud University Medical Center's Reshape Center for Innovation, studied whether Google Glass would work well in operating rooms and ambulances. She discovered that the high caliber of the images and videos would be beneficial for instruction and advice (Khosravi et al., 2022).

The use of technologies in medical education—undergraduate, graduate, and continuing—has grown in popularity. There are several benefits related to schooling. Among the benefits are surroundings that are safe and controlled to reduce patient risk; improved and more accurate visually Realistic settings for education and evaluation, recording of student conduct and performance, instruction based on the needs of the group or the individual, student autonomy in the learning process, practice with purpose and repetition, increase the variety of perceptions and synchronization of skills, uniformity in education and evaluation (Guze, 2015). These techniques make it easier to learn new information, boost decision-making, increase perceptual variety, improve coordination of skills, and offer an engaging learning environment that permits learning without putting patients at risk. The ability to evaluate abilities and milestones and give students, regardless of skill level, the resources they need to stay current on medical knowledge, continue to get high-quality care, and become lifelong learners are two more advantages of using computer technologies (Guze, 2015). Virtual, augmented, and mixed reality; robots; artificial intelligence; serious games and gamification; and open educational resources are some of the technologies that are being looked into more and more for use in health professions education (Frenk et al., 2022).

Wu and colleagues examined the application of Glass in a simulation-based training program by contrasting postgraduate trainees with medical students. Every individual using the Glass said it was comfortable to wear. It also didn't interfere with the simulation program, and the recorded simulation aided with participant debriefing by offering comments (Wu et al., 2014). In two studies, Yoshida et al. (Yoshida et al., 2014a, 2014b) found that head-mounted displays, such as the Sony

Corporation's HMZ-T2 and HMM-3000MT (Japan) models, had a legitimate place in healthcare education. During transurethral resection of the prostate procedures, surgeons were able to point regions of interest using these head-mounted displays as a vision-based finger tracking system. These areas can then be transmitted live to trainers and medical learners. This enables surgeons to communicate with, advise, and train colleagues as well as students while maintaining sterility (Yoshida et al., 2014b). A head-mounted display virtual reality simulator has been evaluated, and the results show a strong favorable correlation between the simulator and behind-the-wheel performance in patients recovering from spinal cord injuries (Carlozzi et al., 2013). Nonetheless, the study notes that individuals who used the head-mounted display experienced more "off road" crashes (0.4 vs. 0.25, $p > 0.05$) and stopped for longer periods of time because of the device's limitations. During the simulation, a few participants also complained of severe simulator sickness (Carlozzi et al., 2013).

Wearable Technology For Enhancing Clinical Simulation

"Complex learning" is what medical education entails, which includes learning about physiological systems, cultivating social skills, and making therapeutic judgments that are impacted by a variety of variables within dynamic systems. Whole task learning, which immerse students in actual circumstances, is therefore essential for mastering (Van Merriënboer & Kirschner, 2017).

The purpose of simulation is to imitate actual patients, anatomical locations, clinical tasks, and/or the real-world environments in which healthcare is provided. Several goals for education can be achieved through simulations (Guze, 2015). A wide range of complexity is covered by the use of simulation, from the basic replication of individual body parts to intricate interactions between individuals portrayed by simulated patients or high-fidelity human patient simulators that replicate the appearance of the entire body and changeable physiological variables (Bradley, 2006). When mouth-to-mouth resuscitation procedures were first presented 35 years ago, one of the first simulators was created—a mannequin known as Rescusi Anne (Resusci, 1971). Harvey, a simulator for teaching cardiac assessment techniques, was created around the same period and is being utilized in hospitals and medical institutions all around the world (Gordon et al., 1980). The term "haptic systems" describes simulators that imitate tactile and kinesthetic perception as well as create a sense of resistance when manipulating instruments in a virtual setting (Hamza-Lup et al., 2013). Integrated simulators combine a mannequin, typically consisting of their entire body, with advanced computer controls that may be adjusted to provide a variety of physiological parameter outputs. These outputs can be electrical (seen as monitor readouts) or physical (such as respiration movements or pulse rate) (Lane et al., 2001).

With wearable simulators, clinical professionals can practice a skill while conversing with a simulated or normal patient. Many believe that the best level of fidelity for teaching healthcare learners is when wearable simulators and real people are combined. Some of the items can be used to conduct technical skills while engaging with a human by being worn by a standardized or simulated patient. With the help of the wearable simulators, the healthcare participant can practice: Foleys Catheter, giving birth, placement of a central venous catheter line and/or site care, thoracentesis and ultrasonography-guided puncture using a chest overlay, and parental injections. Certain devices possess inbuilt sensors or wearable wireless sensors that can provide clues to the simulated or standardized patient about when and how to react to them (Gore, n.d.).

Augmented Reality in Anatomy Education and Training

A relatively new technology, augmented reality (AR), has drawn a lot of interest in medical education and training. The foundation of this technology is the combination of real-world environmental inputs with digitally created three-dimensional (3D) representations (Dhar et al., 2021). An online visualization of several anatomical structures is possible for students with this technology. Students may enlarge an image and gain a better understanding of the anatomical structures covered in the lecture by pointing the camera pointer to a specific location while the instructor explains the model and its attributes. To see deeper anatomical features and remove layers, there is also a dissect button. The learner can undo edits and investigate the subject further with the undo command (Bose, 2024).

AR apps for smartphones and tablets offer visual 3D structures and a wealth of information on human anatomy. These pictures are frequently connected to conventional anatomy textbook pages. New developments in technology, like Microsoft's creation of the Hololens glasses, have improved the use of these apps (Palumbo, 2022). Applications based on augmented reality provide interactive 3D brain investigation. Based on MRI data, these applications were created. Neurosurgery, ophthalmology, spine surgery, and ENT surgery are among the various specialty training fields that use Immersive Touch. Another AR simulator for brain tumor removal is called NeuroTouch Cranio. In medical education, AR simulators like RoboSim, ANGIO Mentor, and Anatomical Simulator for Pediatric Neurosurgery are frequently utilized (Bose, 2024).

Virtual Reality for Surgical Training

Virtual Reality (VR) technology is assisting medical professionals in numerous areas. Nonetheless, among the most potential fields of use include ophthalmology, surgery, rehabilitation therapy, and the treatment of psychiatric illnesses. For aspiring surgeons, state-of-the-art virtual reality surgical simulators are an invaluable training tool. They greatly enhance the results of training. Experienced practitioners can also gain from using "virtual surgeons." For instance, using simulators enables medical professionals to practice various surgical scenarios and determine the best course of action for each unique case. With the use of a combination of CT and MRI scans to construct a three-dimensional model of the patient, surgeons may now arrange a surgical treatment for a specific patient in virtual reality thanks to advancements in technology (Linskyi, 2023).

Industry-Academia Collaboration in Technology-Enhanced Educational Opportunities

It is obvious that the sustainability of industry depends on high-quality educational programs. As a result, there has been a thought process change about industry and education collaboration that is currently gaining momentum. Universities can work in a mutually beneficial partnership with industries to increase educational opportunities and value. This can be achieved by developing better resources for postsecondary and advanced education, promoting the transfer of cutting-edge concepts and technologies, and fostering an atmosphere that encourages lifelong learning. Industry-academia partnerships have been growing over the past few years. The Ministry of Education's "Industry-Academia Cooperation (IAC) Project" and other research funds, among other initiatives in China, have further encouraged industry and academia to work together and artificial intelligence is the core component of this partnership with IAC (Esangbedo et al., 2023). The history of artificial intelligence (AI) in education begins with its early incarnations in computer and computer-related technologies. Its influence grew over time to include online and web-based intelligent education systems (Murphy, 2019). Customization is a unique feature of AI-enabled education. Students can have individualized learning paths created for them thanks to the analysis of data and patterns by AI algorithms. This customized strategy raises student engagement and retention rates, which improves the quality of instruction given and the overall learning experience (Chen et al., 2020). Essentially, AI has developed into a key facilitator in the field of education, significantly influencing instructional strategies, administrative procedures, and student performance. The educational industry stands to gain even more from the continued advancement of technology and the growing sophistication of AI capabilities. These developments have the potential to completely transform the ways in which knowledge is obtained, applied, and conveyed in the global quest for academic achievement in the creation of education based on AI (Esangbedo et al., 2023). Peacock et al. (2023) offers numerous instances of possible applications for ChatGPT 3.5, a large language model or generative artificial intelligence (genAI) that was made available for free on November 30, 2022, by OpenAI, in medical education, encompassing research, teaching, development, and learning. The authors emphasize that in order to produce relevant outputs, prompts must be carefully created and refined (Peacock et al., 2023). The findings of educational research can be applied to the teaching contexts of readers through a thorough evaluation and research documenting the use of technologies by educators in a variety of learning designs and environments and their effects on the learning of health professions students (Allen et al., 2022).

Technology should be used in medical education to enhance instruction rather than to take the place of in-person instruction. Teachers still need to concentrate on the fundamentals of education rather than on the newest gadgets. Technologies are one instrument in the toolbox of education. Medical educators' job is to make use of these new tools so that learning becomes a more individualized, interactive, and inspiring process (Guze, 2015).

CHALLENGES

Despite the widespread use of these devices, they require continuous monitoring, regular maintenance and constant updates. Since, these devices have access to personal data, it poses significant risks related to the privacy and security. Here we try to list some of these challenges:

Privacy and Security

Clarity about the storage and handling of data remains a pressing issue. The way this data is shared with third parties and the necessity of patient consent are paramount considerations to avoid a data breach or a theft. Addressing these issues must be of utmost priority to foster trust among users.

Cost

Cost is a critical factor in ensuring the accessibility of these technologies to a wider level. Cheaper devices lack accuracy as well as security, while the brands are not affordable by everyone. It is essential for tech industry to balance the cost implications with technological advancements for their widespread adoption.

Accuracy and Reliability

Sustaining the accuracy of these gadgets should be maintained through continuous R & D process. Some devices are built with substandard machinery and sensors which can compromise the results leading to patients' health at stake. This issue must be taken into direct consideration of health regulatory agencies like FDA to ensure the safety and efficacy of such devices.

Compatibility Issue

compatibility is a main issue while purchasing a good health tracker device. Most companies make their wearable devices compatible to their own mobiles and computer systems. It means if you want to purchase a good health tracker device you must buy other devices too.

Companies should ensure seamless integration of their devices across different platforms. so that a user can use it conveniently and data can be shared effectively within healthcare facilities.

Manufacturing Quality

the durability and performance of these devices is crucial. Manufacturers must uphold stringent quality standards of the machinery and sensors to safeguard user data and device reliability.

Water or sweat penetration is another problem for devices. Devices may go defunct when they are exposed to water. Waterproofing can enhance their functionality, particularly in challenging environments.

Power Management

battery is the life of these devices. User must recharge the battery frequently which is cumbersome. Because power failure can lead to disruption of health monitoring. Enhancing battery capacity and fast charging can mitigate disruption in health monitoring caused by power outage. These aspects are of high priority to improve user experience and device performance.

REFERENCES

K201525-Luke Olson. (n.d.). https://www.fda.gov/medical-

A brief history of wearable computing. (n.d.). Retrieved October 2, 2024, from https://www.media.mit.edu/wearables/lizzy/timeline.html#1268

Adel, A. (2024). The Convergence of Intelligent Tutoring, Robotics, and IoT in Smart Education for the Transition from Industry 4.0 to 5.0. *Smart Cities, 7*(1), 325–369. https://www.mdpi.com/2624-6511/7/1/14

Aderinto, N., Olatunji, G., Abdulbasit, M. O., Edun, M., Aboderin, G., & Egbunu, E. (2023). Exploring the efficacy of virtual reality-based rehabilitation in stroke: a narrative review of current evidence. In *Annals of Medicine* (Vol. 55, Issue 2). Taylor and Francis Ltd. DOI: 10.1080/07853890.2023.2285907

Aeffner, F., Wilson, K., Martin, N. T., Black, J. C., Hendriks, C. L. L., Bolon, B., Rudmann, D. G., Gianani, R., Koegler, S. R., Krueger, J., & Young, G. D. (2017). The gold standard paradox in digital image analysis: Manual versus automated scoring as ground truth. In *Archives of Pathology and Laboratory Medicine* (Vol. 141, Issue 9, pp. 1267–1275). College of American Pathologists. DOI: 10.5858/arpa.2016-0386-RA

Afshari, M. (2014). *Wearable technology in education*. Cloud Tweaks–Connecting Our Planet.

Agarwal, M., Narcisse, D., Khouzam, N., & Khouzam, R. N. (2018). Wearable Cardioverter Defibrillator "The Lifevest": Device Design, Limitations, and Areas of Improvement. *Current Problems in Cardiology*, 43(2), 45–55. DOI: 10.1016/j.cpcardiol.2017.04.002 PMID: 29277221

Al-Alusi, M. A., Ding, E., McManus, D. D., & Lubitz, S. A. (2019). Wearing Your Heart on Your Sleeve: the Future of Cardiac Rhythm Monitorin. In *Current Cardiology Reports* (Vol. 21, Issue 12). Springer. DOI: 10.1007/s11886-019-1223-8

Albogami, Y., Alfakhri, A., Alaqil, A., Alkoraishi, A., Alshammari, H., Elsharawy, Y., Alhammad, A., & Alhossan, A. (2024). Safety and quality of AI chatbots for drug-related inquiries: A real-world comparison with licensed pharmacists. *Digital Health*, 10, 20552076241253523. Advance online publication. DOI: 10.1177/20552076241253523 PMID: 38757086

Alim, A., & Imtiaz, M. H. (2023). Wearable Sensors for the Monitoring of Maternal Health—A Systematic Review. In *Sensors* (Vol. 23, Issue 5). MDPI. DOI: 10.3390/s23052411

Allegra, A., Tonacci, A., Sciaccotta, R., Genovese, S., Musolino, C., Pioggia, G., & Gangemi, S. (2022). Machine Learning and Deep Learning Applications in Multiple Myeloma Diagnosis, Prognosis, and Treatment Selection. In *Cancers* (Vol. 14, Issue 3). MDPI. DOI: 10.3390/cancers14030606

Allen, L. M., Hay, M., & Palermo, C. (2022). Evaluation in health professions education—Is measuring outcomes enough? *Medical Education*, 56(1), 127–136. DOI: 10.1111/medu.14654 PMID: 34463357

Almusawi, H. A., Durugbo, C. M., & Bugawa, A. M. (2021a). Innovation in physical education: Teachers' perspectives on readiness for wearable technology integration. *Computers & Education*, 167, 104185. DOI: 10.1016/j.compedu.2021.104185

Almusawi, H. A., Durugbo, C. M., & Bugawa, A. M. (2021b). Wearable technology in education: A systematic review. *IEEE Transactions on Learning Technologies*, 14(4), 540–554. DOI: 10.1109/TLT.2021.3107459

Alowais, S. A., Alghamdi, S. S., Alsuhebany, N., Alqahtani, T., Alshaya, A. I., Almohareb, S. N., Aldairem, A., Alrashed, M., Bin Saleh, K., Badreldin, H. A., Al Yami, M. S., Al Harbi, S., & Albekairy, A. M. (2023). Revolutionizing healthcare: the role of artificial intelligence in clinical practice. In *BMC Medical Education* (Vol. 23, Issue 1). BioMed Central Ltd. DOI: 10.1186/s12909-023-04698-z

Alvarez, V., Bower, M., de Freitas, S., Gregory, S., & de Wit, B. (2016). *The Use of Wearable Technologies in Australian Universities: Examples from Environmental Science*. Cognitive and Brain Sciences and Teacher Training.

Apple. (n.d.). *Healthcare - Apple Watch* Retrieved September 30, 2024, from https://www.apple.com/healthcare/apple-watch/

Apple Support (BH). (n.d.). *How to use the Blood Oxygen app on Apple Watch* Retrieved October 1, 2024, from https://support.apple.com/en-bh/120358

Askin, S., Burkhalter, D., Calado, G., & El Dakrouni, S. (2023). Artificial Intelligence Applied to clinical trials: opportunities and challenges. In *Health and Technology* (Vol. 13, Issue 2, pp. 203–213). Springer Science and Business Media Deutschland GmbH. DOI: 10.1007/s12553-023-00738-2

Baker, S. B., Xiang, W., & Atkinson, I. (2017). Internet of Things for Smart Healthcare: Technologies, Challenges, and Opportunities. In *IEEE Access* (Vol. 5, pp. 26521–26544). Institute of Electrical and Electronics Engineers Inc., DOI: 10.1109/ACCESS.2017.2775180

Begg, E. J., Barclay, M. L., & Kirkpatrick, C. J. M. (1999). The therapeutic monitoring of antimicrobial agents. *British Journal of Clinical Pharmacology*, 47(1), 23–30. DOI: 10.1046/j.1365-2125.1999.00850.x PMID: 10073735

Bhaltadak, V., Ghewade, B., & Yelne, S. (2024). A Comprehensive Review on Advancements in Wearable Technologies: Revolutionizing Cardiovascular Medicine. *Cureus*. Advance online publication. DOI: 10.7759/cureus.61312 PMID: 38947726

Blanco-González, A., Cabezón, A., Seco-González, A., Conde-Torres, D., Antelo-Riveiro, P., Piñeiro, Á., & Garcia-Fandino, R. (2023). The Role of AI in Drug Discovery: Challenges, Opportunities, and Strategies. In *Pharmaceuticals* (Vol. 16, Issue 6). Multidisciplinary Digital Publishing Institute (MDPI). DOI: 10.3390/ph16060891

Bose, P. (2024). *Augmented Reality in Anatomy Education and Training*. News Medical Life Sciences. https://www.news-medical.net/health/Augmented-Reality-in-Anatomy-Education-and-Training.aspx

Bradley, P. (2006). The history of simulation in medical education and possible future directions. *Medical Education*, 40(3), 254–262. DOI: 10.1111/j.1365-2929.2006.02394.x PMID: 16483328

Brandt, J. D., DuBiner, H. B., Benza, R., Sall, K. N., Walker, G. A., Semba, C. P., Budenz, D., Day, D., Flowers, B., Lee, S., Nguyen, Q., & Wirta, D. (2017). Long-term Safety and Efficacy of a Sustained-Release Bimatoprost Ocular Ring. *Ophthalmology*, 124(10), 1565–1566. DOI: 10.1016/j.ophtha.2017.04.022 PMID: 28528010

Cappon, G., Acciaroli, G., Vettoretti, M., Facchinetti, A., & Sparacino, G. (2017). Wearable continuous glucose monitoring sensors: A revolution in diabetes treatment. *Electronics (Basel)*, 6(3), 65. Advance online publication. DOI: 10.3390/electronics6030065

Carlozzi, N. E., Gade, V., Rizzo, A. S., & Tulsky, D. S. (2013). Using virtual reality driving simulators in persons with spinal cord injury: Three screen display versus head mounted display. *Disability and Rehabilitation. Assistive Technology*, 8(2), 176–180. DOI: 10.3109/17483107.2012.699990 PMID: 22775982

Chen, L., Chen, P., & Lin, Z. (2020). Artificial Intelligence in Education: A Review. *IEEE Access : Practical Innovations, Open Solutions*, 8, 75264–75278. DOI: 10.1109/ACCESS.2020.2988510

Dash, S., Shakyawar, S. K., Sharma, M., & Kaushik, S. (2019). Big data in healthcare: Management, analysis and future prospects. *Journal of Big Data*, 6(1), 54. Advance online publication. DOI: 10.1186/s40537-019-0217-0

Dexcom. (n.d.). *Dexcom Continuous Glucose Monitoring* Retrieved September 30, 2024, from https://www.dexcom.com/

Dhar, P., Rocks, T., Samarasinghe, R. M., Stephenson, G., & Smith, C. (2021). Augmented reality in medical education: Students' experiences and learning outcomes. *Medical Education Online*, 26(1), 1953953. DOI: 10.1080/10872981.2021.1953953 PMID: 34259122

Dimitrov, D. V. (2016). Medical internet of things and big data in healthcare. In *Healthcare Informatics Research* (Vol. 22, Issue 3, pp. 156–163). Korean Society of Medical Informatics. DOI: 10.4258/hir.2016.22.3.156

Ding, L., Hua, H., Zhu, H., Zhu, S., Lu, J., Zhao, K., & Xu, Q. (2020). Effects of virtual reality on relieving postoperative pain in surgical patients: A systematic review and meta-analysis. In *International Journal of Surgery* (Vol. 82, pp. 87–94). Elsevier Ltd., DOI: 10.1016/j.ijsu.2020.08.033

ECRI. (n.d.). *Pulse Oximetry: Choosing the Right Sensor and Applying It Correctly* Retrieved September 30, 2024, from https://www.ecri.org/components/HDJournal/Pages/Pulse-Oximetry-Choosing-the-Right-Sensor.aspx

Esangbedo, C., Zhang, J., Esangbedo, M., Kone, S., & Xu, L. (2023). The role of industry-academia collaboration in enhancing educational opportunities and outcomes under the digital driven Industry 4.0. *Journal of Infrastructure. Policy and Development*, 8(1). Advance online publication. DOI: 10.24294/jipd.v8i1.2569

eSight Eyewear. (n.d.). *How it Works* Retrieved September 30, 2024, from https://www.esighteyewear.com/how-it-works/

Evenson, K. R., Goto, M. M., & Furberg, R. D. (2015). Systematic review of the validity and reliability of consumer-wearable activity trackers. In *International Journal of Behavioral Nutrition and Physical Activity* (Vol. 12, Issue 1). BioMed Central Ltd. DOI: 10.1186/s12966-015-0314-1

Fitbit Sense. (n.d.). *Advanced Health Smartwatch* Retrieved September 30, 2024, from https://www.fitbit.com/global/us/products/smartwatches/sense

Frajhof, L., Borges, J., Hoffmann, E., Lopes, J., & Haddad, R. (2018). Virtual reality, mixed reality and augmented reality in surgical planning for video or robotically assisted thoracoscopic anatomic resections for treatment of lung cancer. *Journal of Visualized Surgery*, 4, 143–143. DOI: 10.21037/jovs.2018.06.02

Frenk, J., Chen, L. C., Chandran, L., Groff, E. O. H., King, R., Meleis, A., & Fineberg, H. V. (2022). Challenges and opportunities for educating health professionals after the COVID-19 pandemic. *Lancet*, 400(10362), 1539–1556. DOI: 10.1016/S0140-6736(22)02092-X PMID: 36522209

Gibson, W., Wand, B. M., & O'Connell, N. E. (2017). Transcutaneous electrical nerve stimulation (TENS) for neuropathic pain in adults. In *Cochrane Database of Systematic Reviews* (Vol. 2017, Issue 9). John Wiley and Sons Ltd. DOI: 10.1002/14651858.CD011976.pub2

Gordon, M. S., Forker, A. D., Gessner, I., McGuire, C., Mayer, J. W., Sajid, D. P. A., Ewy, G. A., Felner, J. M., & Waugh, R. A. (1980). Teaching bedside cardiologic examination skills using "Harvey", the cardiology patient simulator. *The Medical Clinics of North America*, 64(2), 305–313. DOI: 10.1016/S0025-7125(16)31620-0 PMID: 6155573

Gore, T. (n.d.). *How Wearable Simulators Can Improve Clinical Simulation Realism*. https://www.healthysimulation.com/how-wearable-simulators-can-improve-clinical-simulation-realism/#:~:text=Wearable%20simulators%20allow%20clinical%20professional,available%20to%20educate%20healthcare%20learners

Gupta, A., Scott, K., & Dukewich, M. (2018). Innovative technology using virtual reality in the treatment of pain: Does it reduce pain via distraction, or is there more to it? *Pain Medicine*, 19(1), 151–159. DOI: 10.1093/pm/pnx109 PMID: 29025113

Gurcan, M. N., Boucheron, L. E., Can, A., Madabhushi, A., Rajpoot, N. M., & Yener, B. (2009). Histopathological Image Analysis: A Review. *IEEE Reviews in Biomedical Engineering*, 2, 147–171. DOI: 10.1109/RBME.2009.2034865 PMID: 20671804

Guze, P. A. (2015). Using Technology to Meet the Challenges of Medical Education. *Transactions of the American Clinical and Climatological Association*, 126, 260–270. PMID: 26330687

Haghi, M., Thurow, K., & Stoll, R. (2017). Wearable devices in medical internet of things: Scientific research and commercially available devices. *Healthcare Informatics Research*, 23(1), 4–15. DOI: 10.4258/hir.2017.23.1.4 PMID: 28261526

Hamza-Lup, F. G., Popovici, D. M., & Bogdan, C. M. (2013). Haptic feedback systems in medical education. *Journal of Advanced Distributed Learning Technology*, 1(2), 7.

Hee Keum, D., Kim, S.-K., Koo, J., Lee, G.-H., Jeon, C., Won Mok, J., Ho Mun, B., Jae Lee, K., Kamrani, E., Joo, C.-K., Shin, S., Sim, J.-Y., Myung, D., Hyun Yun, S., Bao, Z., & Kwang Hahn, S. (2020). *Wireless smart contact lens for diabetic diagnosis and therapy*. https://www.science.org

Hooda, M., Rana, C., Dahiya, O., Rizwan, A., & Hossain, M. S. (2022). Artificial intelligence for assessment and feedback to enhance student success in higher education. *Mathematical Problems in Engineering*, 2022(1), 5215722. DOI: 10.1155/2022/5215722

Horvers, A., Tombeng, N., Bosse, T., Lazonder, A. W., & Molenaar, I. (2021). Detecting emotions through electrodermal activity in learning contexts: A systematic review. *Sensors (Basel)*, 21(23), 7869. Advance online publication. DOI: 10.3390/s21237869 PMID: 34883870

Hosny, A., Parmar, C., Quackenbush, J., Schwartz, L. H., & Aerts, H. J. W. L. (2018). Artificial intelligence in radiology. In *Nature Reviews Cancer* (Vol. 18, Issue 8, pp. 500–510). Nature Publishing Group. DOI: 10.1038/s41568-018-0016-5

Hossain, M. I., Zahid, M. S., Chowdhury, M. A., Hossain, M. M. M., Hossain, N., Islam, M. A., & Mobarak, M. H. (2024). Smart bandage: A device for wound monitoring and targeted treatment. In *Results in Chemistry* (Vol. 7). Elsevier B.V., DOI: 10.1016/j.rechem.2023.101292

Hu, J. R., Martin, G., Iyengar, S., Kovell, L. C., Plante, T. B., van Helmond, N., Dart, R. A., Brady, T. M., Turkson-Ocran, R. A. N., & Juraschek, S. P. (2023). Validating cuffless continuous blood pressure monitoring devices. In *Cardiovascular Digital Health Journal* (Vol. 4, Issue 1, pp. 9–20). Elsevier Inc. DOI: 10.1016/j.cvdhj.2023.01.001

Hu, J. R., Park, D. Y., Agarwal, N., Herzig, M., Ormseth, G., Kaushik, M., Giao, D. M., Turkson-Ocran, R. A. N., & Juraschek, S. P. (2023). The Promise and Illusion of Continuous, Cuffless Blood Pressure Monitoring. *Current Cardiology Reports*, 25(10), 1139–1149. DOI: 10.1007/s11886-023-01932-4 PMID: 37688763

Ijeh, S., Okolo, C. A., Arowoogun, J. O., Adeniyi, A. O., & Omotayo, O. (2024). Predictive modeling for disease outbreaks: a review of data sources and accuracy. *International Medical Science Research Journal*, 4(4), 406–419. DOI: 10.51594/imsrj.v4i4.999

Ikegami, H., Hiromine, Y., & Noso, S. (2022). Insulin-dependent diabetes mellitus in older adults: Current status and future prospects. In *Geriatrics and Gerontology International* (Vol. 22, Issue 8, pp. 549–553). John Wiley and Sons Inc. DOI: 10.1111/ggi.14414

Iqbal, S. M. A., Mahgoub, I., Du, E., Leavitt, M. A., & Asghar, W. (2021). Advances in healthcare wearable devices. In *npj Flexible Electronics* (Vol. 5, Issue 1). Nature Research. DOI: 10.1038/s41528-021-00107-x

Irvine, U. C. (n.d.). *UCI School of Medicine first to integrate Google Glass into curriculum – wearable computing technology will transform training of future doctors. UCIrvine News.* https://news.uci.edu/press-releases/uci-school-of-medicine-first-to-integrate-google-glass-into-curriculum/

Jamison, R. N., Wan, L., Edwards, R. R., Mei, A., & Ross, E. L. (2019). Outcome of a High-Frequency Transcutaneous Electrical Nerve Stimulator (hfTENS) Device for Low Back Pain: A Randomized Controlled Trial. *Pain Practice*, 19(5), 466–475. DOI: 10.1111/papr.12764 PMID: 30636101

Kanschik, D., Bruno, R. R., Wolff, G., Kelm, M., & Jung, C. (2023). Virtual and augmented reality in intensive care medicine: a systematic review. In *Annals of Intensive Care* (Vol. 13, Issue 1). Springer Science and Business Media Deutschland GmbH. DOI: 10.1186/s13613-023-01176-z

Kennedy, H. L. (2013). The evolution of ambulatory ECG monitoring. *Progress in Cardiovascular Diseases*, 56(2), 127–132. DOI: 10.1016/j.pcad.2013.08.005 PMID: 24215744

Khosravi, S., Bailey, S. G., Parvizi, H., & Ghannam, R. (2022). Wearable Sensors for Learning Enhancement in Higher Education. *Sensors (Basel)*, 22(19), 7633. Advance online publication. DOI: 10.3390/s22197633 PMID: 36236732

Kim, I., Kang, K., Song, Y., & Kim, T. J. (2022). Application of Artificial Intelligence in Pathology: Trends and Challenges. In *Diagnostics* (Vol. 12, Issue 11). Multidisciplinary Digital Publishing Institute (MDPI). DOI: 10.3390/diagnostics12112794

Koh, A., Kang, D., Xue, Y., Lee, S., Pielak, R. M., Kim, J., Hwang, T., Min, S., Banks, A., Bastien, P., Manco, M. C., Wang, L., Ammann, K. R., Jang, K.-I., Won, P., Han, S., Ghaffari, R., Paik, U., Slepian, M. J., . . . Rogers, J. A. (2016). *A soft, wearable microfluidic device for the capture, storage, and colorimetric sensing of sweat*. https://stm.sciencemag.org/

Korn, E. L., Moscow, J. A., & Freidlin, B. (2023). Dose optimization during drug development: whether and when to optimize. In *Journal of the National Cancer Institute* (Vol. 115, Issue 5, pp. 492–497). Oxford University Press. DOI: 10.1093/jnci/djac232

Kuzu Demir, E., & Demir, K. (2017). *Enhancing Learning with Wearable Technologies In and Out of Educational Settings*. DOI: 10.4018/978-1-5225-1692-7.ch006

Kvedar, J., Coye, M. J., & Everett, W. (2014). Connected health: A review of technologies and strategies to improve patient care with telemedicine and telehealth. *Health Affairs*, 33(2), 194–199. DOI: 10.1377/hlthaff.2013.0992 PMID: 24493760

Lan, L., Sikov, J., Lejeune, J., Ji, C., Brown, H., Bullock, K., & Spencer, A. E. (2023). A Systematic Review of using Virtual and Augmented Reality for the Diagnosis and Treatment of Psychotic Disorders. In *Current Treatment Options in Psychiatry* (Vol. 10, Issue 2, pp. 87–107). Springer Science and Business Media Deutschland GmbH. DOI: 10.1007/s40501-023-00287-5

Lane, J. L., Slavin, S., & Ziv, A. (2001). Simulation in medical education: A review. *Simulation & Gaming*, 32(3), 297–314. DOI: 10.1177/104687810103200302

Lee, D. A. of, D., & Higginbotham, E. J. (n.d.). Clinical Review Glaucoma and its treatment: A review. In *Am J Health-Syst Pharm*.

Li, L. R., Du, B., Liu, H. Q., & Chen, C. (2021). Artificial Intelligence for Personalized Medicine in Thyroid Cancer: Current Status and Future Perspectives. In *Frontiers in Oncology* (Vol. 10). Frontiers Media S.A. DOI: 10.3389/fonc.2020.604051

Lin, Y. R., Hung, C. C., Chiu, H. Y., Chang, B. H., Li, B. R., Cheng, S. J., Yang, J. W., Lin, S. F., & Chen, G. Y. (2018). Noninvasive glucose monitoring with a contact lens and smartphone. *Sensors (Basel)*, 18(10), 3208. Advance online publication. DOI: 10.3390/s18103208 PMID: 30249021

Linskyi, A. (2023). *Virtual Reality for Surgical Training: Theory & Practice*. https://smarttek.solutions/blog/vr-for-surgery/

Lopez-Jimenez, F., Attia, Z., Arruda-Olson, A. M., Carter, R., Chareonthaitawee, P., Jouni, H., Kapa, S., Lerman, A., Luong, C., Medina-Inojosa, J. R., Noseworthy, P. A., Pellikka, P. A., Redfield, M. M., Roger, V. L., Sandhu, G. S., Senecal, C., & Friedman, P. A. (2020). Artificial Intelligence in Cardiology: Present and Future. In *Mayo Clinic Proceedings* (Vol. 95, Issue 5, pp. 1015–1039). Elsevier Ltd. DOI: 10.1016/j.mayocp.2020.01.038

Manov, A. E., Chauhan, S., Dhillon, G., Dhaliwal, A., Antonio, S., Donepudi, A., Jalal, Y. N., Nazha, J., Banal, M., & House, J. (2023). The Effectiveness of Continuous Glucose Monitoring Devices in Managing Uncontrolled Diabetes Mellitus: A Retrospective Study. *Cureus*. Advance online publication. DOI: 10.7759/cureus.42545 PMID: 37637581

Maples-Keller, J. L., Bunnell, B. E., Kim, S. J., & Rothbaum, B. O. (2017). The use of virtual reality technology in the treatment of anxiety and other psychiatric disorders. In *Harvard Review of Psychiatry* (Vol. 25, Issue 3, pp. 103–113). Lippincott Williams and Wilkins. DOI: 10.1097/HRP.0000000000000138

Medtronic. (n.d.). *Nellcor™ SpO_2 Forehead Sensor with OxiMax™ Technology* Retrieved September 30, 2024, from https://www.medtronic.com/covidien/en-us/products/pulse-oximetry/nellcor-spo2-forehead-sensor.html

Mihai, D., Stefan, D., Stegaru, D., Bernea, G., Vacaroiu, I., Papacocea, T., Lupu o-ru, M., Nica, A., Stiru, O., Dragos, D., & Olaru, O. (2021). Continuous glucose monitoring devices: A brief presentation [Review]. *Experimental and Therapeutic Medicine*, 23(2), 174. Advance online publication. DOI: 10.3892/etm.2021.11097 PMID: 35069855

Mubin, O., Stevens, C. J., Shahid, S., Al Mahmud, A., & Dong, J.-J. (2013). A review of the applicability of robots in education. *Journal of Technology in Education and Learning, 1*(209–0015), 13.

Murphy, R. (2019). *Artificial Intelligence Applications to Support K–12 Teachers and Teaching: A Review of Promising Applications.* Challenges, and Risks., DOI: 10.7249/PE315

MyHeart. (n.d.). *Your Guide to Heart Health* Retrieved September 30, 2024, from https://myheart.net/

Nosta, J. (n.d.). *How Google Glass is changing medical education. Forbes Technology.*https://www.forbes.com/sites/johnnosta/2013/06/27/google-glass-teach-me-medicine-how-glass-is-helping-change-medical-education/

Palanivel, K. (2020). Emerging Technologies to Smart Education. *International Journal of Computer Trends and Technology*, 68(2).

Palumbo, A. (2022). Microsoft HoloLens 2 in Medical and Healthcare Context: State of the Art and Future Prospects. *Sensors (Basel)*, 22(20), 7709. https://www.mdpi.com/1424-8220/22/20/7709. DOI: 10.3390/s22207709 PMID: 36298059

Panula, T., Sirkia, J. P., Wong, D., & Kaisti, M. (2023). Advances in Non-Invasive Blood Pressure Measurement Techniques. *IEEE Reviews in Biomedical Engineering*, 16, 424–438. DOI: 10.1109/RBME.2022.3141877 PMID: 35015647

Patel, R. H., Foltz, E. A., Witkowski, A., & Ludzik, J. (2023). Analysis of Artificial Intelligence-Based Approaches Applied to Non-Invasive Imaging for Early Detection of Melanoma: A Systematic Review. In *Cancers* (Vol. 15, Issue 19). Multidisciplinary Digital Publishing Institute (MDPI). DOI: 10.3390/cancers15194694

Patreon. (n.d.). *Functional Vision To Legally Blind Individuals - Interview With eSight's Director Of Product Marketing, Roland Mattern* Retrieved September 30, 2024, from https://www.patreon.com/posts/78398589

Peacock, J., Austin, A., Shapiro, M., Battista, A., & Samuel, A. (2023). Accelerating medical education with ChatGPT: An implementation guide. *MedEdPublish*, ●●●, 13. PMID: 38440148

Pelletier, S. G. (2014). *Technology in academic medicine: Medicine takes a closer look at Google Glass*. AAMC Reporter.

Philips. (n.d.). *DreamStation Go* Retrieved September 30, 2024, from https://www.usa.philips.com/c-e/hs/sleep-apnea-therapy/i-currently-use-sleep-apnea-therapy/dreamstation-go.html

Pratley, R. E., Kanapka, L. G., Rickels, M. R., Ahmann, A., Aleppo, G., Beck, R., Bhargava, A., Bode, B. W., Carlson, A., Chaytor, N. S., Fox, D. S., Goland, R., Hirsch, I. B., Kruger, D., Kudva, Y. C., Levy, C., McGill, J. B., Peters, A., Philipson, L., & Miller, K. M. (2020). Effect of Continuous Glucose Monitoring on Hypoglycemia in Older Adults With Type 1 Diabetes: A Randomized Clinical Trial. *Journal of the American Medical Association*, 323(23), 2397–2406. DOI: 10.1001/jama.2020.6928 PMID: 32543682

Rana, M., & Bhushan, M. (2023). Machine learning and deep learning approach for medical image analysis: Diagnosis to detection. *Multimedia Tools and Applications*, 82(17), 26731–26769. DOI: 10.1007/s11042-022-14305-w PMID: 36588765

Razdan, S., & Sharma, S. (2022). Internet of Medical Things (IoMT): Overview, Emerging Technologies, and Case Studies. In *IETE Technical Review (Institution of Electronics and Telecommunication Engineers, India)* (Vol. 39, Issue 4, pp. 775–788). Taylor and Francis Ltd. DOI: 10.1080/02564602.2021.1927863

REAL® System. (n.d.). *The REAL® System, Virtual Reality Rehabilitation Tool* Retrieved September 30, 2024, from https://www.realsystem.com/

Resusci, A. (1971). *Advanced Cardiac Life Support Training guides*. American Heart Association.

Rickards, E., Miller, S., Hampson, O., & Wat, D. (2021). Evaluating the accuracy of pulse oximetry: Finger probe versus earlobe probe in patients with confirmed or suspected COVID-19 within the community setting. *The European Respiratory Journal*, 58(suppl 65), PA456. Advance online publication. DOI: 10.1183/13993003.congress-2021.PA456

Rodgers, M. M., Alon, G., Pai, V. M., & Conroy, R. S. (2019). Wearable technologies for active living and rehabilitation: Current research challenges and future opportunities. *Journal of Rehabilitation and Assistive Technologies Engineering*, 6, 205566831983960. DOI: 10.1177/2055668319839607 PMID: 31245033

Saghir, N., Aggarwal, A., Soneji, N., Valencia, V., Rodgers, G., & Kurian, T. (2020). A comparison of manual electrocardiographic interval and waveform analysis in lead 1 of 12-lead ECG and Apple Watch ECG: A validation study. *Cardiovascular Digital Health Journal*, 1(1), 30–36. DOI: 10.1016/j.cvdhj.2020.07.002 PMID: 35265871

Siren Care. (n.d.). *What is the Siren Care Diabetic Sock and Foot Monitoring System?* Retrieved September 30, 2024, from https://www.siren.care/blog/what-is-siren-diabetic-sock-system

Steiner, D. F., Macdonald, R., Liu, Y., Truszkowski, P., Hipp, J. D., Gammage, C., Thng, F., Peng, L., & Stumpe, M. C. (2018). *Impact of Deep Learning Assistance on the Histopathologic Review of Lymph Nodes for Metastatic Breast Cancer*. www.ajsp.com

Stellarix. (n.d.). *Challenges in Wearable Technology In Healthcare* Retrieved October 2, 2024, from https://stellarix.com/wearable-health-technology-benefits-challenges-types/

Steve Jobs Was Real Inspiration Behind Creation Of Apple Watch To Improve Health Care. (n.d.). Retrieved October 2, 2024, from https://www.techtimes.com/articles/157379/20160510/steve-jobs-real-inspiration-behind-creation-apple-watch-improve-healthcare.htm

Stock, R., Gaarden, A. P., & Langørgen, E. (2024). The potential of wearable technology to support stroke survivors' motivation for home exercise–Focus group discussions with stroke survivors and physiotherapists. *Physiotherapy Theory and Practice*, 40(8), 1795–1806. DOI: 10.1080/09593985.2023.2217987 PMID: 37246716

Tamargo, J., Le Heuzey, J. Y., & Mabo, P. (2015). Narrow therapeutic index drugs: A clinical pharmacological consideration to flecainide. *European Journal of Clinical Pharmacology*, 71(5), 549–567. DOI: 10.1007/s00228-015-1832-0 PMID: 25870032

Teja Nallamothu, P., & Bharadiya, J. P. (2023). Artificial Intelligence in Orthopedics: A Concise Review. In *Asian Journal of Orthopaedic Research* (Vol. 9, Issue 1). https://www.sdiarticle5.com/review-history/98595

Thomas, B. (2021). Artificial Intelligence: Review of Current and Future Applications in Medicine. *Federal Practitioner*, 38(11). Advance online publication. DOI: 10.12788/fp.0174 PMID: 35136337

Tian, T. (2016). *Proceedings, 2016 IEEE International Conference on Bioinformatics and Biomedicine :* Dec 15-18, 2016, Shenzhen, China. IEEE.

Tiger Medical. (n.d.). *Covidien Specialty SpO2 Sensors* Retrieved October 1, 2024, from https://tigermedical.com/products/specialty-non-adhesive-spo2-sensors-covsc-pr-oo?srsltid=AfmBOopn6zXnthlui9BQqG0z_E1eeyVivw71xSoHZOrQQcnHHnTdq3-o

Turner Medical. (n.d.). *Ear-Clip Sensor Pulse Oximeters* Retrieved October 1, 2024, from https://www.turnermedical.com/EAR-PULSE-OXIMETER-s/74.htm

UAE Government. (n.d.). *The Emirati Genome Programme* Retrieved October 1, 2024, from https://u.ae/en/information-and-services/health-and-fitness/research-in-the-field-of-health/the-emirati-genome-programme

Van Merriënboer, J. J. G., & Kirschner, P. A. (2017). *Ten steps to complex learning: A systematic approach to four-component instructional design*. Routledge. DOI: 10.4324/9781315113210

Vator TV. (n.d.). *When Fitbit was young: the early years* Retrieved October 2, 2024, from https://vator.tv/news/2017-02-21-when-fitbit-was-young-the-early-years

Wang, Z., Ong, C. L. J., & Fu, Z. (2022). AI Models to Assist Vancomycin Dosage Titration. *Frontiers in Pharmacology*, 13, 801928. Advance online publication. DOI: 10.3389/fphar.2022.801928 PMID: 35211014

Wellue. (n.d.). *O2Ring Smart Ring Pulse Oximeter. Overnight Tracking* Retrieved September 30, 2024, from https://getwellue.com/pages/o2ring-oxygen-monitor

Wong, K. P., Tse, M. M. Y., & Qin, J. (2022). Effectiveness of Virtual Reality-Based Interventions for Managing Chronic Pain on Pain Reduction, Anxiety, Depression and Mood: A Systematic Review. In *Healthcare (Switzerland)* (Vol. 10, Issue 10). MDPI. DOI: 10.3390/healthcare10102047

Wu, T., Dameff, C., & Tully, J. (2014). Integrating Google Glass into simulation-based training: Experiences and future directions. *Journal of Biomedical Graphics and Computing*, 4(2), 49. DOI: 10.5430/jbgc.v4n2p49

Xu, F., Sepúlveda, M.-J., Sepúlveda, S., Jiang, Z., Wang, H., Li, J., Liu, Z., & Yin, Y. (2020). Effect of an Artificial Intelligence Clinical Decision Support System on Treatment Decisions for Complex Breast Cancer. *JCO Clinical Cancer Informatics*, 4(4), 824–838. DOI: 10.1200/CCI.20.00018 PMID: 32970484

Yamauchi, M., Moritsuchi, Y., & Hamada, E. (2024). Novel concept of personalized therapy with continuous positive airway pressure. In *Sleep and Breathing*. Springer Science and Business Media Deutschland GmbH., DOI: 10.1007/s11325-024-03095-0

Yapici, M. K., & Alkhidir, T. E. (2017). Intelligent medical garments with graphene-functionalized smart-cloth ECG sensors. *Sensors (Basel)*, 17(4), 875. Advance online publication. DOI: 10.3390/s17040875 PMID: 28420158

Yoo, J. H., & Kim, J. H. (2023). Advances in Continuous Glucose Monitoring and Integrated Devices for Management of Diabetes with Insulin-Based Therapy: Improvement in Glycemic Control. In *Diabetes and Metabolism Journal* (Vol. 47, Issue 1, pp. 27–41). Korean Diabetes Association. DOI: 10.4093/dmj.2022.0271

Yoshida, S., Kihara, K., Takeshita, H., & Fujii, Y. (2014a). A head-mounted display-based personal integrated-image monitoring system for transurethral resection of the prostate. *Wideochirurgia i Inne Techniki Malo Inwazyjne*, 9(4), 644–649. DOI: 10.5114/wiitm.2014.43040 PMID: 25562008

Yoshida, S., Kihara, K., Takeshita, H., & Fujii, Y. (2014b). Instructive head-mounted display system: Pointing device using a vision-based finger tracking technique applied to surgical education. *Wideochirurgia i Inne Techniki Malo Inwazyjne*, 9(3), 449–452. DOI: 10.5114/wiitm.2014.44132 PMID: 25337172

Zarella, M. D., Bowman, D., Aeffner, F., Farahani, N., Xthona, A., Absar, S. F., Parwani, A., Bui, M., & Hartman, D. J. (2019). A practical guide to whole slide imaging a white paper from the digital pathology association. In *Archives of Pathology and Laboratory Medicine* (Vol. 143, Issue 2, pp. 222–234). College of American Pathologists. DOI: 10.5858/arpa.2018-0343-RA

Zhao, A. P., Li, S., Cao, Z., Hu, P. J. H., Wang, J., Xiang, Y., Xie, D., & Lu, X. (2024). AI for science: Predicting infectious diseases. In *Journal of Safety Science and Resilience* (Vol. 5, Issue 2, pp. 130–146). KeAi Communications Co. DOI: 10.1016/j.jnlssr.2024.02.002

Chapter 9
MindWatch Leveraging Wearable Technology for Real-Time Monitoring and Support of Student Mental Health

Fathima Mehnaz
https://orcid.org/0009-0006-1066-3686
Sahyadri College of Engineering and Management, India

Mustafa Basthikodi
https://orcid.org/0000-0001-7443-4774
Sahyadri College of Engineering and Management, India

Habeeb Ur Rahiman
https://orcid.org/0000-0002-8042-7375
Kingdom University, Bahrain

Rashmi Kodikal
https://orcid.org/0000-0001-7026-6654
Graphic Era University, India

ABSTRACT

The increasing demands of academics and societal issues are leading to a rise in health concerns, like anxiety and stress among college students affecting their performance and overall well-being significantly. This chapter delves into the progressions in gadgets such as smartwatches and biosensors that monitor vital signs like heart rate variability as well as stress and sleep patterns instantly. These innovations are

DOI: 10.4018/979-8-3693-7817-5.ch009

boosted by intelligence (AI) allowing for tailored guidance and interventions to support well-being effectively. Using gadgets has proven to be effective, in offering observations into the mental well-being of students by recognizing problems early on and promoting self-care practices in educational environments. However, these devices face challenges related to safeguarding data privacy accurately and ensuring accessibility, which must be addressed for their broader adoption in schools and universities.

1. INTRODUCTION

Mental health has become one of the most significant aspects that encompasses emotional, psychological, and social well-being of an individual that influence their well-being. Studies identified that it plays a vital role in individuals both psychological and physical well-being(Jiangxin & Zhiguo, 2021). In an educational environment, multiple incidents and situation reminds importance of mental health for the wellness and academic performance of the students who encounter unique challenges such as academic pressures, social transitions, and career uncertainties. Frequent reports and studies noticed that Covid pandemic has contributed severely to increase the stress and disrupted their mental health severely. This stress often led to mental health issues, including anxiety, depression, and sleep disturbances, which can adversely affect both personal well-being and academic performance (Agnafors et al., 2021). The studies further explored that the younger population in facing higher challenges due to uncertain and traditional coping mechanism. In many situations, due to lack of awareness and lacunas in the system, many students in educational institution students are reported to be committing suicide. Considering these circumstances and in response, educational institutions are increasingly exploring proactive approaches to address mental health concerns within academic settings (Lazarou & Exarchos, 2024). However, the rising prevalence of mental health issues among university students underscores the inadequacy of traditional support systems, such as counselling services and wellness programs, in meeting the growing demand. Many students face barriers to accessing professional help due to stigma, lack of awareness, or logistical challenges, leading to untreated mental health issues that can contribute to decreased academic performance, lower retention rates, and higher dropout rates. In today's fast facet digitally equipped world, technology has emerged as a viable solution to address these challenges with wearable devices initially developed for physical health tracking now being increasingly used to monitor mental health. Despite their potential, the application of wearable technology for mental health support in university settings remains underexplored, necessitating further investigation into their effectiveness and implementation.

Wearable technology offers a notably transformative approach to address mental health care by introducing various innovative technologies and platforms. This system provides non-intrusive, real-time monitoring of key indicators such as heart rate variability (HRV), sleep patterns, smartwatches etc. that can measure the various aspects that contributes the stress levels. With the help of multiple advanced technologies institutions are now leveraging advancements in sensors, algorithms, and artificial intelligence (AI) that possibly provide valuable insights for early detection and intervention of mental health challenges (Bloomfield et al., 2024a). For university students, wearable devices serve as a bridge between traditional mental health support and the increasing need for accessible, preventative care. By enabling students to autonomously monitor their mental health, these devices facilitate timely interventions that could prevent more severe mental health issues, fostering a healthier and more supportive academic environment (Debard et al., 2020).

Although previous studies explored the role of wearable devices to identify the mental health of an individual, the study was more focused on general medical terms and several gaps were identified. The observation notifies that most of the data were comes from developing countries and the existing research restricted in only limited set of area. Similarly, the research explores that gender-specific differences in the area are unexplored. Another important observation in the study is that integration is an educational ecosystem and is not must explored. The current research lacks framework in linking the wearable device and educational metal health related issues. There are hardly few studies explored the possibility of implying a technology-based solution in addressing the student's mental health related problem.

Considering the importance of adapting technology in preventing the disruption of mental health the current chapter intended to examine how technology possibly address the mental health challenges faced by university students and explores possible framework to address the metal health related issues through wearable technology in real-time monitoring and early intervention, evaluate its ethical implementation and effectiveness.

The structure of the chapter begins with exploring the introduction of wearable devices and mental health related issues. The initial part of the chapter explored how sever mental health related issues are impacting students study environments and their academic performance. The study notifies the role of technology as a coping strategy and predicting the issues more accurately by proposing technology and framework. The further part of the study explored the research gap and exploring the past studies in this domain. The literature review explores the major cluster and based on the literature; the study proposed framework to address implementing technological solutions.

The chapter further explores the major solutions and ways to attain the implementation of technology elaborating its implementation. The chapter comprehensively addresses the wearable device and how real time monitoring system works in this domain. The key observation of the study notifies comprehensive framework to tackle this metal health problem amongst students' community.

Finally, the chapter concludes with recommending the limitation and future directions in wearable technology and student mental health, proposing strategies for integrating wearables into educational settings, advancing technological innovations, and fostering equitable access. The future proposals are addressing the key stakeholders like academic institutions, universities, government, and policy makers. This section provides a comprehensive framework for understanding the application and implications of wearable technology in student mental health monitoring.

2. CONCEPTUAL FRAMEWORKS AND LITERATURE REVIEW

Although multiple research address role of technology in addressing stress and mental health of the students, the existing literature lacks comprehensive coverage of technology, psychology, education and ethical issues. The literature review explored the association between wearable technology and mental health in educational ecosystem.

2.1 Literature Review

Wearable technology and metal health are two different dimensions in an academic arena. There are a couple of studies conducted by various disciplines on role of technology in addressing the psychological well-being of individuals in various organizations(Christmann et al., 2018; Hovnanyan et al., 2022; Morash et al., 2008). However, studies like P. Zhang et al. (2024) and Xia & Lin (2021) notified how influence role of AI in coping the mental health of college students and role of data mining technology as a coping strategy. Such studies explored the effective implementation of AI in students' activities and monitoring them through technology like IoT, AI and ML to predict possible stress that disrupts mental peace of students(Rahiman & Kodikal, 2024). The study further explored the role of Electrodermal Activity (EDA) that identifies the changes of skin conductance and correlates the stress level(Ji, 2024).

Real-time monitoring is another essential and effective system that enhance detection level of mental health of students. Continuous monitoring system in an educational institution enable educators to understand the student stress and causes of stress. This predictive approach helps to identify the most suitable mechanism to propose effective and productive strategies for most of the cases(Xie, 2021). One of the most potential data the institution identifies is the data related to students at risk. Continues technology enables the educator the flag the students at risk much advance(Liu & Mai, 2022).

One of the most challenging aspects of adapting technology in education or any other sector is privacy issues. The studies like Sokolovskaia et al. (2022) explores the data privacy is one of the major concerns in adapting technology in educations. The further studies like Rahiman & Kodikal (2024) also raised concerns in data privacy issues in higher education equipped with AI in various academic activities and roles. The current research also lacks comprehensive solution to address privacy concerns in technological adaption.

The bibliographic map explores the comprehensive literature and with the heal of five cluster and associated key words. The figure notifies that as per the Scopus database, the key words related to wearable device and mental health are notably lesser in compared to other key words. The study on wearable devices and allied in risk assessment, anxiety, psychology, classification of information in higher education, depressions are the other important studies which has been carried out in this subject. Therefore, this bibliometric analysis clearly explores the existing research gap on wearable devices and mental health (Figure 1).

Figure 1. Bibliometric analysis of wearable device

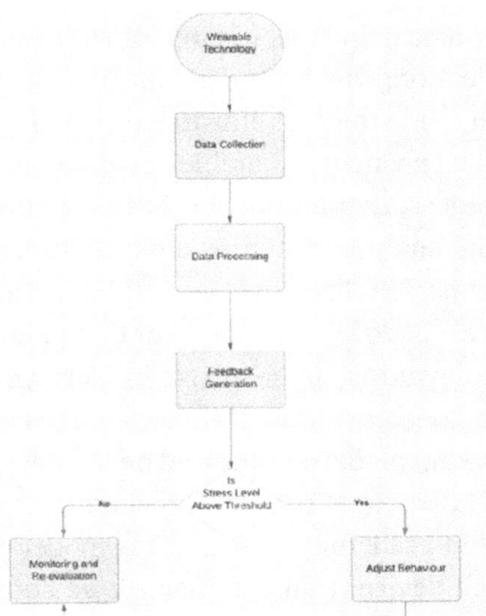

3. CONCEPTUAL FARMWORK

3.1 Wearable Technology in Mental Health Monitoring

The role of wearable technology in managing mental health related task has advanced rapidly, offering a promising method for monitoring mental health and providing real-time data that can shed light on an individual's emotional and psychological well-being (Sokolovskaia et al., 2022). The studies notified that devices such as fitness trackers, smartwatches, and medical-grade wearables come equipped with various sensors that used by various premier institutions to monitor vital physiological and behavioural markers important for assessing mental health (Escobar-Linero et al., 2023). Further studies identified that such technologies play a crucial role in tracking key mental health indicators like stress, anxiety, and emotional states (Hickey et al., 2021a). By measuring physiological signals such as heart rate variability (HRV), sleep patterns, and levels of physical activity, wearables offer a non-intrusive way to monitor changes in mental health. The studies like Jiang (2020) notified that HRV, which measures the variation between heartbeats, which is commonly used as an indicator of stress, as a lower HRV often correlates with higher stress and anxiety levels (Zhang et al., 2020).

One of the major challenges in this process is availability of data. The continuous collection these process data allows for early intervention when signs of mental health deterioration appear, a critical factor in high-stress environments such as academic settings. Gaining timely data helps specifically for students who can greatly benefit from real-time data collection, which supports the proactive identification of issues like anxiety, depression, and stress, as illustrated in **Figure 2** (Garg et al., 2020). These studies explore that enhancing wearable devices with data analysis techniques improves their effectiveness, making mental health monitoring more precise. AI models on the other hand analyzes patterns in physiological data to predict mental health crises and suggest personalized behavioural interventions that promote better mental well-being (Liu et al., 2021). With help of these algorithms institutional can be able to detect subtle deviations in heart rate or sleep patterns that may indicate emerging mental health concerns, prompting timely intervention.

The proposed conceptual model emphasised on the integration of wearable technologies into daily life holds great promise for monitoring and improving mental health, particularly within academic settings. This model prosed to provide a real-time insight, so that such devices make it effective and empower the users to take a proactive approach to mental health management, making it more personalized and timelier. This seamless integration of technology into everyday routines offers valuable support for individuals experiencing mental health challenges, with early detection and intervention as the ultimate goal.

Figure 2. Framework for leveraging wearable technology

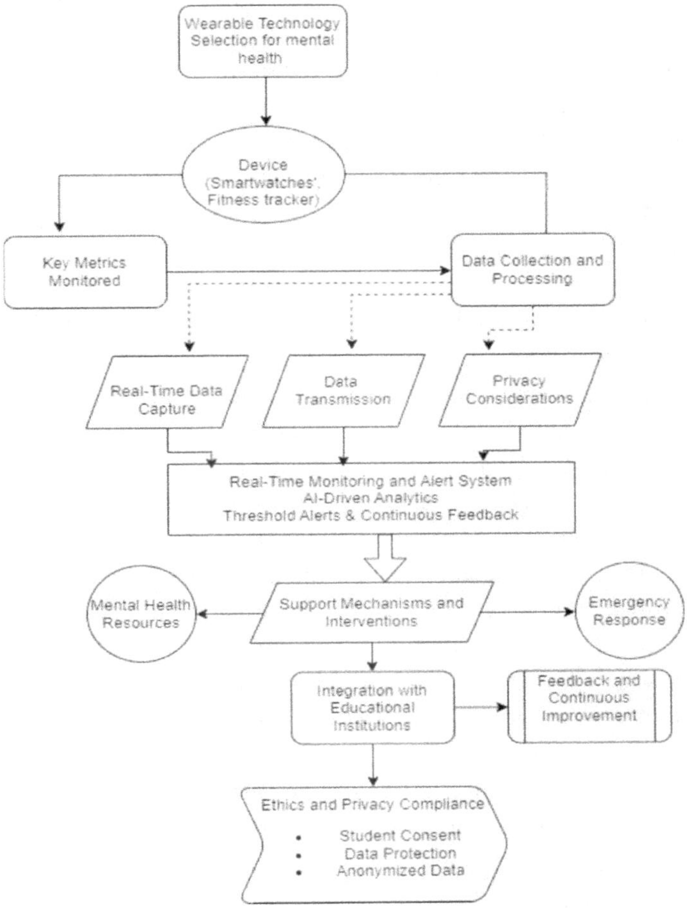

Source: Developed by Authors

3.2 Real-Time Data Collection and Analysis

One of the most effective approaches in to manage mental health challenges is Real-time data collection. This process involves continues data collection related to psychological, and behavioural to that helps to generate actionable input. This is the central to the effectiveness of wearable technology in mental health monitoring which can be used to make continues decision making to institutions to reach effective outcome(X. Song et al., 2024). Wearable devices seamlessly capture physiological data such as heart rate, skin temperature, physical activity, and sleep patterns, which serve as indicators of mental health(Wu & Feng, 2023). This continuous stream of data enables immediate detection of anomalies or trends, making these devices valuable for early intervention in mental health care. Data transmission occurs through secure channels to smartphones or cloud-based systems, where it undergoes analysis using advanced algorithms and machine learning models(Lattie et al., 2019). The continuous nature of real-time monitoring eliminates the reliance on intermittent, self-reported data typical of traditional assessments, allowing for a more nuanced understanding of an individual's mental state. Real-time monitoring of HRV and sleep patterns enable the user to quickly identify stress or depression indicators, prompting timely action (Aquilina et al., 2020). Similarly effective implementation of AI enhances the analytical capability of wearable devices by identifying patterns in the data. Machine learning models can detect deviations that correlate with mental health issues (Panakaje et al., 2024). Prolonged decreases in HRV or consistent sleep disturbances may signal heightened stress or emerging depressive symptoms. These insights can enable healthcare providers or academic counsellors to intervene before these conditions worsen (Vinod Kumar Dhiman, 2024). To visualize the role of AI and ML in wearable technology for mental health monitoring, the following flowchart outlines the key steps involved in data collection, analysis, early detection, personalized interventions, and integration with health systems.

The proposed model shown in **Figure 3**, wearable devices begin with data collection through sensors, followed by AI and ML-driven data analysis to uncover patterns. These steps enable early detection of mental health issues, paving the way for tailored interventions and seamless integration with broader health support systems.

In universities, wearable technology has been particularly effective in monitoring students' mental well-being during high-pressure periods, such as exams or project deadlines. Real-time data allows support systems to respond to stress peaks, reducing the risk of burnout. For example, institutions that integrate wearables with counselling services have reported success in managing stress through personalized feedback and intervention (Hamza et al., 2023).

Real-time feedback from wearables empowers students to take control of their mental health. Insights into stress levels or sleep quality can prompt individuals to engage in mindfulness exercises or modify their study schedules. Such interventions are supported by evidence showing that self-monitoring enhances awareness and encourages healthier behaviours (Hovsepian et al., 2015). By enabling continuous data collection and actionable insights, real-time monitoring addresses limitations of traditional mental health assessments and fosters a proactive approach to student well-being.

Figure 3. Wearable technology for mental health management

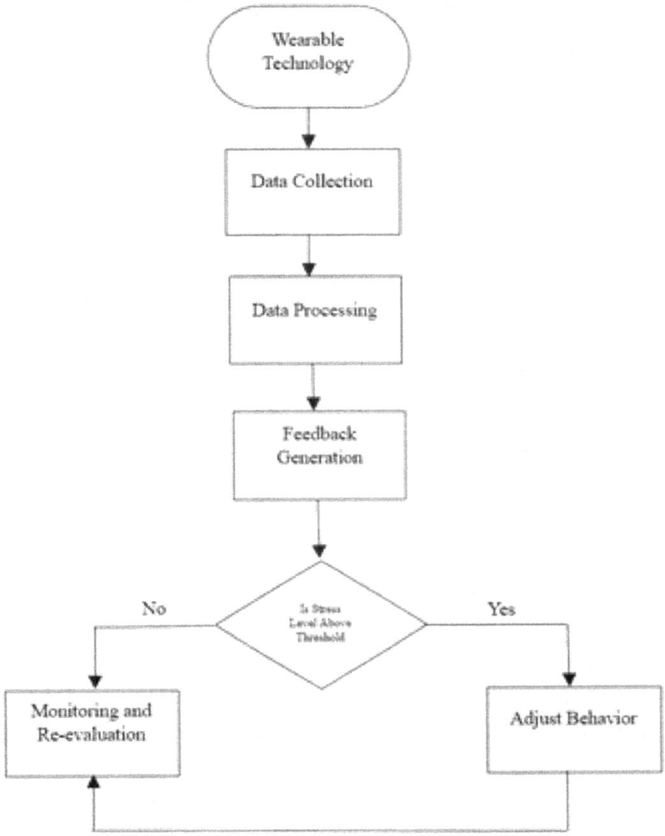

Source: Developed by authors

4. PRACTICAL APPLICATIONS OF WEARABLE TECHNOLOGY IN MENTAL HEALTH MONITORING

4.1 Real-Time Stress Monitoring in College Students

The research explores the internal study conducted by involved 54 college students who used wrist-worn sensors, specifically the Apple Watch, to monitor physiological data such as heart rate and hand acceleration over a period of 40 days. The research notifies that all the participants made to self-report their stress levels using a mobile health application. The data collected was analyzed using machine learning algorithms, with the XGBoost model achieving an accuracy of 84.5% and an AUC of 0.64. The outcome of the study demonstrates the effectiveness of wearables in detecting stress patterns and highlights their potential for real-time stress monitoring in academic settings.

4.2 The cStress Initiative: Continuous Stress Tracking

The continues stress tacking is another impotent aspect in large number of populated environments. The research explored that as per cStress project, wearable technology has been used for continuous monitoring of stress through sensors embedded in smart devices. The results indicates that this initiative tracks key physiological indicators, such as HRV and electroencephalography (EEG) patterns, providing real-time feedback on stress levels. This result of this approach allows individuals or an institution to make continues monitor their stress dynamically and make proactive adjustments to their routines. The cStress project underscores the importance of continuous monitoring and its potential to empower individuals to manage their mental health more effectively.

4.3 Integrating AI and Wearables for Anxiety Prediction

AI application is an effective tool to predict the anxiety. The effective AI embedded application helps the institutions to make systematic prediction to reach proper prediction. The research initiative conducted in Malaysia combined wearable devices with AI and social media data to predict anxiety levels in students. According to this study involved collecting physiological data from smartwatches, including HRV and heart rate, and integrating it with behavioral insights from students' social media activity. By using machine learning algorithms processed this data to identify patterns and accurately predict anxiety levels. The outcome of this study highlights the potential of combining wearable technology with AI to enable continuous monitoring, early detection, and proactive interventions, bridging the gap between technology and psychological support.

5. ETHICS AND PRIVACY CONCERNS

Ethics is another major concern in the integration of wearable technology for mental health monitoring. It holds great promise for supporting students' well-being by providing real-time insights into their physical and emotional states. It is essential to institution to make continuously tracking data such as heart rate variability, sleep patterns, and physical activity levels, these devices can enable early interventions and promote proactive mental health management. However, the continuous collection of sensitive data also introduces important ethical and privacy considerations. Balancing the benefits of these devices with concerns regarding data security, informed consent, and user autonomy is essential. Ensuring robust privacy protections and transparent data-sharing practices can help mitigate potential risks, fostering trust and ensuring that the technology serves students' best interests.

5.1 Privacy and Data Security

Multiple research projects raised concern about privacy and data security of the data collected from various users. It is essential to ensure the data protection in uncompromised way to ensure that all the users are safe when they be part of the process. Wearable devices that continuously collect real-time data, such as heart rate variability, sleep patterns, and stress levels, offer tremendous potential for early intervention in mental health crises(Dalledonne, 2020). However, the storage and transmission of such data raise concerns about privacy and security, as unauthorized access could expose sensitive personal information, leading to potential stigma or discrimination (Wang & Gao, 2019). To mitigate these risks, transparent data-sharing practices and strong security measures are essential to protect students' mental health data (Vijayan et al., 2021). Additionally, many wearable devices rely on third-party companies for data storage and analysis, which may not always adhere to the same stringent privacy standards as healthcare providers. This highlights the need for clearer regulatory guidelines to ensure users maintain control over their data, ensuring transparency and privacy in the data-sharing process (Bloomfield et al., 2024).

5.2 Informed Consent and Autonomy

Informed consent is essential in wearable technology deployment, ensuring that users fully understand what data is being collected and how it will be used. Users' needs to give their consent before institution make them part of this approach. This transparency helps users make informed decisions about their participation in mental health monitoring (Yllaconza et al., 2024). Wearables should offer control over data collection, allowing students to determine what information they share and how it's utilized. By providing clear, accessible information about data usage, universities can empower students to engage with wearables in ways that feel both safe and voluntary (Paroi, n.d.).

5.3 Stigma, Discrimination, and Over-Surveillance

Wearable devices in mental health monitoring could inadvertently contribute to stigma and discrimination. If collected data is improperly managed or anonymized, individuals might face adverse consequences based on their mental health data. If a student is flagged as experiencing high levels of stress, this information could be shared with their academic institution or health professionals without their consent, potentially leading to unnecessary interventions or unfair treatment (Shahul et al., 2023). Furthermore, as wearable devices become increasingly integrated into public health systems, there is a growing risk of over-surveillance. Continuous monitoring may create a society where individuals feel compelled to maintain a certain standard of mental health, exacerbating feelings of anxiety or inadequacy among those who struggle with mental health challenges(Meng et al., 2021). While over-surveillance and the potential for discrimination raise serious concerns about the impact on students, ensuring the ethical use of data is equally critical in mitigating these risks. By establishing clear guidelines for responsible data handling, we can ensure that the collected information supports students' mental health without contributing to harm or exploitation.

5.4 Ethical Use of Data

Ethical usage of data un an educational institution is highly essential. While these devices can provide valuable insights into mental health, the potential for misuse of personal data such as its use for marketing or insurance purposes must be addressed(Ruotsalainen & Blobel, 2020). The study has proposed clear ethical frameworks are necessary to ensure that data is used solely for supporting students' mental health and not for exploitative practices. The study stated that effective ethical usage can be achieved by establishing guidelines that prioritize users' well-being and ensure that their data is not used in ways that could disadvantage them (Jovanovic & Kay, 2023).

6. POTENTIAL CHALLENGES AND LIMITATIONS

While wearable technology holds significant promise for mental health monitoring, several challenges and limitations need to be addressed to ensure its effectiveness and ethical implementation. These challenges encompass technical issues, user engagement, and the integration of wearables into existing healthcare systems.

6.1 Technical Challenges

Ensuring accuracy and reliability is one of the primary technical challenges with wearable devices for mental health monitoring. The technology typically relies on sensors that track physiological signals, such as heart rate, skin temperature, and sleep patterns, to infer mental health status (M. Aldiabat et al., 2014). However, these metrics can be influenced by various external factors such as physical activity, environmental conditions, and even device malfunctions. There is certain major concern noticed in the process. For example, increase in the heart rate due to exercise might be considered or misinterpreted as a sign of mental health distress, potentially leading to false alarms or unnecessary interventions. Moreover, the data collected by these devices may not fully capture the complex nature of mental health conditions like anxiety or depression, which cannot be accurately diagnosed solely based on physiological signals (Ruby Dhar et al., 2023). This limitation raises concerns about the over-reliance on wearable devices to make critical health decisions without appropriate context or additional clinical inputs.

6.2 User Engagement and Compliance

It is also important to notice that sustaining user engagement is another considerable challenge, as wearable devices are most effective when showed consistently(Darejeh & Salim, 2016). However, user compliance can be hampered by noticeable technical factors such as discomfort, battery life limitations, or a lack of perceived benefit. Research suggests that continuous monitoring might not always be felt as relevant or necessary to the user, particularly in the case of mental health, where the immediate benefits are not always clear (Banaee et al., 2013). To overcome from this disengagement, it is essential that users are educated about the device's capabilities and how to interpret their data. If users fail to understand how to properly use the device or misinterpret the readings, it could lead to unnecessary anxiety or missed warning signs.

6.3 Data Overload and Interpretation

Another challenge associated with wearable devices is data overload. A vast amount of data that must be processed and interpreted accurately. If institution unable to maintain effective data management systems, users or clinicians may struggle to extract meaningful insights. It is essential to note that the volume of data generated possibly overwhelm users or clinicians, potentially causing important information to be missed (Vijayan et al., 2021). The sheer quantity of data may also complicate the identification of meaningful patterns, requiring advanced analytics or artificial intelligence (AI) algorithms to discern useful information.

Moreover, another essential aspect here is interpreting wearable data is highly individualized, so there are possible chances of same physiological signals may have different implications for different users. A spike in heart rate might signify distress in one individual but be a normal response to exercise in another. Additionally, integrating data from multiple devices or platforms adds another layer of complexity, as these systems may not always be compatible with one another, leading to fragmented or incomplete datasets (Saddique & Seng Choi, 2023).

5.4 Integration with Existing Healthcare Systems

Integrating wearable technology into established healthcare systems presents numerous challenges. Although wearables provide valuable data, incorporating this information into clinicians' workflows is often difficult. Many healthcare professionals are not trained to interpret wearable data, and the lack of interoperability between wearable devices and healthcare platforms can result in fragmented care. A clinician might struggle to integrate real-time data from wearables with existing medical records, leading to potential gaps in care (Chung et al., 2020). Additionally, traditional healthcare systems may not be ready to handle the continuous data streams generated by wearables. This could result in difficulties in responding to urgent mental health concerns or tracking long-term trends in mental health over time.

5.5 Ethical and Legal Considerations

As wearable devices become more integrated into healthcare, ethical and legal considerations surrounding their use become increasingly critical(Gerber et al., 2021). The accuracy of data collected by wearables is paramount, as inaccuracies could lead to misdiagnosis or inappropriate treatment plans, with potentially serious consequences for students' mental health(Yaokumah, 2020). More importantly, the collection and use of personal health data raise privacy and security concerns. Wearable devices collect sensitive data that must be protected to avoid misuse, particularly for commercial purposes or by third parties such as insurance companies (Zamani et al., 2023). Informed consent must be an integral part of the deployment of wearables, ensuring that students are fully aware of how their data will be used and have the ability to opt-out if desired. For students in particular, ethical considerations must be given to ensure that they are not coerced into participation and that their autonomy is respected (Hickey et al., 2021b). Additionally, data security remains a key issue, especially in the context of wearable IoT (WIoT) devices generating personal health data. The integration of cloud computing, fog computing, and big data platforms presents potential vulnerabilities that could be exploited, leading to unauthorized access or misuse of sensitive information (Kapoor et al., n.d.). It is essential that these devices be designed with robust security features to protect against data breaches and unauthorized access.

5.6 Bias and Accessibility

Adapting technology in routine assessment is to avoid the traditional approach and generating unbiased results. However, in some cases inefficient application of resources make leads to raise ethical concern that requires attention is the bias inherent in the design and distribution of wearable devices(Feng et al., 2022). Students from different backgrounds for example, certain demographic groups, particularly those in lower-income or rural areas, may face significant barriers to accessing wearable technologies(Sarraf et al., 2021). These accessibility gaps possibly worsen existing disparities in mental health care, as individuals without access to such devices may miss out on the benefits of real-time health monitoring and intervention (Seng et al., 2023). Therefore, ensuring equitable access to wearable devices, and considering alternative low-cost options for underserved communities, is critical for addressing these biases (Kraiss, 2017).

7. FUTURE DIRECTIONS IN WEARABLE TECHNOLOGY AND STUDENT MENTAL HEALTH

The finding and proposed framework stressed the importance of adapting effective wearable technology in educational institution to develop effective students' engagement and academic performance(Hong, 2023). Since wearable technology have a potential to support mental health monitoring among students presents new opportunities and challenges, it is essential that future advancements are likely to focus on improving the accuracy of mental health detection, enhancing real-time interventions, and ensuring the integration of wearables into holistic mental health care frameworks(Qu, 2022). The study proposes upcoming devices needs to identify the lacunas in the system especially major concern regarding redundancy, data misinterpretation, and privacy concern. Although various research and technological is taking place across the world and some of the major concerns are still unaddressed due to some reasons. Therefore, incorporating more advanced resources, by consulting key stakeholders and taking users feedback is crucial.

The study technically proposes that incorporating the effective sensors that capable of detecting a wider range of physiological and psychological markers that are directly linked to mental health(L. Song, 2022). For example, wearables could integrate sensors that measure brain activity using EEG or similar technologies, more precise heart rate variability measures, or biomarkers linked to conditions like depression, anxiety, or stress These innovations could lead to more accurate real-time detection of mental health fluctuations, allowing for personalized interventions that are timely and targeted (Yee Chung et al., 2021).

The research further opines that adapting AI and ML algorithms are expected to play a significant role in the future of wearable mental health technology. Since AI-powered systems analyzes vast amounts of data collected by wearables, identify patterns, and provide personalized insights for students(Hu, 2020). AI might predict mental health episodes, such as anxiety or depressive episodes, offering early warning signs and suggesting coping mechanisms or interventions (Kustimah et al., n.d.). Institution can tailor the system as per the nature and requirement of their requirement and demographic(Yang, 2018). The individual's unique psychological and physiological state, wearables could enhance their usefulness and accessibility for students. Future wearables may seamlessly integrate with existing campus health systems, enabling continuous data flow between students and healthcare providers(-Jafarlou et al., 2023). This would help clinicians track students' mental health over time and intervene proactively, potentially reducing the need for in-person appointments. Wearable data could alert counsellors or mental health professionals about signs of anxiety or depression, allowing for timely interventions or targeted support (Chung et al., 2020). Such integration could normalize mental health monitoring as an ongoing part of student health care.

Also, wearable devices might focus not just on monitoring mental health but also on empowering students to manage their well-being. Through real-time feedback mechanisms, such as notifications or guided breathing exercises, wearables could teach students to recognize stress and anxiety triggers and offer strategies for self-regulation(Huang et al., 2016). This proactive approach could help students gain more control over their mental health, making wearables a tool for both prevention and intervention (Banaee et al., 2013). As wearable technology becomes more commonplace, ethical considerations will remain crucial. Future designs must prioritize students' privacy, ensure transparent data collection practices, and clearly communicate the intended use of data. Moreover, the normalization of wearable technology in mental health care could help reduce the stigma associated with mental health challenges. By making mental health monitoring an everyday part of student life, wearables could encourage more students to seek support and engage in proactive mental health management (Park et al., 2021). Wearable devices in the future could foster a sense of community by facilitating peer-to-peer support networks. Wearables could integrate with social platforms or campus mental health networks, allowing students to share their mental health data with trusted peers or family members. This could encourage openness and support among students, making it easier for them to discuss mental health struggles (Park et al., 2021). Additionally, wearables might connect students to wellness communities or offer group activities aimed at improving mental health, further promoting collective well-being. For wearable technology to have a widespread impact, it must be accessible to all students, regardless of socioeconomic status. The future of wearable mental health technology will likely focus on reducing costs, making devices more affordable and accessible. As competition in the wearable market grows, prices may decrease, and more budget-friendly options may emerge. Moreover, universities or policy programs could help subsidize or provide wearables for students who cannot afford them, ensuring equitable access to mental health monitoring.

8. CONCLUSION

The objective of this chapter is to identify and propose the effective framework on application of wearable device in an educational eco-system. The chapter examined the some of the key role of wearable technology in monitoring and supporting student mental health. As mental health challenges among students continue to rise, the potential of integrating wearable devices into wellness programs presents a promising approach for proactive care.

The chapter explored the how the technology helps to predict the mental health and well-being of the students by applying in educational framework. With the help of advanced analytics and artificial intelligence, study proposed effective means to monitor physiological and behavioural data, enabling timely, individualized interventions. The study also raised concern over certain key aspects like ensuring data accuracy, data privacy, ethical usage of data and how to be maintaining real-time monitoring capabilities and promoting user engagement were identified as essential for the successful implementation of these technologies. The main takeaway is that wearable technology has the potential to significantly enhance mental health care for students. To realize this potential, continued research and collaboration among researchers, developers, and stakeholders are vital. By integrating these insights into practical applications, wearable technologies can be a key component in creating healthier, more supportive environments for students.

REFERENCES

Aldiabat, K. M., Matani, N. A., & Navenec, C. L. (2014). Mental health among undergraduate university students: A background paper for administrators, educators and healthcare providers. *Universal Journal of Public Health*, 2(8), 209–214.

Bloomfield, L. S. P., Fudolig, M. I., Kim, J., Llorin, J., Lovato, J. L., McGinnis, E. W., McGinnis, R. S., Price, M., Ricketts, T. H., Dodds, P. S., Stanton, K., & Danforth, C. M. (2024). Predicting stress in first-year college students using sleep data from wearable devices. *PLOS Digital Health*, 3(4), e0000473. DOI: 10.1371/journal.pdig.0000473 PMID: 38602898

Christmann, C. A., Hoffmann, A., Zolynski, G., & Bleser, G. (2018). Stress-mentor: Linking gamification and behavior change theory in a stress management application. In, S. C. (Ed.), *20th International Conference on HCI, HCI International 2018* (Vol. 851, pp. 387–393). Springer Verlag. DOI: 10.1007/978-3-319-92279-9_52

Dalledonne, S. (2020). *Remote sensing data access policy, data products regulatory framework and intellectual property rights challenges in an era of environmental protection urgency. 2020-Octob.* https://www.scopus.com/inward/record.uri?eid=2-s2.0-85100923787&partnerID=40&md5=200d351cdc4f11e90f366bf237a40ff7

Darejeh, A., & Salim, S. S. (2016). Gamification Solutions to Enhance Software User Engagement—A Systematic Review. *International Journal of Human-Computer Interaction*, 32(8), 613–642. DOI: 10.1080/10447318.2016.1183330

Debard, G., De Witte, N., Sels, R., Mertens, M., Van Daele, T., & Bonroy, B. (2020). Making wearable technology available for mental healthcare through an online platform with stress detection algorithms: The CareWear project. *Journal of Sensors*, 2020, 1–15. Advance online publication. DOI: 10.1155/2020/8846077

Dhar, R. (2023). Smart wearable devices for real-time health monitoring. *Asian Journal of Medical Sciences*, 14(12), 1–3. DOI: 10.3126/ajms.v14i12.58664

Dhiman, V. K. (2024). The emergence of AI in mental health: A transformative journey. *World Journal of Advanced Research and Reviews*, 22(1), 1867–1871. DOI: 10.30574/wjarr.2024.22.1.1298

Escobar-Linero, E., Muñoz-Saavedra, L., Luna-Perejón, F., Sevillano, J. L., & Domínguez-Morales, M. (2023). Wearable Health Devices for Diagnosis Support: Evolution and Future Tendencies. In *Sensors* (Vol. 23, Issue 3). MDPI. DOI: 10.3390/s23031678

Feng, S., Chong, Y., Yu, H., Ye, X., & Li, G. (2022). Digital financial development and ecological footprint: Evidence from green-biased technology innovation and environmental inclusion. *Journal of Cleaner Production*, 380, 135069. https://doi.org/https://doi.org/10.1016/j.jclepro.2022.135069. DOI: 10.1016/j.jclepro.2022.135069

Gerber, K., Flores, I. G., Ruiz, A. C., Ali, I., Ginsberg, N. L., & Schenberg, E. E. (2021). Ethical Concerns about Psilocybin Intellectual Property. *ACS Pharmacology & Translational Science*, 4(2), 573–577. DOI: 10.1021/acsptsci.0c00171 PMID: 33860186

Hamza, M. A., Abdalla Hashim, A. H., Alsolai, H., Gaddah, A., Othman, M., Yaseen, I., Rizwanullah, M., & Zamani, A. S. (2023). Wearables-Assisted Smart Health Monitoring for Sleep Quality Prediction Using Optimal Deep Learning. *Sustainability (Basel)*, 15(2), 1084. Advance online publication. DOI: 10.3390/su15021084

Hickey, B. A., Chalmers, T., Newton, P., Lin, C. T., Sibbritt, D., McLachlan, C. S., Clifton-Bligh, R., Morley, J., & Lal, S. (2021a). Smart devices and wearable technologies to detect and monitor mental health conditions and stress: A systematic review. In *Sensors* (Vol. 21, Issue 10). MDPI AG. DOI: 10.3390/s21103461

Hickey, B. A., Chalmers, T., Newton, P., Lin, C. T., Sibbritt, D., McLachlan, C. S., Clifton-Bligh, R., Morley, J., & Lal, S. (2021b). Smart devices and wearable technologies to detect and monitor mental health conditions and stress: A systematic review. In *Sensors* (Vol. 21, Issue 10). MDPI AG. DOI: 10.3390/s21103461

Hong, L. (2023). Study on the influence of social network for college students' mental health through big data technology. *ACM International Conference Proceeding Series*, 147–153. DOI: 10.1145/3584871.3584892

Hovnanyan, A., Mastromatteo, L. Y., Rubaltelli, E., & Scrimin, S. (2022). Stress and Emotional Intelligence Shape Giving Behavior: Are There Different Effects of Social, Cognitive, and Emotional Stress? *Frontiers in Psychology*, 13, 800742. Advance online publication. DOI: 10.3389/fpsyg.2022.800742 PMID: 35282189

Hovsepian, K., Al'absi, M., Ertin, E., Kamarck, T., Nakajima, M., & Kumar, S. (2015). CStress: Towards a gold standard for continuous stress assessment in the mobile environment. *UbiComp 2015 - Proceedings of the 2015 ACM International Joint Conference on Pervasive and Ubiquitous Computing*, 493–504. DOI: 10.1145/2750858.2807526

Hu, Y. (2020). Improve the mental health of college students by means of computer network technology. *Proceedings - 2020 International Conference on Computer Engineering and Intelligent Control, ICCEIC 2020*, 5–9. DOI: 10.1109/ICCEIC51584.2020.00008

Huang, Y., Xiong, H., Leach, K., Zhang, Y., Chow, P., Fua, K., Teachman, B. A., & Barnes, L. E. (2016). Assessing social anxiety using GPS trajectories and point-of-interest data. *UbiComp 2016 - Proceedings of the 2016 ACM International Joint Conference on Pervasive and Ubiquitous Computing*, 898–903. DOI: 10.1145/2971648.2971761

Jafarlou, S., Lai, J., Azimi, I., Mousavi, Z., Labbaf, S., Jain, R. C., Dutt, N., Borelli, J. L., & Rahmani, A. (2023). Objective Prediction of Next-Day's Affect Using Multimodal Physiological and Behavioral Data: Algorithm Development and Validation Study. *JMIR Formative Research*, 7, e39425. Advance online publication. DOI: 10.2196/39425 PMID: 36920456

Ji, D. (2024). Research on Intelligent Multimedia Technology Path of College Students' Mental Health Education in Higher Vocational Colleges and Universities. *Applied Mathematics and Nonlinear Sciences*, 9(1), 20243379. Advance online publication. DOI: 10.2478/amns-2024-3379

Jiang, L. (2020). Research on the Construction of New Mode of Mental Health Education for Higher Vocational College Students Based on Computer Big Data Technology. *Journal of Physics: Conference Series*, 1578(1), 012113. Advance online publication. DOI: 10.1088/1742-6596/1578/1/012113

Jiangxin, H., & Zhiguo, X. (2021). Research on Mental Health Assessment and Crisis Intervention Mechanism of University Students with Data Technology. *Proceedings - 2021 2nd International Conference on Education, Knowledge and Information Management, ICEKIM 2021*, 451–454. DOI: 10.1109/ICEKIM52309.2021.00105

Jovanovic, P., & Kay, R. (2023). Exploring Ethical Issues with Using Wearable Technology in K-12 Classrooms: A Review of the Literature. *Journal of Digital Life and Learning*, 2(2), 3–11. DOI: 10.51357/jdll.v2i2.203

Kapoor, V., Singh, R., Reddy, R., & Churi, P. (n.d.). *International Conference on Innovative Computing and Communication (ICICC 2020) Privacy Issues in Wearable Technology: An Intrinsic Review*. https://ssrn.com/abstract=3566918

Kraiss, K. N. (2017). *Evaluating the Effects of Wearable Technologies to Improve Evaluating the Effects of Wearable Technologies to Improve Physical Activity Levels for College Students with Intellectual and Physical Activity Levels for College Students with Intellectual and Developmental Disabilities Developmental Disabilities*. https://trace.tennessee.edu/utk_graddiss

Kustimah, K., Devy Kumalasari, A., & Meindy, N. (n.d.). *The Open Psychology Journal When do College Students Seek Mental Health Services?* https://doi.org/ DOI: 10.2174/18743501-v16-e230420-2022-112

Lattie, E. G., Lipson, S. K., & Eisenberg, D. (2019). Technology and college student mental health: Challenges and opportunities. *Frontiers in Psychiatry*, 10(APR), 246. Advance online publication. DOI: 10.3389/fpsyt.2019.00246 PMID: 31037061

Liu, F., & Mai, Y. (2022). Application of Computer Technology in College Students' Mental Health Education. *Proceedings - 2022 3rd International Conference on Education, Knowledge and Information Management, ICEKIM 2022*, 1019–1024. DOI: 10.1109/ICEKIM55072.2022.00222

Meng, X., Zhang, J., & Ren, G. (2021). The Evaluation Model of College Students' Mental Health in the Environment of Independent Entrepreneurship Using Neural Network Technology. *Journal of Healthcare Engineering*, 2021, 1–9. Advance online publication. DOI: 10.1155/2021/4379623 PMID: 34608410

Morash, M., Kwak, D.-H., Hoffman, V., Lee, C. H., Cho, S. H., & Moon, B. (2008). Stressors, coping resources and strategies, and police stress in South Korea. *Journal of Criminal Justice*, 36(3), 231–239. https://doi.org/https://doi.org/10.1016/j.jcrimjus.2008.04.010. DOI: 10.1016/j.jcrimjus.2008.04.010

Panakaje, N., Ur Rahiman, H., Parvin, S. M. R., Shareena, P., Madhura, K., Yatheen, M., & Irfana, S. (2024). Revolutionizing pedagogy: Navigating the integration of technology in higher education for teacher learning and performance enhancement. *Cogent Education*, 11(1), 2308430. Advance online publication. DOI: 10.1080/2331186X.2024.2308430

Paroi, S. (n.d.). *LEVEL AND SOURCES OF STRESS: EVIDENCE FROM COLLEGE STUDENTS*. https://www.researchgate.net/publication/378306172

Qu, Z. (2022). Analysis of the Effect of Martial Arts on Students' Physical and Mental Health Based on Internet of Things Technology. *Security and Communication Networks*, 2022, 1–9. Advance online publication. DOI: 10.1155/2022/8663157

Rahiman, H. U., & Kodikal, R. (2024). Revolutionizing education: Artificial intelligence empowered learning in higher education. *Cogent Education*, 11(1), 2293431. Advance online publication. DOI: 10.1080/2331186X.2023.2293431

Ruotsalainen, P., & Blobel, B. (2020). Health information systems in the digital health ecosystem—Problems and solutions for ethics, trust and privacy. *International Journal of Environmental Research and Public Health*, 17(9), 3006. Advance online publication. DOI: 10.3390/ijerph17093006 PMID: 32357446

Saddique, A., & Seng Choi, C. (2023). *Systematic Review Provides Evidence That Anxiety and Negative Thoughts are Prevalent among Medical Students in Malaysia*. https://doi.org/DOI: 10.6007/IJARBSS/v13-i7/16938

Sarraf, D., Vasiliu, V., Imberman, B., & Lindeman, B. (2021). Use of artificial intelligence for gender bias analysis in letters of recommendation for general surgery residency candidates. *American Journal of Surgery*, 222(6), 1051–1059. DOI: 10.1016/j.amjsurg.2021.09.034 PMID: 34674847

Seng, K. P., Ang, L. M., Peter, E., & Mmonyi, A. (2023). Machine Learning and AI Technologies for Smart Wearables. In *Electronics (Switzerland)* (Vol. 12, Issue 7). MDPI. DOI: 10.3390/electronics12071509

Shahul, A., Hovan George, A. S., & George, A. S. (2023). *Partners Universal International Innovation Journal (PUIIJ) Wearable Sensors: A New Way to Track Health and Wellness*. DOI: 10.5281/zenodo.8260879

Sokolovskaia, S., Nezhkina, N., Orlova, E., Karpycheva, M., Bocharin, I., Kolokoltsev, M., Limarenko, O., Romanova, E., & Martusevich, A. (2022). Pedagogical online technology to increase the level of students' conscious attitude to physical and mental health. *Journal of Physical Education and Sport*, 22(10), 2380–2385. DOI: 10.7752/jpes.2022.10303

Song, L. (2022). An Intelligent Method for Determining the Mental Health of University Students Based on IoT Technology. In Y. Pei, C. J.-W., & J. C. Hung (Eds.), *Lecture Notes in Electrical Engineering: Vol. 935 LNEE* (pp. 1283–1289). Springer Science and Business Media Deutschland GmbH. DOI: 10.1007/978-981-19-4132-0_175

Song, X., Zhang, Y., & Chang, X. (2024). Development of a Micro-course App for College Students' Mental Health Education Courses Based on Computer Software Technology. In Y. Zhang & N. Shah (Eds.), *Lecture Notes of the Institute for Computer Sciences, Social-Informatics and Telecommunications Engineering, LNICST: Vol. 583 LNICST* (pp. 572–581). Springer Science and Business Media Deutschland GmbH. DOI: 10.1007/978-3-031-63139-9_59

Vijayan, V., Connolly, J., Condell, J., McKelvey, N., & Gardiner, P. (2021). Review of wearable devices and data collection considerations for connected health. In *Sensors* (Vol. 21, Issue 16). MDPI AG. DOI: 10.3390/s21165589

Wang, Q.-Z., & Gao, L. (2019). Neural network for processing privacy-protected data. *Journal of Cryptologic Research*, 6(2), 258–268. DOI: 10.13868/j.cnki.jcr.000300

Wu, Y., & Feng, R. (2023). Application of virtual reality technology in promoting mental health of college students. *Chinese Journal of School Health, 44*(7), 1095-1098and1102. DOI: 10.16835/j.cnki.1000-9817.2023.07.031

Xia, Y., & Lin, Z. (2021). College Students' Mental Health Education System Based on Data Mining Technology. *Proceedings - 2021 International Symposium on Advances in Informatics, Electronics and Education, ISAIEE 2021*, 106–109. DOI: 10.1109/ISAIEE55071.2021.00033

Xie, L. (2021). Practice of Big Data Technology and Improved BP Neural Network in College Students' Mental Health Education. In V. Sugumaran, Z. Xu, & H. Zhou (Eds.), *Advances in Intelligent Systems and Computing: Vol. 1384 AIST* (pp. 831–837). Springer Science and Business Media Deutschland GmbH. DOI: 10.1007/978-3-030-74811-1_120

Yang, M. (2018). Application of Virtual Reality Technology in College Students' Mental Health Education. *Journal of Physics: Conference Series*, 1087(6), 062056. Advance online publication. DOI: 10.1088/1742-6596/1087/6/062056

Yaokumah, W. (2020). Predicting and explaining cyber ethics with ethical theories. *International Journal of Cyber Warfare & Terrorism*, 10(2), 46–63. DOI: 10.4018/IJCWT.2020040103

Yee Chung, J. W., Fuk So, H. C., Tak Choi, M. M., Man Yan, V. C., & Shing Wong, T. K. (2021). Artificial Intelligence in education: Using heart rate variability (HRV) as a biomarker to assess emotions objectively. *Computers and Education: Artificial Intelligence*, 2, 100011. Advance online publication. DOI: 10.1016/j.caeai.2021.100011

Yllaconza, J. E. C., Elescano, R. A. V., Cuevas, T. L. V., Orellana, M. V., Campos, J. H. T., Montenegro, J. S., Livia, D. P. R., & Loayza, E. F. P. (2024). Use of wearable technologies in health promotion in Human Medicine students. *EAI Endorsed Transactions on Pervasive Health and Technology*, 10. Advance online publication. DOI: 10.4108/eetpht.10.5701

Zamani, A. S., Hashim, A. H. A., Akhtar, M. M., Samdani, F., Siddiqui, A. T., Alluhayb, A., Hamza, M. A., & Ahmad, N. (2023). The prediction of sleep quality using wearable-assisted smart health monitoring systems based on statistical data. *Journal of King Saud University. Science*, 35(9), 102927. Advance online publication. DOI: 10.1016/j.jksus.2023.102927

Zhang, P., Han, W., & Liu, Q. (2024). Research on Predicting the Mental Health of College Students with Prediction Models based on Big Data Technology. *IEIE Transactions on Smart Processing and Computing*, 13(4), 393–401. DOI: 10.5573/IEIESPC.2024.13.4.393

Chapter 10
Integrating AI/ML With Wearable Devices for Monitoring Student Mental Health

Uddalak Mitra
JIS College of Engineering, India

Shafiq Ul Rehman
https://orcid.org/0000-0003-2266-218X
Kingdom University, Bahrain

ABSTRACT

The rise of wearable technologies has transformed how we monitor and support student mental health. This chapter explores the exciting potential of Artificial Intelligence (AI) and Machine Learning (ML) to further enhance wearables for mental health monitoring in educational settings. By integrating AI and ML, wearable devices can evolve from simple data collectors to intelligent assistants that provide personalized mental health support, improve student well-being, and empower educators. We will delve into the core concepts of AI and ML, their role in wearable technology, and how they can be harnessed to create a more supportive and effective learning environment. This chapter will explore specific applications of AI-powered wearables in mental health monitoring, analyze the benefits for both students and educators, and address potential challenges and considerations for responsible implementation.

DOI: 10.4018/979-8-3693-7817-5.ch010

1. INTRODUCTION

Education is undergoing a profound transformation (Aldowah, H et al. 2017), driven by technological advancements that are reshaping student learning and well-being. The intricacies of contemporary student life cannot be adequately addressed by traditional approaches to student mental health care (Wang et al. 2023, Wiedermann et al. 2023, Gonzo et al. 2023, Dayal et al. 2023). Wearable technologies have become a key innovation in this shift, providing new ray-of-hope to understand, monitor, and support mental health (Robinson et al. 2023, Masri et al. 2024, Lagera et al. 2023, Fucarino et al. 2024). From fitness trackers and smartwatches to sophisticated biometric monitors, wearable technology records data in real time on a variety of physiological and behavioral patterns, including heart rate, sleep quality, physical activity, and emotional states. When monitored over time, these data provide educational institutions with vital information about the mental health and general well-being of their students. For example, heart rate variability can indicate stress or anxiety (Immanuel et al. 2023, Gullett et al. 2023), poor sleep patterns can suggest mental health issues like depression (Corbo et al. 2023, Yasugaki et al. 2023), and shifts in physical activity levels can reflect changes in mood or motivation (Budnick et al. 2023, Flack et al. 2023). This real-time data collection offers an unprecedented opportunity for proactive mental health support, enabling educators and support staff to intervene early.

With the addition of artificial intelligence (AI) and machine learning (ML), wearables have possibilities beyond data collecting. Their combinations become an intelligent system that can actively monitor, interpret, and respond to students' mental health in real time (Ravi Prakash et al. 2023, Sharma et al. 2023). AI and ML algorithms process extensive data from wearables, identifying patterns that may not be immediately visible to human observers, enabling timely, targeted interventions.This predictive power transforms wearables in education. For example, AI algorithms can detect physiological markers—such as increased heart rate during inactivity or irregular sleep patterns—signaling the onset of anxiety or depression. Recognizing these early signs allows AI-driven systems to alert educators or mental health professionals, who can then provide personalized support. This proactive approach contrasts with traditional methods, which rely on students self-reporting or waiting until symptoms are severe.

Additionally, AI and ML systems continually improve through *adaptive learning*, enhancing accuracy and effectiveness over time. As data accumulates, algorithms refine their pattern recognition and prediction, leading to higher-quality mental health interventions. For students, this means more personalized, timely support, reducing the risk of unnoticed or untreated mental health issues. In educational settings, where early intervention is crucial, this capability is particularly valuable.

The chapter's contributions can be summarized as:

1. **Synergy of Wearables and AI/ML**: Discusses how integrating wearable technologies with AI and ML can enhance mental health support for students, enabling real-time stress detection and early identification of conditions like anxiety and depression.
2. **Current State and Limitations of Wearables**: Examines the strengths and limitations of wearables in educational contexts, highlighting challenges such as discomfort with constant monitoring, incomplete mental health data capture, and the need for optimized implementation practices.
3. **Applications of AI in Wearables for Mental Health**: Provides examples of AI-driven applications, such as detecting physiological signs of acute stress (e.g., sudden heart rate spikes) and identifying patterns indicating mental health concerns through trends like reduced activity or poor sleep.
4. **Case Studies and Real-world Applications**: Includes case studies showcasing practical uses of AI-integrated wearables, such as monitoring student stress during exams and enabling counselors to create personalized mental health plans based on real-time data.
5. **Ethical and Practical Considerations**: explains significant practical and ethical issues, including as consent, data protection, and the possibility of false positives. It highlights the need of handling sensitive health data carefully and avoiding needless interventions.
6. **Guidelines for Responsible Use**: Provides actionable guidelines for educators, policymakers, and developers on ensuring data privacy, obtaining informed consent, and fostering transparency with students and parents. Emphasizes the importance of testing and validating AI algorithms to ensure prediction accuracy and reliability.
7. **Future Directions**: Envisions future advancements in AI, wearable technology, and educational applications, suggesting expanded uses beyond mental health to areas like academic performance and physical health, aiming to create supportive, inclusive learning environments.

In conclusion, combining wearable technology with AI and ML is a big step forward for helping students' mental health. Educational institutions may address specific mental health issues and set the stage for future learning environments where students can flourish academically and emotionally by utilizing these technologies to deliver more intelligent, individualized treatment.

2. BACKGROUND

2.1 Wearable Technologies in Mental Health Monitoring

Wearable technologies are becoming increasingly prominent in the field of mental health monitoring (Robinson et al. 2023, Alhejaili et al. 2023, Gomes et al. 2023, Ahmed et al. 2023), particularly in educational settingsInnovative approaches to tracking and analysing physiological and psychological data are provided by gadgets like smartwatches (like the Apple Watch and Samsung Galaxy Watch), fitness trackers (like the Fitbit and Garmin Vivosmart), and brain-computer interfaces (BCIs) like the Muse Headband and EmotivEpoc+. These devices provide several key benefits for monitoring and supporting mental health.

- **Data Collection:**Wearables capture real-time physiological data like heart rate variability, brain activity, and physical activity**(Table 1).** Smartwatches monitor heart rate and skin temperature, correlating with stress, while brain-computer interfaces track brainwaves to assess cognitive states. Fitness trackers provide insights into activity and sleep patterns, offering educators and mental health professionals a comprehensive view of students' emotional and physiological states.
- **Personalized Support:**Integrated with AI, wearables deliver tailored mental health support by analyzing individual data patterns. For instance, consistent heart rate spikes before exams may prompt AI to recommend breathing exercises or breaks, ensuring interventions suit each student's needs.
- **Early Detection:**Wearables excel at identifying early signs of mental health issues like depression or anxiety by monitoring subtle physiological changes. For example, poor sleep paired with reduced activity may signal depression, enabling timely interventions to support students before issues escalate.

In summary, wearable technologies hold great promise in transforming mental health monitoring through real-time data collection, personalized interventions, and early detection of potential issues. As AI continues to evolve, the integration of wearable devices in educational mental health support systems offers the potential to significantly improve outcomes for students.

Table 1. Wearable devices for mental health monitoring and their key features

Device Type	Key Features	Mental Health Monitor	Example Devices
Smartwatches	Heart rate, skin temperature, activity tracking	Abnormalities indicates stress	Apple Watch, Samsung Galaxy Watch
Fitness Trackers	Physical activity, sleep pattern monitoring	Depression can be detected from sleep disorder	Fitbit, Garmin Vivosmart
Brain-Computer Interfaces	Cognitive state assessment, brain activity tracking	Can be used to capture the cognitive abilies and hence personalized education	Muse Headband, EmotivEpoc+

2.2 Understanding AI and Machine Learning

- **Artificial Intelligence (AI)** is the capacity of machines to carry out operations that normally call for human intelligence. Learning, thinking, problem-solving, and comprehending natural language are some of these activities. AI systems receive information and make judgments based on it in order to mimic human cognitive processes. In educational settings, AI can be used to analyze vast amounts of data generated by students through various sources, such as wearables, to provide insights into their mental health, learning behaviors, and overall well-being. By leveraging AI, schools can develop intelligent systems that support mental health interventions in real-time, enhancing the educational experience for students.
- **Machine Learning (ML)** is a branch of artificial intelligence (AI) (Figure 1) that focuses on creating algorithms that can learn from data without being specifically built for a given job. Unlike traditional programming, where explicit rules and logic are required, ML algorithms identify complex patterns, relationships, and trends in data autonomously. These algorithms improve over time as they process more data, becoming more accurate in their predictions and classifications. In the context of mental health monitoring, machine learning (ML) is essential for evaluating the vast amounts of data produced by wearable technology, including activity levels, sleep patterns, and heart rate variability, in order to spot any mental health problems before they worsen. For example, ML models can analyze a student's heart rate and activity levels over time to detect abnormal patterns, such as prolonged periods of inactivity combined with disrupted sleep, which may indicate the onset of stress or anxiety. The ability to identify these patterns allows educators and mental health professionals to intervene early, offering targeted support tailored to individual needs. The use of AI and ML in this way holds transformative potential

for personalizing mental health care and improving outcomes in educational environments.

Figure 1. Relationship between AI and ML

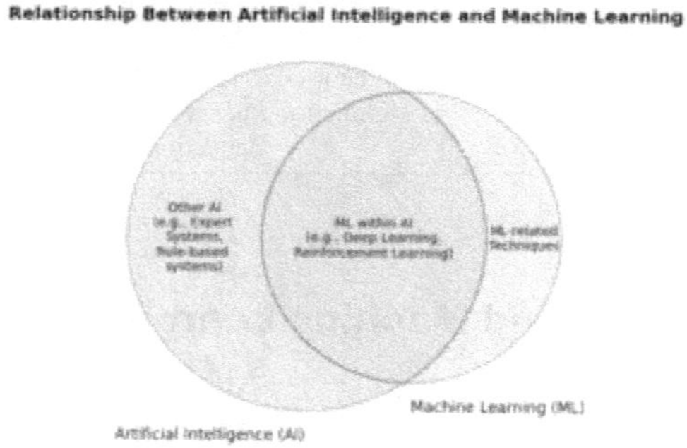

2.3 Integrating AI/ML with Wearables in Mental Health Monitoring

An important development in mental health monitoring is the combination of wearable technology with artificial intelligence (AI) and machine learning (ML). AI/ML algorithms may be used to improve wearable devices that gather data from users in real time. This data can then be processed, analysed, and interpreted to produce actionable insights (Rehman, S. U et al, 2024). Either the wearable devices themselves or cloud-based solutions that compile data from several devices can accomplish this integration. These qualities make it possible to monitor mental health in school settings in a number of significant ways.

- **Real-time-Feedback:** Giving pupils real-time input on their mental health condition is one of the most potent benefits made possible by AI/ML integration with wearables (Roos et al., 2023). For example, wearable technology can instantly analyse data and provide students feedback if it detects high heart rates or erratic breathing patterns, which are classic physiological indicators of stress or worry. This might be as straightforward as sending them a notification advising them to seek help, take a break, or engage in relaxation techniques. By delivering instant feedback, students can make immediate adjustments to their behavior, potentially mitigating the impact of stress or anxiety before it escalates (Fucarino et al.2024). This real-time support mechanism also allows students to become more aware of their mental and emotional states, promoting self-regulation and proactive mental health management.

- **Adaptive-Interventions:** Another significant advantage of AI/ML integration is the ability to deliver adaptive, personalized mental health interventions(Oyebode et al. 2023). Unlike traditional one-size-fits-all approaches to mental health care, AI systems analyze the data collected by wearables and generate interventions tailored to the individual. For example, if a student consistently shows signs of stress during certain times of day or after specific activities (like exams or presentations), the AI system could suggest targeted coping strategies such as mindfulness exercises, breathing techniques, or scheduling breaks during these high-stress periods. These interventions adapt to the evolving needs of the student, ensuring they receive the appropriate support at the right time (Immanuel et al. 2023). This personalized, data-driven approach enhances the effectiveness of mental health interventions, making them more relevant to the student's unique experiences and stressors.
- **Early-Intervention:** One of the most critical roles AI/ML plays in mental health monitoring is early detection and intervention(Diaz-Ramos et al. 2023). Wearable devices continuously monitor physiological and behavioral patterns such as sleep quality, heart rate variability, and activity levels, which are often indicative of mental health status. AI algorithms analyze these patterns over time, identifying subtle changes that may indicate the onset of mental health challenges, such as stress, anxiety, or depression. For instance, a sudden decrease in physical activity or poor sleep quality over several days may trigger the AI system to flag the student as potentially at risk of developing depression. Educators, mental health professionals, or the students themselves can then intervene promptly, preventing the condition from worsening (Gullett et al. 2023). Early detection is particularly valuable in educational settings where mental health issues can significantly affect academic performance, social interactions, and overall well-being.
- **Scalability-and-Efficiency:** Integrating AI/ML with wearables also allows mental health support systems to scale efficiently across large student populations(Roos et al. 2023). AI systems can process data from thousands of students simultaneously, identifying those who may require immediate attention while continuously monitoring others. This reduces the burden on mental health professionals, allowing them to focus their efforts on the students who need the most support. Additionally, AI-driven systems can operate 24/7, providing continuous monitoring and interventions outside of traditional school hours, which is crucial for detecting issues that may arise during evenings or weekends.

- **Data-Driven-Insights-for-Educators-and-Mental-Health-Professionals:** In addition to providing instant feedback and interventions, the data gathered and examined by AI/ML systems may give educators and mental health specialists important information on the general patterns in student mental health. Aggregate data can reveal patterns such as increased anxiety during exam periods or higher stress levels in certain demographic groups, allowing schools to implement broader mental health strategies. Additionally, these insights can guide the development of programs and initiatives aimed at improving student well-being.
- **Ethical-and-Privacy-Considerations:** While the integration of AI/ML with wearables in mental health monitoring offers many benefits, it also raises ethical and privacy concerns (Aldowah, H et al. 2021). Schools must ensure that students' data is securely stored and processed, with clear consent obtained from both students and guardians. Furthermore, AI-driven interventions must be transparent and interpretable, so that students and educators understand how decisions are being made and what data is being used. To guarantee the appropriate use of these technologies, ethical issues pertaining to data privacy, consent, and bias in AI algorithms must be properly addressed.

In conclusion, integrating AI/ML with wearable devices in mental health monitoring has the potential to transform how educational institutions approach student well-being. From real-time feedback and adaptive interventions to early detection and scalable support systems, these technologies offer personalized and timely mental health care that can significantly improve student outcomes. However, as schools adopt these systems, they must navigate the associated ethical and practical challenges to ensure their responsible use.

Table 2. AI/ML contributions to wearables

Feature	Description	Example Use Case
Real-time Feedback	Alerts users of stress/anxiety signs for immediate action.	Notification for relaxation when high stress is detected.
Adaptive Interventions	Provides personalized coping strategies based on activity and time patterns.	Suggests breathing exercises during exam preparation.
Early Detection	Identifies subtle signs of mental health issues before they escalate.	Detects poor sleep and inactivity, flags potential depression.
Scalability and Efficiency	Enables monitoring of large student populations with minimal resource strain.	Tracks thousands of students simultaneously, highlighting at-risk cases.
Data-Driven Insights	Aggregates data to guide mental health strategies and programs.	Identifies increased anxiety during exam periods across student groups.

3. APPLICATIONS OF AI-POWERED WEARABLES IN MENTAL HEALTH MONITORING

3.1 Personalized Mental Health Support

Personalized mental health support through wearable technologies, such as smartwatches, fitness trackers, and brain-computer interfaces (BCIs), is transforming how educators and mental health professionals approach student well-being(Li, J et al. 2024). These devices monitor physiological and neurological indicators, offering individualized insights into a student's mental health. When paired with AI/ML algorithms, wearables can provide tailored interventions that address specific needs, allowing students to receive real-time guidance for improving their mental health.

Smartwatches and Fitness Trackers

Fitness trackers and smartwatches, like the Fitbit and Apple Watch, have developed from fitness aids to vital tools for mental health monitoring. These wearables provide information on mental health by gathering physiological data such as heart rate, HRV, sleep patterns, and activity levels (Sadek, I. et al. 2019). Stress, anxiety, or depression might be indicated by symptoms like high heart rates, disturbed sleep, or low activity levels (Rehman, S. U. et al., 2024). These patterns are analysed by integrated AI/ML algorithms, which then offer timely, tailored treatments, such as breathing exercises or mindfulness exercises, to aid with stress management.

These devices also promote better sleep hygiene and enable early detection of mental health issues. Sleep tracking features identify restlessness or poor-quality sleep and offer tailored advice to improve sleep habits. Over time, AI/ML models learn an individual's typical physiological patterns, detecting subtle changes that may indicate risks like depression. By flagging these early warning signs, smartwatches and trackers empower users to seek support proactively, making them invaluable tools for maintaining mental health.

Proposed prototype: To integrate AI/ML into smartwatches and fitness trackers for mental health monitoring, we can develop a prototype model using the Fitbit dataset available on Kaggle(https://www.kaggle.com/datasets/arashnic/fitbit/data). In order to extract important physiological markers like heart rate variability (HRV), activity levels, and sleep patterns, the procedure starts with data preparation, which includes feature engineering and cleaning. AI/ML algorithms, such as Random Forest or Support Vector Machines, can then be trained to predict mental health status by labeling deviations from normal physiological patterns (e.g., elevated heart rate and poor sleep) as potential signs of stress or mental health issues. Once trained, the model can be deployed to analyze real-time data from wearables, offering personalized interventions like breathing exercises or relaxation tips when stress or anxiety is detected. This approach allows for continuous monitoring and early detection of mental health risks, providing proactive support for users.

Brain-Computer Interfaces (BCIs)

Brain-Computer Interfaces (BCIs) offer a groundbreaking approach to personalized mental health support by measuring brain activity through techniques like electroencephalography (EEG). Devices such as the Muse Headband and EmotivEpoc+ monitor cognitive functions like attention, focus, and stress levels, providing real-time feedback. For instance, BCIs can detect brainwave patterns indicating high stress or distraction and suggest relaxation techniques such as deep breathing or mindfulness exercises. This immediate feedback helps users manage stress effectively and maintain focus, especially in high-pressure academic environments.

Beyond stress reduction, BCIs enhance focus and support mental health management. By monitoring brainwave patterns, these devices can guide users through exercises to improve attention and cognitive performance, reducing distractions during study or lessons. BCIs also detect abnormal brain activity linked to conditions like anxiety or depression, alerting users to seek support early. With applications like neuro-feedback therapy, BCIs enable users to self-regulate their brain activity, making them valuable tools for fostering mental health and resilience in educational settings.

Proposed prototype: To integrate AI/ML with Brain-Computer Interfaces (BCIs) for personalized mental health support, we can develop a prototype model using EEG data from devices like the Muse Headband or EmotivEpoc+, with pre-recorded dataset available at *(*https://www.kaggle.com/c/eeg-based-brain-computer-interface*)*. The first step involves preprocessing the EEG data to extract relevant features such as brainwave patterns linked to stress, attention, and focus. AI/ML algorithms like Random Forest or Neural Networks can then be trained to detect deviations from baseline brain activity, which may indicate stress, distraction, or early signs of anxiety or depression. Once trained, the model can provide real-time feedback, suggesting relaxation techniques or focus-enhancing exercises based on the detected patterns. This approach offers immediate, data-driven interventions, enabling users to self-regulate their mental health and improve cognitive performance, particularly in stressful academic environments.

In conclusion, integrating AI/ML with smartwatches, fitness trackers, and BCIs offers transformative potential for mental health support. By analyzing real-time physiological and neurological data, these technologies provide personalized interventions, promote stress management, improve focus, and enable early detection of mental health issues. The proposed prototypes demonstrate how AI-driven models can empower users to proactively manage their well-being, highlighting the vital role of technology in addressing mental health challenges.

3.2 Early Detection and Intervention

Early detection and intervention are crucial for mental health support, particularly in educational settings where students may hesitate to seek help (Adeghe, E. P. et al., 2024). Wearable devices, integrated with AI and Machine Learning (ML), provide effective tools to identify early signs of emotional and behavioral distress, enabling timely, personalized interventions that prevent escalation.

Emotion Recognition

AI-powered wearables, such as smartwatches and fitness trackers, can detect subtle physiological changes like heart rate, skin conductance, and voice tone, assessing emotional states in real-time. These devices identify signs of distress such as anxiety, frustration, or sadness, even if the student is unaware or unwilling to express them. Additionally, wearables equipped with cameras or apps can analyze facial expressions and voice patterns, detecting emotions like anger, fear, or sadness. If a student's voice shows signs of tension, the system may flag it as a potential indicator of stress or depression.

In educational environments, where students face emotional pressures from academic and social challenges, these insights are invaluable. For example, if a wearable detects heightened anxiety, it can alert a school counselor or suggest relaxation techniques to prevent distress from escalating into more severe mental health issues.

Behavioral Monitoring

Wearables can track behavioral changes that signal potential mental health issues. Patterns like increased sedentary behavior, irregular sleep, social withdrawal, or reduced physical activity are signs of depression or anxiety. AI-driven wearables continuously monitor these behaviors, offering a view of a student's mental well-being over time. If irregular sleep patterns and decreased activity are detected, the wearable can recommend interventions such as counseling or improved sleep habits.

Subtle changes, like increased isolation, can be flagged by wearables, signaling early signs of social withdrawal, a symptom of depression or anxiety. The wearable can alert educators or mental health professionals, allowing for timely support.

AI/ML integration with wearables allows for continuous, passive monitoring, collecting valuable data even when educators are not directly observing a student. This ensures early detection of issues in students who may be reluctant to seek help.

In practice, wearables have shown promise for early intervention. Studies have used them to monitor heart rates and skin conductance during exams, detecting extreme stress and enabling timely intervention. Similarly, wearables that track sleep patterns can identify students at risk for mental health issues and promote healthier sleep habits.

Incorporating AI/ML with wearables offers a powerful solution for early detection and intervention in mental health, supporting students' well-being and helping them succeed academically and personally.

3.3 Enhanced Support for Students with Special Needs

Wearable technologies integrated with AI and Machine Learning (ML) have transformative potential for supporting students with special needs (Alkhawaldeh et al., 2023). These students often face unique challenges that require tailored interventions. AI-powered wearables provide real-time feedback, personalized interventions, and continuous monitoring to improve their learning experiences and well-being.

Neurofeedback

Neurofeedback is a promising area, especially for students with Attention-Deficit/Hyperactivity Disorder (ADHD), anxiety, or other neurodevelopmental conditions. Wearables such as EEG headbands or brain-computer interfaces (BCIs) monitor brain activity and provide real-time data on mental states like focus and relaxation. This data is processed by AI to give immediate feedback, helping students self-regulate their attention or anxiety.

For students with ADHD, neurofeedback can enhance focus by visualizing brainwave patterns, allowing them to practice better self-control. Combined with behavioral therapies, this biofeedback offers a holistic approach to managing ADHD symptoms. For students with anxiety, neurofeedback can help them monitor stress levels and engage in mindfulness exercises, promoting emotional regulation.

Assistive Technologies

AI-driven wearables also offer assistive technologies that enhance learning for students with sensory processing or communication difficulties. Devices like smart glasses overlay digital information onto the physical environment, aiding students with visual impairments or processing challenges. These glasses help students with autism interpret social cues, such as facial expressions, facilitating better social interaction.

Wearables with haptic feedback, like wristbands or vests, support students with sensory sensitivities by providing gentle vibrations or pressure in response to stimuli. These devices help students manage sensory overload, allowing them to stay focused in the classroom.

Assistive communication devices equipped with voice recognition or text-to-speech functionalities are valuable for students with speech delays or non-verbal conditions. AI-enabled wearables can convert text or symbols into speech, helping these students communicate more effectively with peers and teachers, fostering independence and engagement in the learning process.

3.4 Empowering Educators and Mental Health Professionals

The integration of AI/ML with wearables is not just beneficial for students, but also for educators and mental health professionals(Masri et al 2024). By providing access to real-time data and advanced analytics, AI-driven wearables offer educators a deeper understanding of student well-being and performance, empowering them to tailor their teaching methods and interventions more effectively. The capacity to monitor and evaluate data over time is particularly advantageous to mental health practitioners, since it enables more proactive and individualized mental health treatment.

Real-Time Student Monitoring

Wearables in education offer real-time student monitoring. AI algorithms analyze data from devices, such as heart rate and facial expressions, providing insights into a student's emotional state. For example, if a student shows anxiety during an exam or disengagement in class, educators can be alerted to provide immediate support, like offering a break. Over time, data can reveal trends in student well-being, allowing educators to proactively address stress or disengagement.

Personalized Support Plans

AI-driven analytics help mental health professionals create tailored support plans. By analyzing continuous wearable data, AI identifies specific triggers for stress and effective interventions. For instance, if a student becomes anxious before exams, professionals can introduce coping strategies. AI can also suggest personalized interventions, like mindfulness exercises for chronic stress, ensuring support is customized to each student's needs.

Formative Assessment and Feedback

Wearables enhance formative assessments by including mental and physical well-being data, enabling holistic evaluations. AI can highlight areas where students need additional support, academically or emotionally. For example, if a student shows stress in math, AI might suggest alternative teaching methods. This data-driven approach personalizes learning, helping students thrive both academically and emotionally.

AI-driven wearables empower educators and mental health professionals to offer proactive support, addressing issues early and ensuring every student can succeed.

4. BENEFITS OF AI-POWERED WEARABLES IN MENTAL HEALTH MONITORING

The integration of AI-powered wearables in mental health monitoring offers significant benefits for students, educators, and mental health professionals alike(-Komalasari, R et al. 2024, Negi, R. et al. 2024). These devices not only collect vast amounts of physiological and behavioral data but also utilize AI to interpret this data in meaningful ways, providing real-time feedback and enabling personalized interventions. The following key benefits illustrate the transformative potential of these technologies in improving mental health outcomes and fostering more supportive educational environments.

Improved Mental Health Outcomes

AI-powered wearables enhance mental health by providing personalized support. By tracking data like heart rate and sleep patterns, AI can detect stress or anxiety and recommend tailored interventions like mindfulness or relaxation exercises. This early, targeted support helps reduce mental health crises, fostering resilience and improved outcomes.

Enhanced Student Well-being

With real-time feedback, AI wearables help students manage stress from academics or social challenges. Detecting physiological changes, these devices prompt students to practice calming techniques, empowering them to regulate emotions and build long-term self-care habits, which improves well-being and enhances learning.

Empowered Educators

AI wearables offer educators data-driven insights into student stress and engagement, allowing them to adjust teaching methods and provide personalized support. Recognizing trends like pre-exam anxiety, teachers can implement preventive strategies, creating a responsive, supportive classroom environment.

Early Identification of Mental Health Issues

Wearables enable early detection of mental health concerns by monitoring behavior and physiological data. AI identifies patterns linked to anxiety or depression, prompting timely intervention. This proactive approach helps students manage issues before they escalate, promoting healthier long-term outcomes.

Accessibility and Inclusivity

Wearable technology has promise for increasing accessibility and inclusivity in mental health monitoring, especially for kids with special needs. Students who struggle with communication, sensory processing, or neurodevelopmental disorders like Attention-Deficit/Hyperactivity Disorder (ADHD) or Autism Spectrum Disorder (ASD) sometimes experience particular obstacles that need for specialised assistance. Wearables with AI capabilities, such haptic feedback devices or smart glasses, can offer supportive tools tailored to these kids' unique requirements, increasing access to mental health care for a variety of student demographics.

For example, wearables that provide neurofeedback can help students with ADHD regulate their attention and focus during learning activities, while smart glasses can assist students with visual impairments or sensory processing challenges by overlaying helpful information in real-time. Similarly, haptic feedback devices can provide calming sensory input for students who struggle with sensory overload, helping them remain focused and engaged in the classroom. By offering personalized, real-time support tailored to the individual needs of students, AI-powered wearables contribute to a more inclusive learning environment where all students can thrive, regardless of their abilities or challenges.

In summary, AI-powered wearables offer numerous benefits in the realm of mental health monitoring, from improving individual mental health outcomes to empowering educators with actionable insights. Through real-time feedback, early detection, and personalized interventions, these devices create a more supportive and responsive educational environment. They also foster accessibility and inclusivity by offering specialized tools for students with special needs, ensuring that mental health support is available to all students, regardless of their circumstances. By integrating AI and wearable technology, schools can better meet the diverse mental health needs of their students, creating a foundation for long-term well-being and academic success.

5. CHALLENGES AND CONSIDERATIONS

Wearables with AI capabilities have enormous promise for monitoring and supporting mental health in educational settings, but their deployment presents a number of issues that must be resolved to guarantee their effectiveness, safety, and equity. When integrating these technologies, care must be taken to ensure that educators are properly prepared to utilize them while also giving data protection, equality, and ethical design first priority.

Data Privacy and Security

Making sure the sensitive data collected by AI-powered wearables is secure and private is one of the biggest problems they present (Sadek, I., et al. 2022). Wearables collect a lot of personal data, such as behavioural and physiological information, which might be misused if it is not well secured. To safeguard this data from breaches or unwanted access, schools and other educational institutions must have strong security measures in place, such as data encryption (Manickam, S., et al. 2023). For students and their guardians to understand how their data will be used, who may access it, and how long it will be kept, there must also be clear data privacy regulations in place.

Informed consent is another important consideration. Students and their families should have full transparency regarding the data collected by wearables, including its purpose and potential risks. To guarantee that student data is handled properly, schools must also abide by pertinent data protection legislation, such as the Family Educational Rights and Privacy Act (FERPA) in the US or the General Data Protection Regulation (GDPR) in Europe. Failure to adequately address these privacy concerns could undermine trust in wearable technology and limit its widespread adoption in schools.

Equity and Accessibility

Ensuring that AI-powered wearables are accessible to all students is another critical challenge. These devices can be costly, and not all schools or families may have the resources to afford them. If access to wearables is limited to well-funded institutions or families with higher incomes, there is a risk of exacerbating existing educational inequalities. Students from disadvantaged backgrounds could miss out on the benefits of AI-powered mental health monitoring and support, further widening the gap between those with access to cutting-edge technology and those without.

To address this challenge, policymakers and educational leaders must explore ways to make AI-powered wearables more affordable and accessible. This could include subsidies, partnerships with technology companies, or the development of more cost-effective devices. Additionally, schools must ensure that all students, regardless of socioeconomic status, have equal opportunities to benefit from the personalized mental health interventions and insights provided by wearables. Equitable access to AI-powered technology is essential for fostering inclusive learning environments where every student has the support they need to thrive.

Teacher Training and Support

The successful integration of AI-powered wearables into educational settings also depends on the ability of teachers to use these tools effectively. While wearables can provide valuable insights into student mental health and well-being, these insights are only useful if educators know how to interpret the data and apply it meaningfully in their teaching practices. Therefore, comprehensive training and support for teachers are essential to ensure they can make the most of wearable technology.

Teachers must be equipped with the knowledge and skills to interpret data from wearables, identify early warning signs of mental health issues, and implement appropriate interventions. This may require training programs that focus on the technical aspects of AI and wearables, as well as strategies for integrating mental health monitoring into the classroom. In addition, ongoing support from school administrators and mental health professionals is crucial to help teachers navigate any challenges they encounter when using these technologies. Without adequate training and support, the potential benefits of AI-powered wearables may not be fully realized, and teachers may struggle to use the data effectively.

Ethical Considerations and Algorithmic Bias

Another crucial factor to take into account is the ethical design and implementation of wearables with AI capabilities. Because AI algorithms are not impervious to prejudice, they may unintentionally disfavour particular student groups if they are not properly developed and supervised. Algorithms trained on biassed datasets, for instance, could make incorrect predictions for students from under-represented backgrounds, resulting in differences in how mental health concerns are identified or supported.

When creating AI algorithms for wearables, developers must put equity and inclusion first in order to reduce these dangers. In order to guarantee that the algorithms perform equally effectively for every student, irrespective of their gender, color, socioeconomic status, or other characteristics, this involves use representative and varied datasets throughout the training phase. To find and fix any biases that can develop over time, the algorithms must also be regularly audited and monitored. The use of AI in educational settings also has to be governed by ethical standards, with an emphasis on safeguarding students' rights and making sure the technology is applied responsibly and fairly.

In conclusion, wearables with AI capabilities show promise for monitoring mental health in the classroom, but their effective use necessitates careful consideration of a number of issues. Maximizing the beneficial effects of new technologies requires resolving ethical issues and algorithmic bias, ensuring data privacy and security, fostering fairness and accessibility, and offering sufficient training and support to teachers. Schools may fully utilize wearables with AI capabilities to create safer, more encouraging learning environments for all kids by proactively addressing these issues.

6. CASE STUDY: SUPPORTING STUDENT MENTAL HEALTH WITH AI-POWERED WEARABLE DEVICES IN A UNIVERSITY SETTING

Problem Statement

Imagine Dr. Smith, a university counselor, who faces the challenge of addressing the varied mental health needs of a diverse student body. Students experience a range of issues, such as stress, anxiety, sleep disturbances, and physical inactivity, all of which affect their academic success and overall well-being. While traditional counseling methods offer support, they often fall short in providing personalized, real-time interventions that cater to the unique needs of each student. This gap makes it difficult for Dr. Smith and her team to offer timely, customized care. As a result, some students go without the necessary support, leading to higher stress, disengagement, and a decline in both mental health and academic performance.

Figure 2. Workflow for the hypothetical case study

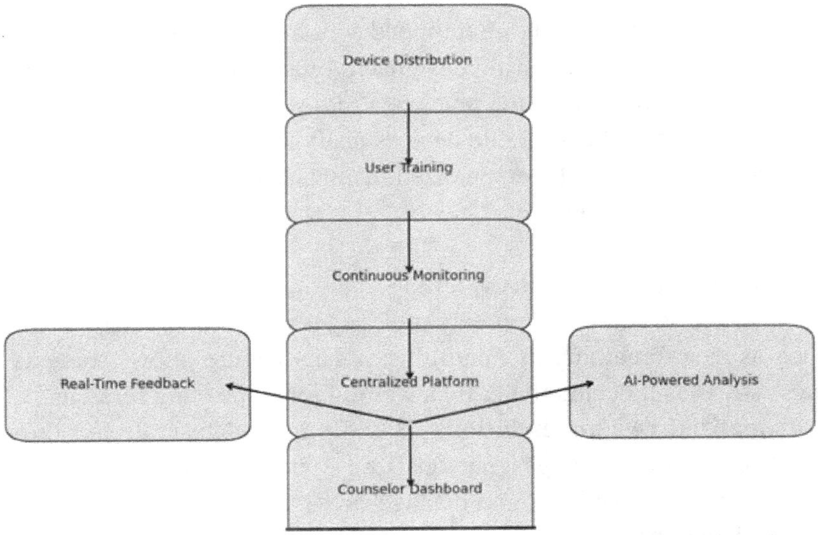

Intervention

To overcome these limitations, Dr. Smith pilots a program that utilizes AI-powered wearable technologies designed to provide continuous mental health support. The program integrates the following technologies:

- **Smartwatches (e.g., Apple Watch)**: These devices monitor physiological indicators such as heart rate variability (HRV), sleep patterns, and physical activity levels. This data provides insights into the physiological impact of stress and anxiety, helping students monitor their overall health.
- **Fitness Trackers (e.g., Fitbit)**: Fitness trackers monitor daily physical activity levels, offering students valuable insights into how physical health and movement impact their mental well-being. These devices encourage students to engage in physical activities that can reduce stress and improve their overall health.
- **Brain-Computer Interfaces (BCIs) (e.g., Muse Headband)**: BCIs are used during relaxation sessions to measure brain activity and provide real-time feedback on stress levels. The BCI devices suggest relaxation techniques like deep breathing or mindfulness exercises, allowing students to manage stress more effectively.

Technology and Approach

In the pilot program, students are equipped with smartwatches and fitness trackers to wear throughout the day. During designated relaxation sessions, they use BCIs to monitor their brain activity and practice stress-reducing techniques. The AI algorithms embedded in the wearable devices analyze physiological and behavioral data to provide students with personalized insights and feedback in real time. Key features of the approach include:

- **Real-time feedback on stress levels**: The smartwatches and BCIs detect rising stress levels and immediately offer suggestions for relaxation techniques such as deep breathing or mindfulness. This feature allows students to address stress as it occurs rather than waiting for counseling sessions.
- **Personalized recommendations**: Based on the data collected throughout the day, the devices suggest personalized strategies to improve sleep quality, increase physical activity, and manage stress. For example, a student experiencing irregular sleep patterns may receive a recommendation to establish a more consistent sleep routine or practice mindfulness before bed.

Implementation

Dr. Smith integrates the AI-powered wearable devices into the university's broader mental health support services. The devices are distributed to participating students, who are encouraged to wear them consistently during their daily activities and relaxation sessions. The platform connected to the devices collects and analyzes data, creating a comprehensive overview of each student's mental health and well-being.

- **Personalized interventions**: The platform suggests mental health interventions that students can complete independently. For instance, a student experiencing high stress levels may be prompted to take a break for a guided mindfulness session, while another may be encouraged to engage in a physical activity to reduce anxiety.
- **Monitoring and support**: Dr. Smith can access a dashboard on the platform, where she monitors student progress and receives alerts when intervention may be necessary. This allows for timely interventions when students are showing signs of significant stress, anxiety, or other mental health challenges. By having access to real-time data, Dr. Smith can offer more personalized and proactive support than traditional counseling methods would allow.

Outcomes

The results of Dr. Smith's pilot program demonstrate the potential of AI-powered wearable devices in enhancing student mental health support.

- **Improved Mental Health:** Students' mental health and general well-being significantly improved after using the wearable AI-powered gadgets. Compared to a control group that did not utilize the devices, students' general mental health improved as a result of the personalized therapies and real-time feedback that helped them better manages stress and anxiety.
- **Reduced Stress and Anxiety**: The data from the BCIs, combined with AI analysis, allowed Dr. Smith to identify students experiencing elevated stress levels early on. By intervening with targeted relaxation techniques and stress management strategies, students experienced a notable reduction in their reported stress and anxiety levels. This proactive approach ensured that mental health issues were addressed before they escalated.
- **Enhanced Engagement in Health Activities**: The continuous feedback from the wearable devices kept students more engaged in health-promoting activities, such as physical exercise and relaxation techniques. Real-time data on physical activity levels and sleep patterns motivated students to make positive lifestyle changes, such as increasing their daily step count or establishing better sleep routines.
- **Personalized Support and Growth**: The AI-powered platform facilitated personalized mental health support for each student. The customized recommendations ensured that students received the help they needed in a timely and effective manner, promoting personal growth and improved well-being.

Challenges

While the outcomes were largely positive, Dr. Smith encountered several challenges during the program's implementation:

- **Initial Learning Curve**: Students required time to adapt to the new technology. Some students initially struggled to use the wearable devices effectively, either forgetting to wear them or misunderstanding the real-time feedback. Over time, however, they became more comfortable with the devices, and usage became more consistent.
- **Data Privacy Concerns**: As with any technology that collects personal data, there were concerns about data privacy, especially regarding the sensitive information collected by BCIs. Dr. Smith had to address these concerns by

ensuring that all data was securely stored and only accessible to authorized individuals. Transparent communication about data privacy policies and the ethical use of BCI data was crucial in building trust with students and their families.

- **Responsible AI Development**: Ensuring that the AI algorithms used in the wearables were unbiased and provided accurate, personalized support was an ongoing challenge. The algorithms required continuous monitoring and adjustments to ensure that they were delivering fair and reliable insights, especially as new data was collected from diverse student populations.

In conclusion, Dr. Smith's pilot program demonstrated that AI-powered wearable devices can play a transformative role in supporting student mental health. Through real-time feedback, personalized recommendations, and proactive interventions, these technologies provide a scalable and effective solution to meet the growing mental health needs of university students. However, addressing challenges related to technology adaptation, data privacy, and responsible AI development remains critical to ensuring the long-term success and ethical implementation of these tools in educational settings.

7. CONCLUSION

Wearable technology combined with artificial intelligence (AI) and machine learning (ML) has the potential to completely transform mental health monitoring and assistance in educational environments. These developments make it possible to design individualized mental health spaces that meet the requirements of each unique student. Wearables with AI capabilities can help students manage stress, anxiety, and other mental health issues by continually evaluating behavioral and physiological data and offering personalized solutions and real-time feedback. This individualized approach makes studying more enjoyable and improves student well-being.

Teachers may modify their teaching methods and give difficult students focused help thanks to the data-driven insights these technologies provide. By giving instructors this kind of empowerment, mental health issues are proactively addressed, enhancing student welfare and academic performance.

However, while the potential benefits are vast, challenges remain. Data privacy and security must be prioritized to protect sensitive student information. Ethical considerations, including the prevention of algorithmic bias and ensuring that AI systems are fair and transparent, are critical. Furthermore, equitable access to AI-powered wearables is essential to avoid exacerbating educational inequalities.

As AI and wearable technologies continue to evolve, their applications in education will likely expand, further transforming how mental health is monitored and supported. Addressing these challenges responsibly will ensure that AI-powered wearables create a more inclusive and supportive learning environment, fostering better mental health outcomes for all students.

REFERENCES

Adeghe, E. P., Okolo, C. A., & Ojeyinka, O. T. (2024). A review of wearable technology in healthcare: Monitoring patient health and enhancing outcomes. *OARJ of Multidisciplinary Studies*, 7(01), 142–148.

Ahmed, A., Aziz, S., Alzubaidi, M., Schneider, J., Irshaidat, S., Serhan, H. A., & Househ, M. (2023). Wearable devices for anxiety & depression: A scoping review. *Computer Methods and Programs in Biomedicine Update*, 3, 100095. DOI: 10.1016/j.cmpbup.2023.100095 PMID: 36743720

Aldowah, H., Rehman, S. U., Ghazal, S., & Umar, I. N. (2017, September). Internet of Things in higher education: A study on future learning. [). IOP Publishing]. *Journal of Physics: Conference Series*, 892(1), 012017. DOI: 10.1088/1742-6596/892/1/012017

Aldowah, H., Ul Rehman, S., & Umar, I. (2021). Trust in iot systems: a vision on the current issues, challenges, and recommended solutions. *Advances on Smart and Soft Computing:Proceedings of ICACIn 2020*, 329-339.

Alhejaili, R., & Alomainy, A. (2023). The use of wearable technology in providing assistive solutions for mental well-being. *Sensors (Basel)*, 23(17), 7378. DOI: 10.3390/s23177378 PMID: 37687834

Alkhawaldeh, M. A., & Khasawneh, M. A. S. (2023). Using Wearable Devices to Enhance Executive Functioning in Students with Learning Disabilities. *Migration Letters : An International Journal of Migration Studies*, 20(S3), 1076–1089.

Budnick, C. J., Stults-Kolehmainen, M., Dadina, C., Bartholomew, J. B., Boullosa, D., Ash, G. I., Sinha, R., Blacutt, M., Haughton, A., & Lu, T. (2023). Motivation states to move, be physically active and sedentary vary like circadian rhythms and are associated with affect and arousal. *Frontiers in Sports and Active Living*, 5, 1094288. DOI: 10.3389/fspor.2023.1094288 PMID: 37143586

Corbo, I., Forte, G., Favieri, F., & Casagrande, M. (2023). Poor sleep quality in aging: The association with mental health. *International Journal of Environmental Research and Public Health*, 20(3), 1661. DOI: 10.3390/ijerph20031661 PMID: 36767029

Dayal, S., & Pratibha, . (2023). Roadblocks in education amidst global crisis—A study based in India. *PLoS One*, 18(10), e0292465. DOI: 10.1371/journal.pone.0292465 PMID: 37847726

Diaz-Ramos, R. E., Noriega, I., Trejo, L. A., Stroulia, E., & Cao, B. (2023). Using wearable devices and speech data for personalized machine learning in early detection of mental disorders: Protocol for a participatory research study. *JMIR Research Protocols*, 12(1), e48210. DOI: 10.2196/48210 PMID: 37955959

Flack, K. D., Stults-Kolehmainen, M. A., Creasy, S. A., Khullar, S., Boullosa, D., Catenacci, V. A., & King, N. (2023). Altered motivation states for physical activity and 'appetite'for movement as compensatory mechanisms limiting the efficacy of exercise training for weight loss. *Frontiers in Psychology*, 14, 1098394. DOI: 10.3389/fpsyg.2023.1098394 PMID: 37187558

Fucarino, A., Fabbrizio, A., Garrido, N. D., Iuliano, E., Reis, V. M., Sausa, M., & Cantoia, M. (2024, July). Emerging Technologies and Open-Source Platforms for Remote Physical Exercise: Innovations and Opportunities for Healthy Population—A Narrative Review. [). MDPI.]. *Health Care*, 12(15), 1466. PMID: 39120170

Gomes, N., Pato, M., Lourenco, A. R., & Datia, N. (2023). A survey on wearable sensors for mental health monitoring. *Sensors (Basel)*, 23(3), 1330. DOI: 10.3390/s23031330 PMID: 36772370

Gonzo, F. (2023). The Impact of Digital Learning Technology on Higher Education Students' Mental Health. *Advances in Hospitality, Tourism, and the Services Industry*.

Gullett, N., Zajkowska, Z., Walsh, A., Harper, R., & Mondelli, V. (2023). Heart rate variability (HRV) as a way to understand associations between the autonomic nervous system (ANS) and affective states: A critical review of the literature. *International Journal of Psychophysiology*, 192, 35–42. DOI: 10.1016/j.ijpsycho.2023.08.001 PMID: 37543289

Immanuel, S., Teferra, M. N., Baumert, M., & Bidargaddi, N. (2023). Heart rate variability for evaluating psychological stress changes in healthy adults: A scoping review. *Neuropsychobiology*, 82(4), 187–202. DOI: 10.1159/000530376 PMID: 37290411

Komalasari, R. (2024). AI-Powered Wearables Revolutionizing Health Tracking and Personalized Wellness Management. *Timor Leste Journal of Business and Management*, 6, 42–50.

Lagera, P. G. D., Chan, S. R., & Yellowlees, P. M. (2023). Asynchronous technologies in mental health care and education. *Current Treatment Options in Psychiatry*, 10(2), 59–71. DOI: 10.1007/s40501-023-00286-6 PMID: 37360962

Li, J., & Washington, P. (2024). A comparison of personalized and generalized approaches to emotion recognition using consumer wearable devices: Machine learning study. *JMIR AI*, 3(1), e52171. DOI: 10.2196/52171 PMID: 38875573

Manickam, S., & Rehman, S. U. (2023). Secured-KDS: Secret key distribution and authentication scheme for resource-constrained devices. *IET Networks*, 12(1), 1–8. DOI: 10.1049/ntw2.12054

Masri, D., Jaber, L., Mashal, R., Albourini, F., Alsaoud, M. A., & Al-Tarawneh, A. M. (2024, February). The Role of Wearables & Technology in Mental Health. In *2024 2nd International Conference on Cyber Resilience (ICCR)* (pp. 1-5). IEEE. DOI: 10.1109/ICCR61006.2024.10532880

Masri, D., Jaber, L., Mashal, R., Albourini, F., Alsaoud, M. A., & Al-Tarawneh, A. M. (2024, February). The Role of Wearables & Technology in Mental Health. In *2024 2nd International Conference on Cyber Resilience (ICCR)* (pp. 1-5). IEEE. DOI: 10.1109/ICCR61006.2024.10532880

Negi, R. (2024). Improving women's mental health through AI-powered interventions and diagnoses. In *Artificial Intelligence and Machine Learning for Women's Health Issues* (pp. 173–191). Academic Press. DOI: 10.1016/B978-0-443-21889-7.00017-8

Oyebode, O., Fowles, J., Steeves, D., & Orji, R. (2023). Machine learning techniques in adaptive and personalized systems for health and wellness. *International Journal of Human-Computer Interaction*, 39(9), 1938–1962. DOI: 10.1080/10447318.2022.2089085

RaviPrakash, H., & Anwar, S. M. (2023). Do It Yourself: Wearable Sensors and AI for Self-Assessment of Mental Health. *AI in Clinical Medicine: A Practical Guide for Healthcare Professionals*, 94-103.

Rehman, S. U., & Manickam, S. (2024). Application of smart sensors for internet of things healthcare environment: study and prospects. In *Next-Generation Smart Biosensing* (pp. 287–305). Academic Press. DOI: 10.1016/B978-0-323-98805-6.00006-3

Rehman, S. U., Sadek, I., Huang, B., Manickam, S., & Mahmoud, L. N. (2024). IoT-based emergency cardiac death risk rescue alert system. *MethodsX*, 13, 102834. DOI: 10.1016/j.mex.2024.102834 PMID: 39071997

Robinson, T., Condell, J., Ramsey, E., & Leavey, G. (2023). Self-Management of Subclinical Common Mental Health Disorders (Anxiety, Depression and Sleep Disorders) Using Wearable Devices. *International Journal of Environmental Research and Public Health*, 20(3), 2636. DOI: 10.3390/ijerph20032636 PMID: 36768002

Robinson, T., Condell, J., Ramsey, E., & Leavey, G. (2023). Self-Management of Subclinical Common Mental Health Disorders (Anxiety, Depression and Sleep Disorders) Using Wearable Devices. *International Journal of Environmental Research and Public Health*, 20(3), 2636. DOI: 10.3390/ijerph20032636 PMID: 36768002

Roos, L. G., & Slavich, G. M. (2023). Wearable technologies for health research: Opportunities, limitations, and practical and conceptual considerations. *Brain, Behavior, and Immunity*, 113, 444–452. DOI: 10.1016/j.bbi.2023.08.008 PMID: 37557962

Roos, L. G., & Slavich, G. M. (2023). Wearable technologies for health research: Opportunities, limitations, and practical and conceptual considerations. *Brain, Behavior, and Immunity*, 113, 444–452. DOI: 10.1016/j.bbi.2023.08.008 PMID: 37557962

Sadek, I., Codjo, J., Rehman, S. U., & Abdulrazak, B. (2022). Security and privacy in the Internet of Things healthcare systems: Toward a robust solution in real-life deployment. *Computer Methods and Programs in Biomedicine Update*, 2, 100071. DOI: 10.1016/j.cmpbup.2022.100071

Sadek, I., Rehman, S. U., Codjo, J., & Abdulrazak, B. (2019). Privacy and security of IoT based healthcare systems: concerns, solutions, and recommendations. In *How AI Impacts Urban Living and Public Health: 17th International Conference, ICOST 2019, New York City, NY, USA, October 14-16, 2019* [Springer International Publishing.]. *Proceedings*, 17, 3–17.

Sharma, A., Arora, S., Kumar, S., & Bansal, A. (2023, January). Ai ML Enabled Wearable Smart Sensors Detecting Psychological Disorders. In *2023 13th International Conference on Cloud Computing, Data Science & Engineering (Confluence)* (pp. 651-656). IEEE. DOI: 10.1109/Confluence56041.2023.10048832

Wang, Y. (2023). The research on the impact of distance learning on students' mental health. *Education and Information Technologies*, 28(10), 12527–12539. DOI: 10.1007/s10639-023-11693-w PMID: 37361750

Wiedermann, C. J., Barbieri, V., Plagg, B., Marino, P., Piccoliori, G., & Engl, A. (2023, May). Fortifying the foundations: A comprehensive approach to enhancing mental health support in educational policies amidst crises. [). MDPI.]. *Health Care*, 11(10), 1423. PMID: 37239709

Yasugaki, S., Okamura, H., Kaneko, A., & Hayashi, Y. (2023). Bidirectional relationship between sleep and depression. *Neuroscience Research*. Advance online publication. DOI: 10.1016/j.neures.2023.04.006 PMID: 37116584

Chapter 11
Measuring Vital Sign: A Case Study of India's Emerging Wearable Technology (EzeCheck) for Mass Screening and Health Monitoring

Zakir Hossen Shaikh
https://orcid.org/0000-0003-4733-4166
Kingdom University, Bahrain

Satya Ratnakaram
https://orcid.org/0000-0002-8908-302X
Bahrain Polytechnic, Bahrain

Mohammad Irfan
https://orcid.org/0000-0002-4956-1170
NSB Academy, India

Joji Abey
https://orcid.org/0000-0003-4176-3023
Kingdom University, Bahrain

Naji Mansour Nomran
https://orcid.org/0000-0003-3024-3057
Kingdom University, Bahrain

Manolo Logrono Anto
https://orcid.org/0009-0009-7682-814X
Kingdom University, Bahrain

ABSTRACT

The emergence of billions of new linked things and their pervasiveness in our daily lives is driving a constant constituent development in technology. Wearables are one of the numerous underlying technologies that may be used to give a truly tailored experience by capturing the rich contextual information supplied by such devices. In recent years, wearable technology has developed rapidly in various fields. The main aim of this paper is to shed light on wearable healthcare monitoring devices EZeCheck. its current application in health care as a quick, easy, and affordable way to screen hemoglobin levels without requiring a single blood sample from the patient, generated in 60 seconds, making it available to a large global population to improve health outcomes, to support the objective of "Anaemia Mukt Bharat

DOI: 10.4018/979-8-3693-7817-5.ch011

(Anaemia free India) and its socioeconomic impact. Further study is necessary to investigate the wearable device's potential uses for health and well-being including heart rate, blood pressure, oxygen saturation, temperature, and continuously measure and monitor vital signs.

1. INTRODUCTION

Technology is always evolving because of the emergence of billions of new, networked things and their pervasiveness in our everyday lives, One of the underlying flexible technologies, Wearables may record the rich contextual data generated by these devices and use it to provide a really customized experience.. The terms "wearables, wearable devices, or wearable technology refer to small electronic and mobile devices, or computers with wireless communications capability that are incorporated into gadgets, accessories, or clothes, which can be worn on the human body, or even invasive and Non-invasive versions such as micro-chips or smart tattoos". Smart wearables have gained considerable attention from Information Systems (IS) academics, business managers, and health practitioners.

In the 13th century the development of wearable technology began when English monk Roger Bacon, who lived in Paris, invented spectacles. In his Opus Majus (c.1266), Bacon described the scientific rationale for the use of corrective lenses. The idea of wearable technology was originally proposed in 1960 by "Edward O Thorp, a mathematics professor at the Massachusetts Institute of Technology in the United States (Amft, O., & Lukowicz, P. (2009)), has developed rapidly in various fields such as health care, education and culture, social networking, and the military, in recent years, with the development of the internet, intelligent hardware, and big data". Wearable devices in the form of portable medical or health electronic devices that can be directly worn on the body can be used to "perceive, record, analyze, regulate, and intervene to maintain health and can even be used to treat diseases with the support of various technologies for identification, sensing, connection, cloud services, and storage(M Haghi, K Thurow, R Stoll Healthcare informatics research, 2017)". Smart medical and wellness wearables have drawn academic researchers' and practitioners' attention as a motivational solution for enhancing individuals' health and well-being (Aliverti, 2017).

With applications ranging from monitoring chronic conditions like diabetes and heart failure to improving fitness tracking and health management, the usage of wearables in healthcare has grown dramatically. With a projected value of US$169 billion by 2029, the worldwide wearable technology in healthcare market was valued at US$42.6 billion in 2023. During the 2024–2029 forecast period, the "wearable technology in healthcare market" is expected to increase at a CAGR of 25.82%. (ResearchAndMarkets.com). The potential of wearable technology to enable continuous, real-time monitoring of vital signs and other health data is one of the main factors driving its acceptance in the healthcare industry. Wearables by continuously measuring and monitoring vital signs and other health factors in real-time with sophisticated sensors allow users to proactively manage their health and well-being such as Heart rate, blood pressure, oxygen saturation, temperature, and activity levels are some of these measurements. A practical and non-intrusive method of keeping an eye on these vital signals is through wearable technology. Vital sign measurement is typically the initial procedure in nearly all medical evaluations. These measures help doctors identify potential health issues and gain a better understanding of how a patient's body works. Anaemia is the medical term for a lower-than-normal hemoglobin level. This results in a reduced capacity for blood to provide organs with oxygen. In turn, the body needs to compensate for this by working harder to keep up with the same requirements such as increasing the breathing rate or heart rate.

This paper presents a comprehensive review of wearable healthcare monitoring systems developed by EzeRx was founded in the year 2018, on the principle of "MedTech for the Masses". India's Most Innovative and Disruptive Healthcare Startup aims in the direction of developing highly progressive medical devices and innovative solutions and making preventive healthcare accessible to people from all walks of life. To be responsible towards society and provide affordable screening solutions for the healthcare problems of the masses and make it accessible even in the country's most remote places to create primary awareness about illnesses and create a healthy society where money and means will not be a constraint for effective healthcare.

A new brand of multidisciplinary care that is evolving around mobile health devices and presents a vision of up-and-coming technology in this space known as **EzeCheck** is main focus of our study, which is a non-invasive, cloud storage, IoT-enabled device that provides a cost-effective, painless, and quick method to screen hemoglobin levels, making it accessible to vast populations worldwide. Operating the EzeCheck app is user-friendly and straightforward, requiring only basic reading and writing skills in English. Anyone familiar with using an Android smartphone can handle it with ease. Despite being stored on a cloud server, the data maintains a high level of security and confidentiality.

Non-invasive hemoglobin tests can be a valuable tool for mass screening in schools and communities, assisting in the identification of individuals affected by anaemia and other blood disorders and improving health outcomes, quality of life, along with socio-economic impact in terms of saving money, time, blood, medical waste, and carbon credit. This device operates seamlessly with the EzeCheck app. While doing hemoglobin tests via this cutting-edge device, doctors can input patient details into the app and start testing which can detect anemia instantly without a single drop of blood from the human body. Results are generated within 60 seconds and stored in EzeCheck's interactive dashboard, benefiting both doctors and patients. By integrating IoT capabilities, EzeCheck has transformed into a smart healthcare tool. The device connects to a secure network, such as Bluetooth, and automatically transmits readings to the EzeCheck dashboard, which is a designated cloud platform. These records are used for data analysis to identify trends, predict potential issues, and guide personalized treatment plans. With EzeCheck, healthcare providers can access patient data remotely, enabling continuous monitoring of hemoglobin levels and improving patient care. This data can then be instantly transmitted to a doctor anywhere in the world using IoT. The doctor can review the data and initiate a tailored treatment plan in a very short time, a process that previously took many days. The World Bank data indicates that one in two Indian women is anemic, which is 20% higher than the global average. Moreover, the device can follow up on a real-time basis and also can be used in remote areas to augment the mission of "Anaemia Mukt Bharat" which is a significant public health challenge in India.

Wearable technology is revolutionizing the healthcare industry by empowering individuals to take charge of their health and well-being. With the ability to track and monitor health metrics in real time, to improve health outcomes and provide preventative healthcare, wearable technology is essential. As wearable technology continues to evolve, the future of healthcare looks brighter than ever, with innovative solutions that have the potential to transform patient care and enhance quality of life. Wearable tech can contribute to **sustainability** by adopting eco-friendly materials and energy-efficient designs. As we look towards 2024, the future of wearable tech promises even more exciting advancements that will enhance our daily lives, improve our health, and offer unprecedented convenience.

To summarize, the study future ICT systems must include wearable technology as a fundamental component. EzeRx addresses the critical gap in health awareness and diagnostics. Many people, due to diverse socioeconomic backgrounds, lack the awareness and resources necessary to maintain a healthy lifestyle. Hemoglobin, a crucial health indicator, often goes unscreened, leading to undiagnosed anemia and related health issues. It is still in its infancy, and several critical challenges from data acquisition and processing, communications, security, privacy aspects, hardware limitations, and user adoption are still to be addressed. The device focuses on the use of hemoglobin screening. It can expand to other vital signs, such as "bio-signals to brain activities or body temperature, heart rate, blood oxygen, blood pressure, or brain functions in muscle movements". However, more research is needed to explore further applications of wearable devices in the medical field.

2. BACKGROUND OF THE STUDY

The term "healthcare AI" describes the application of AI techniques to assess medical data and enhance various aspects of healthcare. It incorporates a variety of technologies, including

Machine Learning (ML). The premise behind this branch of AI is that robots can learn from data, spot patterns, and make judgments with little to no assistance from people.

Natural Language Processing (NLP): NLP enables AI to examine medical literature, doctor-patient talks, and clinical notes to glean insights and automate procedures.

Deep Learning: In the medical field, deep learning is highly effective in interpreting intricate data, such as MRI and X-ray images, to detect anomalies and bolster diagnosis.

Artificial intelligence has unquestionably great potential in the healthcare industry, despite all of its obstacles. But it's crucial to keep in mind that AI cannot take the role of physicians or other medical professionals. These are necessary for medical personnel to effectively manage the emotions of their patients, come up with innovative solutions to difficult problems, and make ethically right and respectable choices.

2.1 Healthcare and AI: Effectiveness

With specialized use cases accessible for all major enterprises, artificial intelligence (AI) is a valuable ally in this era of digital acceleration. Conventional paradigms in a number of industries, including manufacturing, retail, education, healthcare, and finance, are being altered by AI's revolutionary effect. By leveraging machine learning (ML), predictive analytics, and advanced automation, businesses may employ artificial intelligence (AI) to enhance resource efficiency, get insights into consumer behavior and market trends, and streamline processes. Nearly every industry in the economy is seeing a rapid increase in the use of artificial intelligence (AI), including supply chain management, transportation, public safety, retail, HR services, financial services, financial inclusion, robotics, education, and governance. (Shaikh, et al, 2022, 2024, Sheela, P et al, 2024, Selwin et al, 2024, Gupta et al, 2024, Kusuma, K., et al,2024, Panigrahi, Rashmi Ranjan, et al,2023 and Sarea, A.et al, 2021).

Artificial reasoning will have a significant impact on the medical services sector and how AI is used by companies involved in the medical services sector during the next five to ten years. AI is being used by the healthcare industry to diagnose patients more quickly and accurately than humans. It may successfully support physicians in making diagnosis and notify them when a patient's health deteriorates, for example:

1) **Better accessibility:** Many developing nations still have subpar healthcare infrastructure as of 2024. The World Health Organization (WHO) attributes the current 18.1-year difference in life expectancy between the richest and poorest countries to inadequate or nonexistent access to healthcare. These nations can now have an effective healthcare ecosystem thanks to AI involvement, as AI-supported digital technologies make patient diagnosis and treatment easier.
2) **Speedy Data Sharing:** Manually maintaining medical records and data in a hospital is a hectic job. Artificial intelligence can be a great help here. Artificial intelligence in healthcare can locate particular patient data more effectively than conventional care, giving doctors more time to concentrate on medications and treatments. To fully realize the potential of AI and precision medicine, algorithms must be able to analyze large amounts of data quickly.
3) **Early diagnosis:** AI-powered medical technology can assess a person's history and present health issues by analyzing personal data. By comparing the characteristics of the diseases, medical professionals are better able to establish precise diagnosis. Millions of diagnoses and symptoms have been calculated in databases across a number of healthcare-related smartphone apps. More importantly, it can forecast the potential health issues a person might encounter. The Verily app from Google, for instance, is intended to forecast genetic illnesses that are inherited and non-contagious.
4) **Faster and cheaper:** AI has completely transformed healthcare by greatly reducing expenses and expediting patient evaluation and diagnosis. AI, for instance, may identify disease biomarkers more quickly than a human analyst, saving time and effort. Because of this automation, responses can be made faster, saving lives. AI algorithms are also economical since they forecast results using patient data, which minimizes the need for lab visits. As a result, the use of AI in healthcare has significantly increased, with an 88% increase in firms using the technology in the last year.
5) **Improved Care for Patients:** Artificial intelligence (AI) can enhance patient experiences in healthcare by facilitating speedy information navigation, report access, and confusion-free facility navigation. According to study, 83% of patients said that inadequate communication in medical settings bothers them the most. The availability of AI to patients around-the-clock is another important advantage. An example of an app that improves patient experiences is Babylon, which uses sophisticated pharmaceutical side effect checking.

6) **Assistance in surgery:** Significant advancements in robotics and machine learning for surgery have been made using artificial intelligence. The smallest movements may be carried out by dedicated AI surgical systems with 100% accuracy, increasing the effectiveness of difficult procedures and lowering dangers like blood loss and adverse effects. Antibacterial nanorobots that eradicate infections before surgery help patients recover more quickly. AI updates in real time about patient statuses calm patients and help surgeons, especially when general anesthesia is being used.
7) **Robots offering support:** In addition to medical personnel, robots increasingly help patients. Exoskeletons enable paralyzed people to move again on their own. More flexible limbs are possible with AI-driven prosthetics that have sensors. Service robots assist with everyday duties and offer company. Robotic companions monitor vital signs such as blood pressure and sugar levels, and they can even help with medicine administration. Certain robots are made to help people with depression by assessing their emotions and offering emotional assistance to enhance their quality of life.

Source: https://ezerx.in/ezecheck

3. AN INNOVATIVE, IMPACTFUL, AND ECOLOGICALLY RESPONSIBLE JOURNEY OF EZECHECK

Non-invasive technology is revolutionary in today's quickly changing healthcare environment. The first non-invasive hemoglobinometer validated by ICMR-RMRCBB in India is called EzeCheck. The device is in use in more than 19 states in India as well as more than 7 other nations, including Vietnam, South and East Africa, Indonesia, the Philippines, and Myanmar. This benefited more than 23 lakh individuals with significant impact on the environment, the healthcare sector, and society. (https://ezerx.in/ezecheck)

Figure 1. EzeCheck device specification

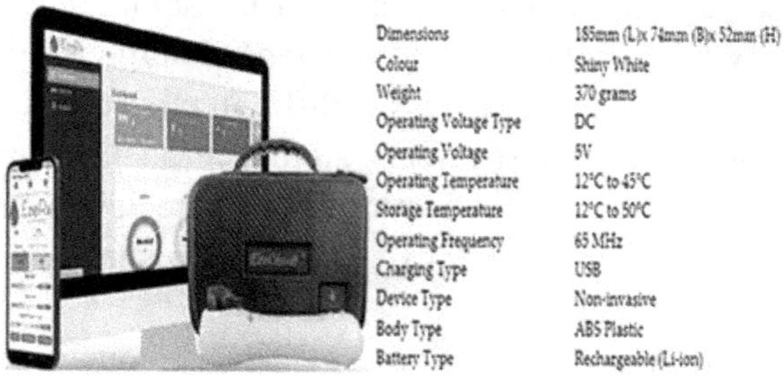

Source: https://www.ezerx.in/ezecheck

Figure 2. How EzeCheck works?

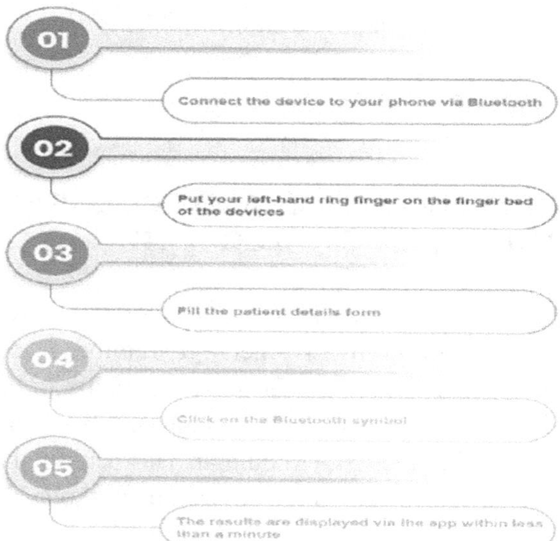

Source: https://www.ezerx.in/ezecheck

3.1 Measurement Process

The EzeCheck gadget uses absorption spectroscopy to function. The LED light in the finger bed area of the device starts to weaken from the reference spectrum to the dark spectrum when the patient places his or her finger on it. Each intensity then gathers spectral signals from the designated range of 300-750 nm (visible spectrum range). Within the region of 300-750 nm, an output is produced in the form of an absorbed signal at around ten different strengths. A specific portion of the signal (450-700 nm) is then extracted for further processing.

Based on the various mathematical permutation combinations of different wavelength points within the specified wavelength range of 530-631 nm, one suitable signal (specifically within the spectral range of 530-631 nm) is chosen from among the ten signals present at varying intensities. The amplitude points of the signals are used to create the Hb output using the ML algorithm (multi-dimensional linear regression) after determining the correct signal and the first and last signal points (at 450 and 700 nm, respectively). The output that was obtained is transmitted to the mobile application so that the person depicted as seen in figure 3 (https://ezerx .in/ezecheck).

Figure 3. Measurement process of the hemoglobin level in the sample

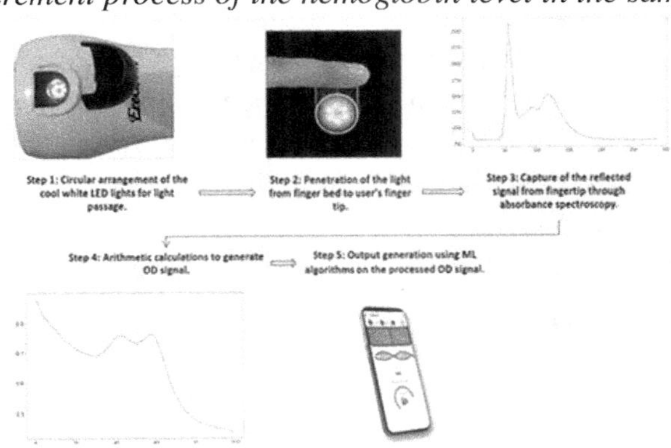

Source: https://www.ezerx.in/ezecheck

3.2 EzeCheck: Using the Internet of Things (IoT) to Revolutionize Anemia Management

A shortage of red blood cells, which carry the protein hemoglobin and are essential for carrying oxygen to the body's tissues, results in anemia. It is a worldwide public health concern because of things like malnutrition, especially in developing nations like India.

The Lancet Haematology estimates that anemia affects around 2 billion individuals globally. The National Family Health Survey, which was done in five rounds between 2019 and 2021, found that anemia rates in India were significantly higher in all age categories. Anemia now affects more Indians than ever before, according to the report, with at least 67% of youngsters suffering from the illness, up from 58.6% in the 2015–16 survey. However, the World Bank data shows that, in nations like India, where malnutrition is a problem, one in two Indian women are anemic, a rate that is 20% higher than the world norm. Early detection is essential to solving this problem. EzeCheck offers this dire scenario a glimmer of hope. This is the first blood-free hemoglobin screening gadget in India that has been verified by the ICMR-RMRC and provides a means to use the Internet of Things (IoT) for effective and easily accessible anemia management, which ensures a healthier and more connected society by revolutionizing the management of anemia and setting the standard for next medical innovations. (https://ezerx.in/ezecheck)

Maternal and Newborn Survival Initiative (MANSI): Benefiting pregnant women and infants in rural regions, the initiative was launched by the Government of India in partnership with the American India Foundation and Tata Group. In India, at least 1.5 million females under the age of eighteen are married each year, according to UNICEF. According to data, India accounts for around 16 million adolescent pregnancies globally, or 11% of all teenage pregnancies. This problem results in anemia and malnutrition in many women, which puts the unborn child and the expectant mother at danger. 52.2% of pregnant women and 57% of women overall experience anemia, according to the National Family Health Survey 5 (2019–21). (https://ezerx.in/ezecheck)

EzeRx is committed to making MANSI staff' life as easy as possible by providing them with all the support they need. MANSI personnel can quickly and easily determine the hemoglobin levels of anyone using this cutting-edge tool. This is especially important for youngsters, since they frequently have difficulties with pricking tests that are traditionally used. Early identification of anemia is made possible by its seamless screening of hemoglobin levels in mothers and children by MANSI personnel. This is a major turning point since it allows for early treatment of the disease in its early stages and promotes preventative healthcare for moms and their children. (https://ezerx.in/ezecheck)

4. RESULTS AND DISCUSSION

Any laboratory or medical test that aids in determining histopathology, biomarkers, diagnosis, and prognosis, typically necessitates the drawing of blood samples or the operation of sample processing machines by professional doctors or trained laboratory personnel. EzeCheck, on the other hand, is a simple and easy-to-use screening device that allows quick-determining Hb results if the person is familiar with using an Android mobile phone.

Many businesses are now able to identify non-invasiveness as a revolutionary criterion and are employing this technology to produce a wide range of products. Blood pressure, step count, and vital signs such as hemoglobin, bilirubin, creatinine, and blood sugar are all measured by these monitors. These are made specifically for everyone. Anyone in need can access and use the technology anywhere at any time thanks to its portability.

Other Hemoglobin Assessment Techniques Employed at Various Sites

Haemocue Portable Photometer Hb-301

A technique for measuring hemoglobin at the point of care is called the Haemocue. Globally, healthcare providers use this compact, lightweight, and user-friendly point-of-care (POC) device extensively. Either a venepuncture or a finger prick is used to collect a blood sample. It is made up of single-use photometers and disposable microcuvettes with reagent. "Sodium deoxycholate haemolyzes the red blood cells inside the microcuvettes, releasing hemoglobin". After that, sodium nitrite changes hemoglobin into methemoglobin, which when combined with sodium azide produces azide methemoglobin. To account for the turbidity in the sample, the absorbance is measured at two wavelengths (570 and 880 mm). The meter then computes and shows the hemoglobin level on the monitor. (Hudson-Thomas, M., Bingham, K. C., & Simmons, W. K. (1994)

Hematology Analyzer (Sysmex XN-1000)

Clinical laboratories can employ the automated multi-parameter blood cell counter known as the Hematology Analyzer for in-vitro diagnostic purposes. This compact, fully automated hematology analyzer is capable of evaluating many parameters at once in both whole blood and capillary modes. This device employs the sodium lauryl sulphate (SLS) cyanide-free hemoglobin detecting technique and the result displayed on the screen. (https://kupdf.net/download/sysmex-xn-series-xn-1000-instruction-for-use-may-2014-english_58fe3037dc0d605513959ebd_pdf)

Sahli's Method

A conventional, non-invasive technique for estimating hemoglobin is Sahli's approach. The main feature is there is a "comparator with a brown glass standard and Sahli's graded hemoglobin tube which is marked in percent and gram". Blood is combined with an acidic solution, which causes hemoglobin to change into the brown-colored acid hematin during the process. After that, the acid hematin is diluted with distilled water until its color resembles that of the brown glass standard. After that, the hemoglobin is estimated by taking the figure straight off the scale. (Patil, P. J., Thakare, G. V., & Patil, S. P. (2013).

One Drop Kit

The app is the One Drop system. The blood glucose, medication, food, and exercise journal is an app that can be downloaded for iOS, Android, and Apple Watch. It allows data sharing with a broader group and has a comprehensive food library as well. It serves as the portal for reminders and support. It monitors sugar levels and gives control input. Because the One Drop Chrome glucometer is Bluetooth-enabled, users may save spending endless hours manually entering blood sugar readings by having the device automatically send BG data to the app. The meter's use of the incredibly precise Aga Matrix test strips from Presto, Jazz, and BGStar is equally significant.

These days, this substitute is becoming a component of preventative healthcare and raising the bar for standards and expectations. There are numerous benefits that EzeCheck has over other devices and traditional techniques are presented in Table 1.

Table 1. Comparison with available hemoglobin estimation methods

Measuring parameters	**EzeCheck**	**Sahli's method**	**One Drop kit**	**Haematology analyzer**
Blood sample quantity	No blood	20-50uL	8-10 uL	200-500 uL
Separate container	Not required	Required	Not Required	Required
Consumables	Not required	Required	Required	Required
Biomedical waste	No	Yes	Yes	Yes
Lab technician	Not required	Required	Required	Required
Time for result generation	60 sec	20-30 min	60 sec	Three hours including sample transportation
Community-level screening	Yes	Time consuming	Yes	Not possible

continued on following page

Table 1. Continued

Measuring parameters	EzeCheck	Sahli's method	One Drop kit	Haematology analyzer
Parent consent in case of age<8 to take blood	Not required	Required	Required	Required
Portability	Yes	Not possible	Yes	Not possible
Immediate data transfer to the next level for medication/ follow-up	available	Not available	Not available	Not available
Operation by any non-technical person	Yes	Not possible	Not possible	Not possible
Any dedicated space or room	Not Required	Dedicated lab	Any where	Dedicated lab
Unit test cost (in INR).	<5	>8	>15	>50
Accuracy	>90%	>80%	>90%	>95%
Training for diagnosis	One hour only to any semiskilled manpower	>1 month to lab technician	>1 day to skilled manpower	>1 month to lab technician

Source: Sahoo et al, 2024

The main benefits of utilizing a non-invasive Hb screening method are its simplicity and ease of use, ability to be carried out in a non-medical environment, and ease of operation for any semi-skilled labor force. It is an easy and practical way to screen big groups of people. Non-invasive hemoglobin screening for anemia or other blood disorders in large populations is less expensive than traditional invasive methods, which can cost up to 5-7 Indian rupees (INR) per test (including the cost of the device, (EzeCheck). Traditional invasive methods, on the other hand, require expensive laboratory equipment, consumables, and trained medical personnel, and can be quite expensive. Furthermore, the latter produces no biomedical waste.

The results of Das Mahapatra et al.'s 2024 examination are shown in Figure 4, which shows how " EzeCheck and the Sysmex XN 1000 hemato-analyzer" concur. It was found that the data from both instruments did not differ statistically substantially, and that there was a strong correlation between the two.

Figure 4. "Scatter plots of EzeCheck with Hematolyzer (A), Hemocue (B) and Sahli's Method (C)"

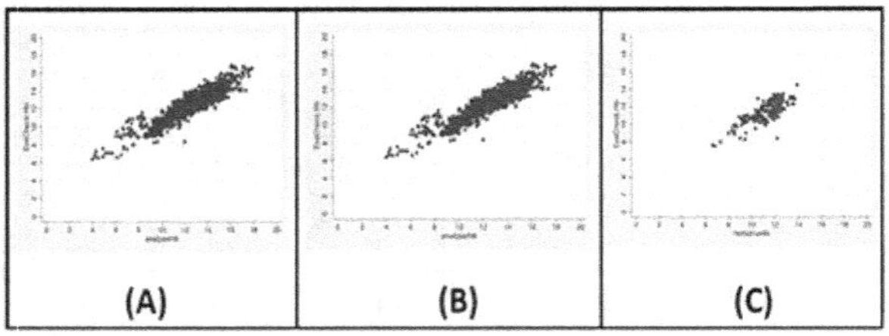

Source: Das Mahapatra et al., 2024doi: https://doi.org/10.1371/journal.pdig.0000500.g001

According to same research by Das Mahapatra et al. (2024), "there was a statistically significant constant and proportionate bias for EzeCheck in respect to Hemocue". Comparing EzeCheck to Sahli's approach revealed statistically significant constant bias but no proportional bias. Shown in figure 5.

Figure 5. Bland-Altman plots of EzeCheck with Hematolyzer (A), Hemocue (B) and Sahli's Method (C)

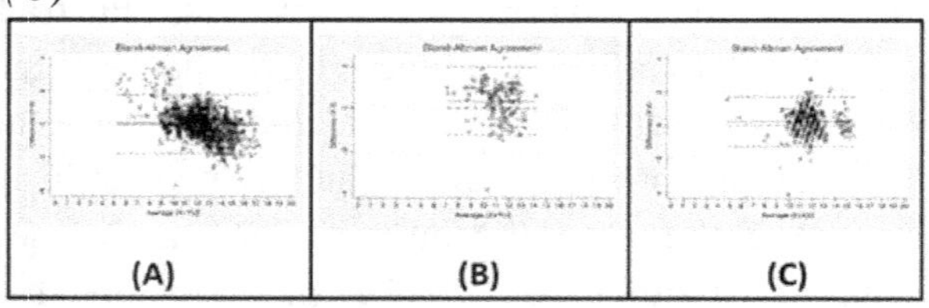

Source: Das Mahapatra et al., 2024doi: https://doi.org/10.1371/journal.pdig.0000500.g001

Das Mahapatra et al. (2024) evaluated three distinct Hgb assessment methods: the whole blood hemato-analyzer (Sysmex XN 1000) and the diagnostic performance, which was regarded as the "gold standard, the conventional Sahli's approach, and the portable invasive (finger-prick) hemoglobinometer HemoCue (Hb-301)". EzeCheck demonstrated compelling sensitivity, specificity, concordance, and accuracy across all age and gender groups are the main findings of their study.

EzeRx: Current and Prospective Outcomes

Leading developer of non-invasive diagnostic technologies, EzeRx, has announced a significant 167% increase in net revenue from FY 21–22 to FY 22–23, going from INR 2.56 crore to INR 6.85 crore. The company's remarkable growth was fueled by the widespread use of EzeCheck, its main product, in government offices, hospitals, schools, and non-governmental organizations. The business intends to use its upcoming revenues to propel innovation and further solidify its position in the market. Regulatory clearances will cost INR 1.62 crore, research and development will cost INR 3.7 crore, clinical trials will cost INR 1.71 crore, and new market growth will cost INR 3.32 crore presented in figure 6. https://www.ezerx.in/ezecheck.

Figure 6. EzeRx growth in year-on-year net revenue

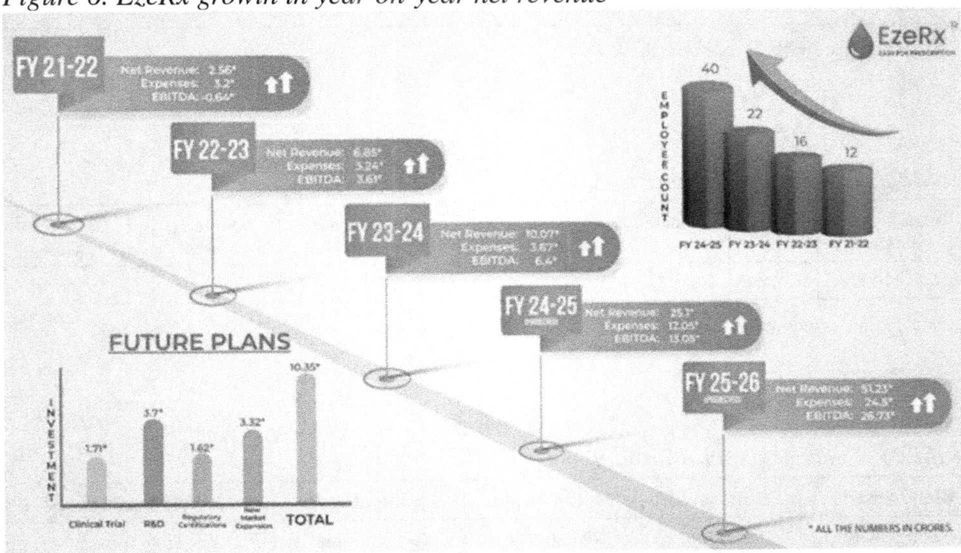

Source: https://biovoicenews.com/ezerx-registers-a-167-growth-in-year-on-year-net-revenue/

EzeCheck Anemia Detection Initiatives in Different States of India and Achievements

Recent findings using the EzeCheck device have uncovered an alarming truth - anemia is more prevalent than we thought, and it's affecting communities across various states, its achievements and social impact shown in Table 2, 3 and 4.

Table 2. EzeCheck state wise anemia detection

State	Screening Percentage %
Gujrat	60
Madhya Pradesh	57.5
Karnataka	58
Uttar Pradesh	66
Bihar	65
Assam	51
Meghalaya	61
Manipur	51
West Bengal	60
Jharkhand	62
Odisha	56

Source: https://www.ezerx.in/ezecheck

Table 3. Ezecheck testing achievements

Total Screening	Male	Female	Devices deployed	Organization covered	State covered	Global Presence
24,49,210	82,0486	16.28,724	5700+	275+	21+	7+ Countries

Source: https://www.ezerx.in/ezecheck

Table 4. Ezecheck social impact

Blood Saved	Time Saved	Money Saved	Carbon Credit	Medical Waste Saved
6,075.00L+	**1064+years**	**$97.50 M+**	**46+Kilotones**	**15+ Tons**

Source: https://www.ezerx.in/ezecheck

5. THE FUTURE OF HEMOGLOBIN TESTING AND ITS SOCIO-ECONOMIC IMPACT: INDIA'S EZECHECK

In India, the main point of interaction between the public and the healthcare system is the health sub-center, primary health center, or Ayushman Bharat Health and Wellness Center. The provision of high-quality services by these centers is critical to the success of national health programs. A tool for mass screening by frontline staff members like "auxiliary nurse midwives (ANM) and accredited social health activists (ASHA)" might be introduced at these centers to address the anemia problem.

The goal of the Ayushman Bharat "School Health Program" is to encourage wellness and illness prevention in public and government-funded educational institutions. It is advised that schoolchildren undergo routine anemia screenings devoid of blood samples. The government should regularly examine teenage girls for anemia as it has recognized women's empowerment as a critical development effort. Basic healthcare services including nutrition instruction, contraceptive counseling, and preschool activities are offered at anganwadi centers, which are rural childcare facilities. The tool may be used to continuously screen children in rural areas at the Anganwadi level.

In the past, blood draws and other intrusive procedures were necessary for hemoglobin testing. But the advent of non-invasive techniques, such as those invented by EzeRx, will make hemoglobin testing more widely available, more pleasant, and more accurate in the future. Methods for measuring hemoglobin levels without intrusive procedures are referred to as non-invasive hemoglobin testing. Advances in optical sensors have made it possible to quickly and easily acquire measurements. The field of hemoglobin testing has a bright future ahead of it, because of the unending advancements in technology.

Due to its ability to provide rapid findings, prompt diagnosis and action, increase awareness of anemia, facilitate preventative treatment, and gather essential health data, the gadget has already reached even the most remote regions of India, including tribal communities. Through these initiatives, not only does it treat short-term health issues but also makes a positive impact on long-term health, demonstrating a dedication to improving the well-being of India's different tribal groups irrespective of their geographical barriers, financial issues, inadequate healthcare infrastructure, lack of awareness and education. A significant social benefit is that it even lessens medical waste, cost and time saving (saves over 1.35 million USD and over 1064 years in terms of testing) shown in figure 7.

Figure 7. EzeCheck Social impact

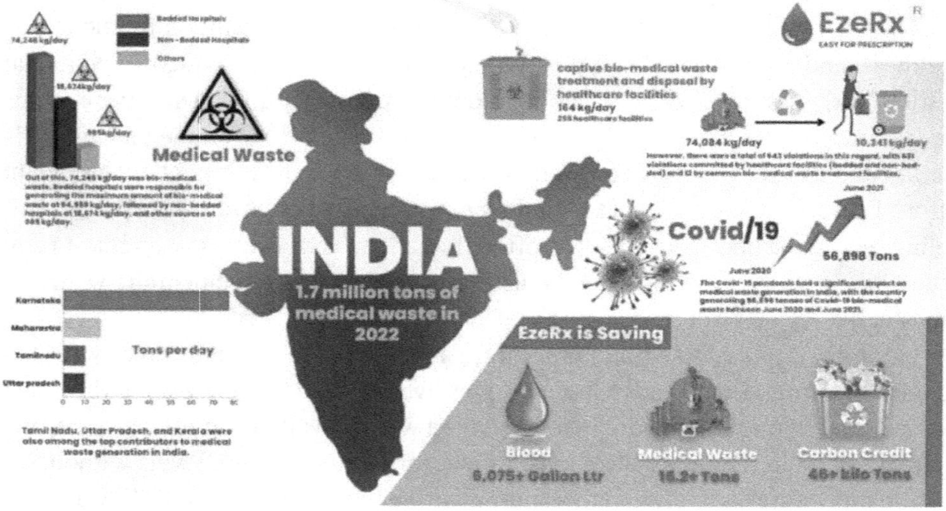

Source: https://www.ezerx.in/ezecheck

The 2012 UN **Sustainable Development Goals (SDGs)** include the prevention and prompt treatment of anemia. This prevalence appears to have changed little globally in the past 20 years and the SDG targets for anaemia reduction might still be beyond reach by 2030. (Daru, Jahnavi The Lancet Global Health, Volume 10, Issue 5, e586 - e587)

In order to minimize anemia in vulnerable age groups, particularly in women, children, and adolescents, the **"Anaemia Mukt Bharat"** (AMB) plan, also known as "Anaemia-Free India," was introduced in 2018.(Saini et al., 2022). It uses a life cycle approach that incorporates preventative and therapeutic processes. Community anaemia screening is important because it helps identify and treat anemia early on, especially in communities who are more susceptible. This can lead to better overall health outcomes and lessen the burden of this common ailment. Mass screening in schools and communities using non-invasive hemoglobin testing will assist discover anaemic individuals for early diagnosis and bring patients for prompt treatment, which can be employed in distant places to support 'Anaemia Mukt Bharat (K, Sinha A, Sahoo R, et al., January 24, 2024).

The non-invasive devices, such as Masimo Pronto and the innovative smartphone apps, were seen as very user-friendly and acceptable. This cutting-edge technique might revolutionize anemia detection in regions with limited resources. (Young MF, Raines K, Jameel F, et al,2021)

6. FINDINGS AND CONCLUSION

EzeRx addresses the important gap in health awareness and diagnosis. Many people don't have the knowledge or resources needed to sustain a healthy lifestyle because of their varied socioeconomic circumstances. By providing thorough screening services for essential health components, such as hemoglobin, this study attempts to fill the gap. A portable, non-invasive gadget that doesn't require a single drop of human blood to measure blood sugar, anemia, or forecast kidney, liver, or lung problems in real time. However, this study's disadvantage is that the wearable technology was limited to anemia. The future perspective is to measure bilurvin,creatine,blood sugar and oxygen saturation for monitoring of health issues.

In communities and schools, non-invasive Hb testing device EzeCheck may prove to be a useful tool for widespread screening. Non-invasive hemoglobin testing can help identify patients with anemia and other blood abnormalities so that treatment and intervention can begin as soon as possible. Additionally, the identification of populations that are more susceptible to anemia, such as those from low-income households or those with limited access to healthcare, can be aided by the use of this device in mass screening programs. To address these discrepancies, specific treatments and programs may be developed using this data. All things considered, non-invasive hemoglobin testing can be a useful instrument for widespread screening in communities and schools, helping people with anemia and other blood diseases, as well as enhancing their quality of life and health consequences. Furthermore, the devices may be utilized in remote locations to support "Anaemia Mukt Bharat" and can track progress in real time.

Early diagnosis is important since medical treatments are easily accessible and reasonably priced. EzeCheck has enormous potential to combat anemia. enhances patient care and makes managing anemia easier, especially in rural regions. This can prevent over a million deaths annually by enabling faster discovery, better treatment strategies, and earlier detection.

REFERENCES

Aliverti, A. (2017). Wearable technology: Role in respiratory health and disease. *Breathe (Sheffield, England)*, 13(2), e27–e36. DOI: 10.1183/20734735.008417 PMID: 28966692

Amano, I., & Murakami, A. (2013). Use of non-invasive total hemoglobin measurement as a screening tool for anemia in children. *Pediatrics International*, 55(6), 803–805. DOI: 10.1111/ped.12236 PMID: 24330295

Amft, O., & Lukowicz, P. (2009). From backpacks to smartphones: Past, present, and future of wearable computers. *IEEE Pervasive Computing*, 8(3), 8–13. DOI: 10.1109/MPRV.2009.44

Amrutha, A.M., & Sidenur, B., P.S, B. *et al.* (. (2024). Estimation of haemoglobin using non-invasive portable device with spectroscopic signal application. *Scientific Reports*, 14, 8697. DOI: 10.1038/s41598-024-58990-z PMID: 38622231

BrundhaDrM. (2019, July 1). Priyadharshini S. Comparisonof haemoglobin estimationby Sahli's two—time average, Sahli's three—time average methods and automatedanalyzermethod:A different approach in clinical pathology. *Int J Clin Diagn Pathol.*, 2(2), 291–295. DOI: 10.33545/pathol.2019.v2.i2e.118

Chamola, P., & Mittal, P. (2023). A non-invasive optical method for anaemia detection. *Physica Scripta*, 98(6), 065023. DOI: 10.1088/1402-4896/acd5b8

Daru, J. (2022, May). Sustainable Development Goals for anaemia: 20 years later, where are we now? *The Lancet. Global Health*, 10(5), e586–e587. DOI: 10.1016/S2214-109X(22)00127-9 PMID: 35427502

Das Mahapatra, P. P., Roy, C., Agarwal, K., Banerjee, J., & Sharma, S. (2024). Performance of the non-invasive point-of-care device, EzeCheck, for haemoglobin assessment in adults and children in community and institutional care settings. *PLOS Digital Health*, 3(5), e0000500. DOI: 10.1371/journal.pdig.0000500 PMID: 38717987

Das Mahapatra, P. P., Roy, C., Agarwal, K., Sharma, S. D., Chowdhury, S. R., Sharma, S., & Meher, D. (2024). Non-invasive hemoglobin screening device: A promising digital method for reducing anemia prevalence through routine screening and timely intervention. *Hematology*, 29(1), 2365078.

Dimauro, G., De Ruvo, S., Di Terlizzi, F., Ruggieri, A., Volpe, V., Colizzi, L., & Girardi, F. (2020). Estimate of anemia with new non-invasive systems—A moment of reflection. *Electronics (Basel)*, 9(5), 780. DOI: 10.3390/electronics9050780

Gupta, S., Srivastava, R., Shaikh, Z. H., & Irfan, M. (2024). Revolutionizing Rural Finance: Exploring the Impact of FinTech on Financial Inclusion in India. In Irfan, M., Muhammad, K., Naifar, N., & Khan, M. A. (Eds.), *Applications of Block Chain technology and Artificial Intelligence. Financial Mathematics and Fintech*. Springer., DOI: 10.1007/978-3-031-47324-1_3

Haghi, M., Thurow, K., & Stoll, R. (2017, January). Wearable devices in medical internet of things: Scientific research and commercially available devices. *Healthcare Informatics Research*, 23(1), 4–15. https://www.e-hir.org/DOIx.php?id=10.4258/hir.2017.23.1.4. DOI: 10.4258/hir.2017.23.1.4 PMID: 28261526

HemoCueHb301Operating-Manual_US[Internet]. [cited2023Jan 14].Available from: https://www. hemocue.us/wp-content/uploads/2020/08/HB-301_Operating-Manual_US.pdf_https://biovoicenews.com/ezerx-registers-a-167-growth-in-year-on-year-net-revenue/

Hudson-Thomas, M., Bingham, K. C., & Simmons, W. K. (1994). An evaluation of the HemoCue for measuring haemoglobin in field studies in Jamaica. *Bulletin of the World Health Organization*, 72(3), 423.

Sahoo, K. C., Sinha, A., Sahoo, R. K., Suman, S. S., Bhattacharya, D., & Pati, S. (2024). Diagnostic validation and feasibility of a non-invasive haemoglobin screening device (EzeCheck) for'Anaemia Mukt Bharat'in India. *Cureus*, 16(1).

Kusuma, K., Sheela, P., Panigrahi, R. R., & Shaikh, Z. H. (2024). Blockchain Technology and Islamic Finance: A Decade of Growth Trends. In Irfan, M., Kadry, S., Sharif, M., & Khan, H. (Eds.), *Fintech Applications in Islamic Finance: AI, Machine Learning, and Blockchain Techniques* (pp. 59–79). IGI Global., DOI: 10.4018/979-8-3693-1038-0.ch004

Mahapatra, P.P.D., Roy, C., Chowdhury, S.R., Sahoo, P. & Mahapatra, P.D. (2023). A Non-Invasive Approach for Anemia Eradication Through Community Based Mass Screening of Rural Population. International Journal of Social Impact, 8(3), 037-046. DIP: 18.02.005/20230803, DOI: DOI: 10.25215/2455/0803005

Ometov, A., Shubina, V., Klus, L., Skibińska, J., Saafi, S., Pascacio, P., & Lohan, E. S. (2021). A survey on wearable technology: History, state-of-the-art and current challenges. *Computer Networks*, 193, 108074.

Panigrahi, R. R., Dash, M., Shaikh, Z. H., & Irfan, M. (2023). Review of Machine Learning Techniques in the Supply Chain Management of Indian Industry: A Future Research Agenda. In Irfan, M., Elhoseny, M., Kassim, S., & Metawa, N. (Eds.), *Advanced Machine Learning Algorithms for Complex Financial Applications* (pp. 199–219). IGI Global., DOI: 10.4018/978-1-6684-4483-2.ch013

Patil, D. P. (2013, January 1). Variability And Accuracy Of Sahli's Method In Estimation Of Haemoglobin Concentration. *National Journal of Integrated Research in Medicine*, 4, 38.

Ramaswamy, G., Vohra, K., Yadav, K., Kaur, R., Rai, T., Jaiswal, A., & Kant, S. (2021). Point-of-care testing using invasive and non-invasive hemoglobinometers: Reliable and valid method for estimation of hemoglobin among children 6-59 months. *Journal of Tropical Pediatrics*, 67(1), fmaa111. Advance online publication. DOI: 10.1093/tropej/fmaa111 PMID: 33367788

Sahoo, K., Sinha, A., Sahoo, R., Suman, S. S., Bhattacharya, D., & Pati, S. (2024, January 24). Diagnostic Validation and Feasibility of a Non-invasive Haemoglobin Screening Device (EzeCheck) for 'Anaemia Mukt Bharat' in India. *Cureus*, 16(1), e52877. DOI: 10.7759/cureus.52877 PMID: 38406104

Saini, A., Shukla, R., Joe, W., & Kapur, A. (2022). Improving nutrition budgeting in health sector plans: Evidence from India's anaemia control strategy. *Maternal and Child Nutrition*, 18(2), e13253. DOI: 10.1111/mcn.13253 PMID: 35231160

Sarea, A., Subramanian, S., Alareeni, B., Shaikh, Z. H., Hawaldar, I. T., & Elshaker, A. H. (2021). Financial Disclosures by Using Machine Learning Analysis. In Alareeni, B., Hamdan, A., & Elgedawy, I. (Eds.), *The Importance of New Technologies and Entrepreneurship in Business Development: In The Context of Economic Diversity in Developing Countries. ICBT 2020. Lecture Notes in Networks and Systems* (Vol. 194). Springer., DOI: 10.1007/978-3-030-69221-6_5

Selwin, S., Saseendran, H., Vemavarapu, K. R., Shaikh, I., Shaikh, Z. H., & Harris, S. B. (2024). AI-Driven Employee Engagement: Transforming Banking Workforce Dynamics Using the BEULA Model. In Awwad, B. (Ed.), *The AI Revolution: Driving Business Innovation and Research. Studies in Systems, Decision and Control* (Vol. 525). Springer., DOI: 10.1007/978-3-031-54383-8_32

Shaikh, Z., Irfan, M., Sarea, A., & Panigrahi, R. R. (2024). *The Emergence of Islamic Fintech and Bahrain: Prospect for Global Financial Sectors* (Vol. 503). Springer., DOI: 10.1007/978-3-031-43490-7_52

Shaikh, Z. H., Irfan, M., Nomran, N. M., & Lawin, M. L. (2024). The Effect of Strategic Partnership on Innovation and Business: Performance of the Fintech Industry in Bahrain. In Irfan, M., Kadry, S., Sharif, M., & Khan, H. (Eds.), *Fintech Applications in Islamic Finance: AI, Machine Learning, and Blockchain Techniques* (pp. 236–254). IGI Global., DOI: 10.4018/979-8-3693-1038-0.ch016

Shaikh, Z. H., Irfan, M., Nomran, N. M., & Ratnakaram, S. P. (2024). Unlocking the Potential of AI for Efficient Governance: Innovative Approaches of Bahrain. In Irfan, M., Elmogy, M., Gupta, S., Khalifa, F., & Dias, R. (Eds.), *AI-Driven Decentralized Finance and the Future of Finance* (pp. 115–142). IGI Global., DOI: 10.4018/979-8-3693-6321-8.ch005

Shaikh, Z. H., Sarea, A., & Irfan, M. (2022). *Islamic Banking Strategies in the World of Fintech: Success Story of Bahrain* (Vol. 423). Springer., DOI: 10.1007/978-3-030-93464-4_10

Sheela, P., Kusuma, K., Panigrahi, R. R., & Shaikh, Z. H. (2024). Fintech and Islamic Banking: A Systematic View and Future Research Agenda. In Irfan, M., Kadry, S., Sharif, M., & Khan, H. (Eds.), *Fintech Applications in Islamic Finance: AI, Machine Learning, and Blockchain Techniques* (pp. 1–23). IGI Global., DOI: 10.4018/979-8-3693-1038-0.ch001

Stevens, G. A., Paciorek, C. J., Flores-Urrutia, M. C., Borghi, E., Namaste, S., Wirth, J. P., & Rogers, L. M. (2022). National, regional, and global estimates of anaemia by severity in women and children for 2000–19: A pooled analysis of population-representative data. *The Lancet. Global Health*, 10(5), e627–e639.

Young, M. F., Raines, K., Jameel, F., Sidi, M., Oliveira-Streiff, S., Nwajei, P., McGlamry, K., Ou, J., Oladele, A., & Suchdev, P. S. (2021). Non-invasive hemoglobin measurement devices require refinement to match diagnostic performance with their high level of usability and acceptability. *PLoS One*, 16(7), e0254629. DOI: 10.1371/journal.pone.0254629 PMID: 34270591

Chapter 12
Wearable Technologies In Education:
Advantages, Limitations, and Technical Difficulties

Bhabani Shankar Mohanty
https://orcid.org/0000-0001-9130-6221
Central University of Tamil Nadu, India

Zakir Hossen Shaikh
https://orcid.org/0000-0003-4733-4166
Kingdom University, Bahrain

Durga Madhab Mahapatra
https://orcid.org/0000-0003-1272-7285
North-Eastern Hill University, India

Satya Ratnakaranm
Bahrain Polytechnic, Bahrain

Devi Manikeswari
https://orcid.org/0000-0002-5014-9423
Bahrain Polytechnic, Bahrain

ABSTRACT

Education is a cornerstone of societal growth, and technology has greatly enhanced this sector. Wearable technologies offer personalized learning experiences and boost student engagement, yet their implementation faces challenges. High costs, integration with current systems, and data privacy concerns are major barriers. Additionally, issues like battery life, device comfort, connectivity problems, and lack

DOI: 10.4018/979-8-3693-7817-5.ch012

of technical support hinder widespread adoption. Teacher training and equitable access are crucial for effective use. While wearables have immense potential, addressing these limitations is essential to ensure they enhance, rather than detract from, educational outcomes. This chapter examines the benefits and challenges of wearables in education.

1. INTRODUCTION

Any electronic gadget made to be worn on the user's body is considered wearable technology. These gadgets can be jewelry, accessories, medical equipment, or parts of clothes, among other things. Smartwatches, activity trackers like the Fitbit Charge, virtual reality headsets, creative jewelry, web-enabled spectacles, and Bluetooth headsets are all examples of modern wearable technology that falls within a wide range of usage. Wearables function differently depending on their health, fitness, or entertainment category. Wearable technology primarily uses microprocessors, batteries, and internet connectivity to synchronize the gathered data with other electronics, such as computers or mobile devices. Sensors that monitor body motions, offer biometric identification, or help with location monitoring are integrated into wearable technology. The education sector has adopted wearable technology to provide a more integrated and thorough learning environment. Smart gadgets like watches, fitness trackers, smart glasses, and even virtual reality headsets worn on the body are all considered wearable technology. These gadgets are perfect for improving learning experiences since they can track, communicate, and give real-time feedback.

(Al-Emran et al., 2022) developed machine learning models to identify key factors influencing students' adoption of wearable technologies in education. (Al-Emran et al., 2020) explained that students' beliefs and security concerns significantly affect the adoption of smartwatches in higher education. Teacher attitudes toward wearable technology are shaped by various determinants, impacting innovative use in education (Almusawi and Durugbo, 2024). A systematic review reveals the growing role of wearable technology in enhancing educational experiences (Almusawi et al., 2021). Wearable electronic devices can potentially advance nursing education (Araújo et al., 2024). Wearable computers offer diverse human-factor benefits but pose usability challenges (Baber, 2001). Wearable video technology enhances surgical education by creating a point-of-view library (Beard et al., 2016). Wearable technologies offer unique educational affordances, such as immersive and personalized learning (Bower and Sturman, 2015). Pedagogical approaches influence girls' physical activity levels in physical education (Breed et al., 2024). Higher education faculty members' intentions to adopt emerging technologies are predicted by various factors

(Campbell and Frawley, 2023). Multitask learning models improve automated sleep staging and wearable technology integration (Chih et al., 2023). A meta-analysis highlights key factors influencing consumers' adoption of wearable technology (Chiu et al., 2021). Wearable technology facilitates global surgical teleproctoring, improving educational outcomes (Datta et al., 2015). Portable brain technologies offer new opportunities for educational neuroscience research (Davidesco et al., 2021). Wearable sleep technology is increasingly used in clinical and research settings (De Zambotti et al., 2019). Wearable technologies in healthcare research show significant advancements and future potential (De-La-Fuente-Robles et al., 2022). Wearable devices effectively study physical activity among early childhood education students (Díaz-Quesada et al., 2021). Simulation-based physical education using wearable VR equipment enhances learning experiences (Dong and A, 2020). Motion sensors aid physical education training through intelligent models (Du and Nan, 2021). Wearable technology interventions support young adults with intellectual disabilities in postsecondary education (Evmenova et al., 2018). A systematic review suggests significant consumer interaction with wearable technology, setting a research agenda (Ferreira et al., 2021). Wearable sensors are effective in educational settings for data-driven teaching and learning (Ferrier et al., 2022). Wearable virtual reality enhances educational visits to world heritage buildings (Fonseca et al., 2017). Innovative glass technology revolutionizes simulation-based medical education (Frederick and Van Gelderen, 2021). Integrating smart wearable devices into public physical education supports health monitoring (Fu et al., 2021). A systematic survey shows that wearable computing's educational applications rapidly expand (Gao et al., 2022). Privacy concerns are critical when using wearable data in higher education (Hagadone-Bedir et al., 2023). Self-efficacy and cognitive development impact the use of wearable technology in virtual language learning (Hsu et al., 2021). Wearable technology facilitates continuous health monitoring for athletes (Huifeng et al., 2020). Wearable technology has diverse medical applications, improving healthcare delivery and patient outcomes (Iqbal et al., 2016). Older adults' technology anxiety influences their intention to use smart health wearable devices (Jeng et al., 2022). Wearable sensors enhance learning in higher education through real-time feedback and monitoring (Khosravi et al., 2022). Nursing students find wearable display technology beneficial for learning and engagement (Kim et al., 2023). 3D printing education improves fashion design coursework through wearable technologies (Kwon et al., 2017). Sustainable wearables enhance the quality of life by integrating environmental considerations (Lee et al., 2016). Exergames and wearable technology improve engagement in physical education (Lindberg et al., 2016). Augmented reality and wearable devices transform real-time surgical education (Lindeque et al., 2014). A three-party authentication scheme enhances security in wearable device environments (Liu et al., 2016). Micro or nanofiber technologies are essential for developing advanced wearable sensors (Ma et al., 2022). Mobile and wearable technologies are vital for healthcare in the aging population (Malwade et al., 2018). Wearable devices play a growing role in the mining industry for worker safety and monitoring (Mardonova and Choi, 2018). Wireless wearable technology supports individuals with intellectual disabilities in workplace settings (Marelle et al., 2022). Wearable technology in PE offers advantages but faces ideological barriers from teachers (Marttinen et al., 2020). Mobile technologies in medical education enhance learning and offer new pedagogical opportunities (Masters et al., 2016). Wearable sensor technology aids in developing computer-aided education for students with special needs (Mehmood and Lee, 2017). Recent advancements in reconfigurable antennas for wearable technology expand healthcare and education possibilities (Musa et al., 2022). Wearable devices address public health ethics concerns during and after COVID-19 (Nakazawa et al., 2021). Wearable computing platforms enhance computing education with hands-on learning experiences (Ngai et al., 2010). Wearable technology supports inclusive education through innovative learning aids (Nogueira et al., 2020). Wearable textiles foster STEM learning and improve student engagement (Nugent et al., 2019). Various factors influence wearable technology usage for healthcare among Indian adults (Patil et al., 2022). XR technology enhances medical education with new visualization methods (Pennefather and Krebs, 2019). User coping strategies influence technology-enabled assessment satisfaction in higher education (Pillai et al., 2021). Wearable devices are critical in heart monitoring and physical health (Prieto-Avalos et al., 2022). Wearable technology offers great potential but challenges healthcare and medical education (Sultan, 2015). Wearable sweat management technologies represent the future of intelligent wearable materials (Wei et al., 2024). High-performance textile supercapacitors enable future wearable energy storage devices (Wen et al., 2021). Machine learning models improve workout detection using wearable device data (Yoshida and Yuda, 2023)

Wearable technologies in education reap from cost management regarding partners and grant sources. These advantages are that it fits well with integration with current systems and improves data security with new protection measures. As we have seen here, battery consumption and device comfort can also be prioritized in an efficient design. Prioritizing both improving the physical connections, providing a specialized tech help line, and working together on content producers that align with curriculum needs are all vital at this scale. The tool is used successfully since there is practical teacher training, and every student has equal chances. Combining wearables with ordinary forms and methods increases the effectiveness of learning outcomes and provides an opportunity to develop critical thinking and comprehensive learning materials. This chapter examines the potential of wearable technologies in education, highlighting their advantages and applications. It discusses the challenges educators and institutions face, such as cost, integration with infrastructure, privacy concerns, and technical limitations. Wearable technologies, including fitness trackers, smartwatches, VR headsets, and AR glasses, are increasingly used in healthcare, sports, and entertainment for data collection and personalized experiences. They create interactive learning environments and enable real-time feedback, enhancing teaching and learning.

1.1 Objective of Study

This study focuses on the advantages of using wearable technologies in the education industry. It has a significant role in promoting education. In education, wearables provide new opportunities to make learning more engaging, personalized, and interactive. However, there are several technological challenges to the sustainability of wearable technology. This study discusses both limitations and technological difficulties in using wearable devices.

2. IMPORTANCE OF WEARABLE TECHNOLOGY IN EDUCATION

AI has unique use cases available to all big enterprises, making it a valuable ally in this era of digital acceleration. Artificial intelligence (AI) is transforming traditional paradigms in every sector of the economy, including supply chain management, robotics, healthcare, public safety, retail, HR services, financial services, and financial inclusion and governance.(Shaikh, et al, 2022, 2024, Sheela, P et al, 2024, Selwin et al, 2024, Gupta et al, 2024, Kusuma, K., et al,2024, Panigrahi, Rashmi Ranjan, et al,2023 and Sarea, A.et al, 2021).

The integration of wearable technologies in education can transform traditional learning environments. Wearables offer numerous advantages by enhancing student engagement, improving the ability to monitor and support individual learning progress, and fostering collaboration in new, exciting ways. These devices allow for real-time data collection, helping educators to understand student behavior and learning patterns more precisely. Through adaptive learning technologies, wearables can provide personalized learning experiences that adjust content delivery based on a student's performance and engagement level. For instance, students with learning difficulties can receive tailored feedback and additional resources through these devices, while gifted students can be challenged with more advanced materials.

2.1 Overview of Wearable Technologies in Education

Wearable technologies are gaining popularity in healthcare, fitness, and entertainment, particularly education. These devices offer new opportunities for engaging students, personalizing learning, and creating immersive experiences. This section provides an overview of these devices and their use in the classroom.

2.2 Definition and Types of Wearable Technologies

Wearable technologies are electronic devices worn on the body or integrated into clothing and accessories. Their sensors and computing capabilities allow them to collect data, monitor physiological metrics, and provide real-time interactive experiences.

In education, wearable technologies can serve multiple purposes, from tracking student engagement to enhancing learning experiences. These devices often monitor students' physical activity, mental well-being, and cognitive engagement during lessons. Wearables can also facilitate collaborative learning and offer new ways to access and interact with educational content.

Table 1. Description of wearable technologies used in education industries

Category	Description	Applications in Education
Smartwatches and Fitness Trackers	Wearable devices monitor physical activities such as step counts, heart rate, and sleep patterns.	Tracking student health, monitoring activity levels in physical education, and gauging focus during lessons.
Virtual Reality (VR) Headsets	Head-mounted devices create immersive, simulated environments.	Simulating real-world experiences like visiting historical sites, exploring scientific phenomena, or practicing technical skills in virtual spaces.
Augmented Reality (AR) Glasses	Glasses or headsets overlaying digital information onto the real world.	Interactive learning experiences such as visualizing anatomy in biology or displaying historical facts during history lessons.
Wearable Biosensors	Devices worn on the skin or clothing to track physiological data (e.g., stress levels, cognitive load, emotional responses).	Understanding student responses to content or instruction; aiding in educational research.
Head-Mounted Displays (HMDs)	AR/VR headsets like Google Glass provide real-time visual data.	Supporting language translation, presenting guided lessons, and assisting fieldwork by overlaying instructions onto real-world tasks.

2.3 Key Features of Educational Wearables

Several key features make wearable technologies precious in the educational sector:

1. **Real-Time Data Collection**: One of the most significant advantages of wearables is their real-time data collection. Whether monitoring a student's physical activity, tracking cognitive engagement during lessons, or measuring emotional responses, wearables offer educators a continuous stream of data to inform decisions. This real-time feedback can help teachers identify when students are struggling, allowing immediate intervention.
2. **Personalized Learning**: Wearables, such as fitness trackers and biosensors, can enhance learning experiences by detecting fatigue, stress, or disengagement and utilizing VR and AR devices to adapt content based on student performance, ensuring personalized learning.
3. **Interactivity and Engagement**: Wearables enhance learning by enhancing interaction. VR and AR devices enable students to engage with content dynamically, such as historical simulations, 3D models, and virtual conversations, fostering a deeper connection with the material.
4. **Collaboration**: Some wearables, particularly AR glasses, facilitate collaborative learning by enabling students to share experiences and interact in a shared digital environment. For example, AR technology can project a 3D model onto a table, allowing multiple students to explore it from different angles simultaneously.
5. **Health and Well-Being Monitoring**: Wearables like fitness trackers and biosensors can monitor school student health and well-being, particularly during physical education or extracurricular activities. They track steps, heart rates, sleep patterns, and stress levels.
6. **Mobility and Flexibility**: Wearables, unlike traditional classroom tools, are portable and versatile, allowing students to access real-time information on historical landmarks or monitor performance during sports activities, unlike conventional tools.

3. ADVANTAGES OF WEARABLE TECHNOLOGIES IN EDUCATION

Wearable technologies are revolutionizing education by enhancing student engagement, collecting real-time data, personalizing learning experiences, and fostering collaboration. These devices create immersive learning environments, making lessons more interactive and responsive.

Table 2. Comparison of advantages of different wearable devices

Aspect	Enhanced Engagement & Interaction	Real-Time Data Collection & Monitoring	Personalized Learning Experiences	Collaborative Learning Opportunities
Key Features	Interactive and immersive experiences using VR/AR, gamified learning with smartwatches, and enhanced field trips using AR glasses.	Sensors monitor physical and cognitive metrics like heart rate, stress, and engagement, providing actionable insights.	Tailored learning pathways using performance data, real-time adaptive feedback, and support for students with special needs.	Enables teamwork via AR/VR environments and collaborative activities like shared digital workspaces or group challenges.
Technology Examples	VR headsets for 3D environments, AR glasses for real-time overlays, and smartwatches for notifications and challenges.	Fitness trackers, smartwatches, and biosensors to monitor health, engagement, and stress levels.	VR/AR headsets for customized simulations, smartwatches for progress alerts, and biosensors for SEN monitoring.	AR glasses for group tasks, VR for virtual classrooms, and fitness trackers for collaborative physical activities.
Educational Applications	Immersive history lessons (e.g., ancient civilizations); biology models (e.g., 3D cells); fitness gamification in physical education.	Assess physical fitness (e.g., heart rate, calorie tracking); detect boredom or stress; enable responsive teaching methods.	Visual math models; personalized content recommendations; tailored support for ADHD or anxiety management.	Virtual group experiments; cultural exchanges via global classrooms; teamwork in gamified activities like step-count challenges.
Advantages	Promotes active, memorable learning; increases curiosity and creativity; engages students through gamification.	Provides immediate feedback; improves student health monitoring; supports dynamic lesson adjustments.	Enhances focus with adaptive prompts; ensures equitable learning opportunities; improves outcomes for diverse learners.	Strengthens collaboration, problem-solving, and communication skills; supports innovation in group-based learning tasks.

4. TECHNOLOGICAL CHALLENGES IN WEARABLE TECHNOLOGY ADOPTION

Wearable technology in education offers immersive, personalized learning experiences but faces challenges such as battery life, connectivity issues, compatibility issues, data storage and processing issues, and software and firmware updates, impacting effectiveness, sustainability, and scalability.

Table 3. Issues and challenges of wearable devices

Aspect	Challenges	Solutions
Battery Life and Power Consumption	-Frequent recharging interrupts lessons. -Advanced functionality (e.g., 3D visuals) shortens battery life. - Logistical burden of managing multiple devices.	- Optimize software and hardware for energy efficiency. - Develop new battery technologies. - Implement wireless charging or battery-swapping systems.
Connectivity and Network Issues	- Poor Wi-Fi or Bluetooth connections disrupt device performance. - High bandwidth demands (e.g., VR/AR environments). - Digital divide in rural/underserved areas.	- Upgrade school networks (Wi-Fi, routers, mesh systems). - Enable offline capabilities with periodic syncing.
Device Compatibility and Integration	- Lack of standardization across devices and platforms. - Complex integration with LMS or SIS systems. - Limited technical expertise for system management.	- Adopt open standards for data sharing. - Provide SDKs for custom integrations. - Invest in middleware for seamless integration.
Data Storage and Processing Challenges	- Large volumes of data require secure, scalable storage. - Ensuring compliance with data protection laws (e.g., GDPR, FERPA). - Lack of tools to process and interpret raw data effectively.	- Partner with providers for secure cloud storage. - Offer user-friendly analytics platforms for data interpretation.
Software and Firmware Updates	- Managing updates for multiple devices is time-consuming. - Updates may cause compatibility issues. - Inconsistent update schedules disrupt functionality across devices.	- Automate update systems to reduce manual effort. - Establish precise update protocols with providers. - Test updates for compatibility before deployment.

5. LIMITATIONS OF WEARABLE TECHNOLOGIES IN EDUCATION

While wearable technologies have significant potential for revolutionizing education by enhancing engagement, personalizing learning experiences, and providing real-time data, their adoption has substantial limitations. These limitations must be carefully considered to ensure that wearable technology improves education equitably and sustainably. Below, we explore five significant limitations associated with integrating wearable technologies in education: high costs of devices, privacy concerns and data security, accessibility and equity, teacher training and professional development, and potential health risks.

Table 4. Comparisons of different limitations of wearable technologies with proposed solutions

Section	Key Points	Proposed Solutions
High Costs of Devices	- Wearables (e.g., VR headsets, AR glasses, smartwatches) are expensive, limiting access for low-income schools. - Related costs: infrastructure upgrades, software licenses, and technical support. - Inequality between well-funded and under-resourced schools.	- Seek funding through grants, sponsorships, or partnerships. - Advocate for government programs to subsidize wearable technologies in education.
Privacy Concerns and Data Security	- Sensitive data collected (e.g., health metrics, cognitive engagement) risks unauthorized access. - Non-compliance with data protection laws like GDPR and FERPA. - Risks of commercialization of student data by third-party companies.	- Implement robust security measures (encryption, anonymization). - Ensure compliance with relevant data protection laws. - Establish clear consent policies for data usage.
Accessibility and Equity	- High costs widen the digital divide. - Rural schools face infrastructure challenges (e.g., internet access, technical support). - Students with disabilities face usability barriers (e.g., visual or motor impairments).	- Design inclusive wearables for students with disabilities. - Provide funding and resources for equitable access. - Develop alternative solutions for underserved schools.
Teacher Training and Professional Development	- Lack of technical skills among teachers to use wearables effectively. - Need for continuous learning due to evolving technologies. - Teacher resistance due to perceived complexity or limited benefits.	- Invest in comprehensive training and professional development programs. - Provide ongoing technical support and resources. - Focus on both technical and pedagogical aspects during training.
Potential Health Risks	- Risk of eyestrain from prolonged use of VR/AR devices. - Ergonomic issues (e.g., neck strain from head-mounted displays). - Concerns about electromagnetic field (EMF) exposure.	- Establish usage guidelines (e.g., usage time, regular breaks). - Prioritize ergonomically designed devices. - Provide clear communication about device safety standards.
Technical Difficulties	- Complexity in developing user-friendly interfaces for young learners. - Need for customization based on diverse curricula and educational needs. - Scalability challenges in managing multiple devices in large classrooms.	- Simplify user interfaces with intuitive designs and testing. - Build flexible platforms that allow content customization. - Develop centralized management tools for large-scale deployment.
Ethical Dilemmas in Data Usage	- Misuse of sensitive data for unfair assessments or commercialization. - Concerns about constant surveillance through wearable devices. - Ethical issues in balancing data collection with students' privacy and autonomy.	- Design wearables with strong privacy protections. - Establish transparent policies on data collection and usage. - Ensure informed consent from students and parents with opt-out options.

continued on following page

Table 4. Continued

Section	Key Points	Proposed Solutions
Complex User Interface Design	- Wearable UIs are often too complex for younger students. - Devices may overwhelm children with limited technical skills. - Lack of accessibility features for students with diverse cognitive and motor abilities.	- Simplify UIs with larger buttons, voice commands, or touch gestures. - Conduct extensive usability testing with children, including those with disabilities.
Customization for Educational Settings	- Schools need flexible wearables to align with unique curricula and teaching methods. - Challenges in addressing diverse student demographics and special education needs. - Difficulty in adapting wearables to various learning preferences and paces.	- Create modular platforms for content flexibility. - Enable educators to modify applications without excessive technical requirements. - Integrate wearables with existing educational tools like LMSs.
Scalability in Large Classrooms	- Managing multiple devices (e.g., VR headsets, smartwatches) can overwhelm teachers. - Challenges in syncing, charging, and updating devices. - Real-time data management requires robust infrastructure.	- Develop centralized tools for device and data management. - Provide educators with technical support for large-scale deployment. - Optimize wearables for ease of management in large environments.

6. FUTURE TRENDS AND INNOVATIONS IN EDUCATIONAL WEARABLES

As wearable technologies evolve, their potential in educational contexts is expanding rapidly. Innovations in hardware, software, and data analytics are opening new possibilities for how wearables can enhance learning, track student progress, and create more personalized educational experiences. This section explores some key future trends and innovations in educational wearables, including predictions for the evolution of wearable technology, emerging technologies like neurotech and biometric sensors, and the integration of artificial intelligence (AI) and machine learning (ML) into these devices.

6.1 Predictions on the Evolution of Wearable Technology

The future of wearable technology in education will likely see a significant increase in both the sophistication of the devices and the range of applications they support. As the costs of wearables decrease and their capabilities expand, schools and universities will be more likely to adopt these technologies on a larger scale. Some of the key trends that may shape the evolution of educational wearables include:

1. **Greater Affordability and Accessibility**: Adopting wearable technology is expected to lower, accessible, and equitable in education for low-income or underserved areas.
2. **Multimodal Devices**: Wearables are set to become integrated, multimodal devices, combining various functions like VR, AR, fitness tracking, and biometric sensing, enabling students to engage in virtual simulations and receive real-time academic feedback.
3. **Cross-Platform Integration**: Future wearables will integrate with educational technologies like smartboards, LMSs, and cloud platforms, facilitating seamless data flow and enabling teachers to track student progress across multiple touchpoints.
4. **Greater Focus on Student Well-Being**: Wearables are expected to prioritize student well-being alongside academic performance, tracking stress levels, attention span, and emotional states, enabling educators to intervene when needed.

6.2 Integration with AI and Machine Learning

Artificial intelligence and machine learning integration with wearable technologies can revolutionize education by providing scale-based personalized learning experiences. These devices track individual student progress, analyze performance data, and adapt content to suit their needs. This technology can help educators better manage large groups of students and ensure active participation.

7. CONCLUSION AND FUTURE SUGGESTIONS

Wearable technologies are transforming the education landscape by offering many advantages that enhance student engagement, personalize learning experiences, promote collaboration, and support inclusion for students with special needs. From real-time data for adaptive learning to immersive experiences enabled by AR/VR, these technologies are helping to shape the future of education in innovative ways. However, alongside these benefits, wearable technologies also bring several limitations and technical difficulties that educators, institutions, and policymakers must address to maximize their effectiveness in learning environments. A key advantage of wearable technologies is their ability to enhance student engagement and motivation. By creating interactive, gamified learning experiences, wearables make lessons more dynamic and enjoyable, fostering a deeper connection between students and the material they are learning. Furthermore, wearable devices provide real-time data that enable adaptive learning, allowing educators to tailor lessons to individual needs and performance. This personalization, combined with the power of AR/VR technologies, creates immersive learning experiences that help students grasp complex concepts through active participation rather than passive consumption of information. Despite these clear benefits, significant limitations exist in adopting wearables in education. High costs and affordability issues remain challenging for many schools, especially in underserved regions. There is also a risk of over-reliance on technology, which can reduce face-to-face interaction and traditional learning methods. Accessibility and compatibility issues arise when institutions attempt to integrate wearables into existing infrastructure, and some educators may lack the necessary training to incorporate these devices into their teaching effectively. In addition, privacy concerns around student data collection and use raise ethical questions, mainly when wearables track physical, emotional, or cognitive metrics. The technical difficulties associated with wearable technologies, including connectivity issues, battery life, and the durability of devices, further complicate their integration into educational settings. Devices require frequent updates and maintenance and can face challenges with scalability, particularly in larger institutions. Moreover, the lack of standardization across platforms and devices can create additional hurdles for educators to integrate wearables with learning management systems (LMS) or other digital tools. Given the clear advantages, challenges, and limitations, the role of wearable technologies in education must be considered carefully. While these tools can revolutionize learning, educators, institutions, and policymakers must ensure that their implementation is thoughtful, equitable, and designed to complement, not replace, traditional teaching methods. Wearables should be integrated in ways that enhance human interaction and critical thinking while maintaining an emphasis on inclusive and accessible education.

In summary, wearable technologies hold immense potential to transform education, but their adoption must be carefully managed to maximize benefits while minimizing drawbacks. By addressing both the technical and pedagogical challenges, educators can harness the full power of wearables to create engaging, personalized, and inclusive learning experiences for students of all backgrounds and abilities.

REFERENCES

Al-Emran, M., Al-Nuaimi, M. N., Arpaci, I., Al-Sharafi, M. A., & Jnr, B. A. (2022). Towards a wearable education: Understanding the determinants affecting students' adoption of wearable technologies using machine learning algorithms. *Education and Information Technologies*, 28(3), 2727–2746. DOI: 10.1007/s10639-022-11294-z

Al-Emran, M., Granić, A., Al-Sharafi, M. A., Ameen, N., & Sarrab, M. (2020). Examining the roles of students' beliefs and security concerns for using smartwatches in higher education. *Journal of Enterprise Information Management*, 34(4), 1229–1251. DOI: 10.1108/JEIM-02-2020-0052

Almusawi, H. A., & Durugbo, C. M. (2024). Determinants of teacher attitudes and innovative use of wearable technology. *IEEE Transactions on Learning Technologies*, 17, 1428–1439. DOI: 10.1109/TLT.2024.3386464

Almusawi, H. A., Durugbo, C. M., & Bugawa, A. M. (2021). Wearable Technology in Education: A Systematic review. *IEEE Transactions on Learning Technologies*, 14(4), 540–554. DOI: 10.1109/TLT.2021.3107459

Araújo, A. A., Gardim, L., Salma, J., Stephen, T., Santos, S. S. D., Silva, Í. R., De Godoy, S., & Mendes, I. C. (2024). Advancing Nursing Education through Wearable Electronic Devices: A Scoping. *Nurse Education in Practice*, 79, 104032. DOI: 10.1016/j.nepr.2024.104032 PMID: 38964082

Baber, C. (2001). Wearable Computers: A Human Factors review. *International Journal of Human-Computer Interaction*, 13(2), 123–145. DOI: 10.1207/S15327590IJHC1302_3

Beard, H. R., Marquez-Lara, A. J., & Hamid, K. S. (2016). Using wearable video technology to build a Point-of-View surgical education library. *JAMA Surgery*, 151(8), 771. DOI: 10.1001/jamasurg.2016.0976 PMID: 27276102

Bower, M., & Sturman, D. (2015). What are the educational affordances of wearable technologies? *Computers & Education*, 88, 343–353. DOI: 10.1016/j.compedu.2015.07.013

Breed, R., Kay, A., Spittle, M., & Orth, D. (2024). The effect of pedagogical approach on physical activity of girls during physical education. *Research Quarterly for Exercise and Sport*, 95(4), 1–10. DOI: 10.1080/02701367.2024.2329165 PMID: 38771866

Campbell, L. O., & Frawley, C. (2023). An exploration of factors that predict higher education faculty members' intentions to utilize emerging technologies. *Educational Technology Research and Development*, 72(2), 643–659. DOI: 10.1007/s11423-023-10321-1

Chih, H., Ahmed, T., Chiu, A. P., Liu, Y., Kuo, H., Yang, A. C., & Lien, D. (2023). Multitask learning for automated sleep staging and wearable technology integration. *Advanced Intelligent Systems*. Advance online publication. DOI: 10.1002/aisy.202300270

Chiu, W., Oh, G. E., & Cho, H. (2021). Factors Influencing Consumers' adoption of Wearable Technology: A Systematic Review and Meta-Analysis. *International Journal of Information Technology & Decision Making*, 20(03), 933–958. DOI: 10.1142/S0219622021500206

Datta, N., MacQueen, I. T., Schroeder, A. D., Wilson, J. J., Espinoza, J. C., Wagner, J. P., Filipi, C. J., & Chen, D. C. (2015). Wearable technology for global surgical teleproctoring. *Journal of Surgical Education*, 72(6), 1290–1295. DOI: 10.1016/j.jsurg.2015.07.004 PMID: 26276303

Davidesco, I., Matuk, C., Bevilacqua, D., Poeppel, D., & Dikker, S. (2021). Neuroscience Research in the Classroom: Portable Brain Technologies in Education Research. *Educational Researcher*, 50(9), 649–656. DOI: 10.3102/0013189X211031563

De-La-Fuente-Robles, Y., Ricoy-Cano, A., Albín-Rodríguez, A., López-Ruiz, J. L., & Espinilla-Estévez, M. (2022). Past, Present and Future of Research on Wearable Technologies for Healthcare: A bibliometric analysis using Scopus. *Sensors (Basel)*, 22(22), 8599. DOI: 10.3390/s22228599 PMID: 36433195

De Zambotti, M., Cellini, N., Goldstone, A., Colrain, I. M., & Baker, F. C. (2019). Wearable sleep technology in clinical and research settings. *Medicine and Science in Sports and Exercise*, 51(7), 1538–1557. DOI: 10.1249/MSS.0000000000001947 PMID: 30789439

Díaz-Quesada, G., Bahamonde-Pérez, C., Giménez-Egido, J. M., & Torres-Luque, G. (2021). Use of wearable devices to study physical activity in early childhood education. *Sustainability (Basel)*, 13(24), 13998. DOI: 10.3390/su132413998

Dong, P., & A, L. (2020). RETRACTED: Simulation of physical education teaching based on FPGA and wearable VR equipment. *Microprocessors and Microsystems*, 81, 103773. DOI: 10.1016/j.micpro.2020.103773

Du, J., & Nan, Z. (2021). RETRACTED: Research on the intelligent model of progress in physical education training based on motion sensor. *Microprocessors and Microsystems*, 82, 103903. DOI: 10.1016/j.micpro.2021.103903

Evmenova, A. S., Graff, H. J., Motti, V. G., Giwa-Lawal, K., & Zheng, H. (2018). Designing a wearable technology intervention to support young adults with intellectual and developmental disabilities in inclusive postsecondary academic environments. *Journal of Special Education Technology*, 34(2), 92–105. DOI: 10.1177/0162643418795833

Ferreira, J. J., Fernandes, C. I., Rammal, H. G., & Veiga, P. M. (2021). Wearable technology and consumer interaction: A systematic review and research agenda. *Computers in Human Behavior*, 118, 106710. DOI: 10.1016/j.chb.2021.106710

Ferrier, B., Lee, J., Mbuli, A., & James, D. A. (2022). Translational Applications of Wearable Sensors in Education: Implementation and efficacy. *Sensors (Basel)*, 22(4), 1675. DOI: 10.3390/s22041675 PMID: 35214578

Fonseca, D., Navarro, I., De Renteria, I., Moreira, F., Ferrer, Á., & De Reina, O. (2017). Assessment of wearable virtual reality Technology for visiting world Heritage Buildings: An Educational approach. *Journal of Educational Computing Research*, 56(6), 940–973. DOI: 10.1177/0735633117733995

Frederick, J., & Van Gelderen, S. (2021). Revolutionizing simulation education with smart glass technology. *Clinical Simulation in Nursing*, 52, 43–49. DOI: 10.1016/j.ecns.2020.12.007

Fu, D., Chen, L., & Cheng, Z. (2021). Integration of Wearable Smart Devices and Internet of Things Technology into Public Physical Education. *Mobile Information Systems*, 2021, 1–10. DOI: 10.1155/2021/6740987

Gao, W., Wei, T., Huang, H., Chen, X., & Li, Q. (2022). Toward a systematic survey on wearable computing for education applications. *IEEE Internet of Things Journal*, 9(15), 12901–12915. DOI: 10.1109/JIOT.2022.3168324

Gupta, S., Srivastava, R., Shaikh, Z. H., & Irfan, M. (2024). Revolutionizing Rural Finance: Exploring the Impact of FinTech on Financial Inclusion in India. In Irfan, M., Muhammad, K., Naifar, N., & Khan, M. A. (Eds.), *Applications of Block Chain technology and Artificial Intelligence. Financial Mathematics and Fintech*. Springer., DOI: 10.1007/978-3-031-47324-1_3

Hagadone-Bedir, M., Voithofer, R., & Kulp, J. T. (2023). Contextualizing privacy with wearable data in higher education. *British Journal of Educational Technology*, 54(6), 1619–1635. DOI: 10.1111/bjet.13378

Hsu, C., Chen, Y., Lin, C., & Lien, W. (2021). Cognitive development, self-efficacy, and wearable technology use in a virtual reality language learning environment: A structural equation modeling analysis. *Current Psychology (New Brunswick, N.J.)*, 41(3), 1618–1632. DOI: 10.1007/s12144-021-02252-y

Huifeng, W., Kadry, S. N., & Raj, E. D. (2020). Continuous health monitoring of sportsperson using IoT devices based wearable technology. *Computer Communications*, 160, 588–595. DOI: 10.1016/j.comcom.2020.04.025

Iqbal, M. H., Aydin, A., Brunckhorst, O., Dasgupta, P., & Ahmed, K. (2016). A review of wearable technology in medicine. *Journal of the Royal Society of Medicine*, 109(10), 372–380. DOI: 10.1177/0141076816663560 PMID: 27729595

Jeng, M., Pai, F., & Yeh, T. (2022). Antecedents for older adults' intention to use smart Health Wearable Devices-Technology Anxiety as a moderator. *Behavioral Sciences (Basel, Switzerland)*, 12(4), 114. DOI: 10.3390/bs12040114 PMID: 35447686

Khosravi, S., Bailey, S. G., Parvizi, H., & Ghannam, R. (2022). Wearable sensors for learning enhancement in higher education. *Sensors (Basel)*, 22(19), 7633. DOI: 10.3390/s22197633 PMID: 36236732

Kim, J., Heo, N., & Kang, H. (2023). Validity and reliability of positive attitudes toward and perceived importance of wearable display technology as an effective learning tool among nursing students. *Nurse Education in Practice*, 73, 103812. DOI: 10.1016/j.nepr.2023.103812 PMID: 37922737

Kusuma, K., Sheela, P., Panigrahi, R. R., & Shaikh, Z. H. (2024). Blockchain Technology and Islamic Finance: A Decade of Growth Trends. In Irfan, M., Kadry, S., Sharif, M., & Khan, H. (Eds.), *Fintech Applications in Islamic Finance: AI, Machine Learning, and Blockchain Techniques* (pp. 59–79). IGI Global., DOI: 10.4018/979-8-3693-1038-0.ch004

Kwon, Y. M., Lee, Y., & Kim, S. J. (2017). Case study on 3D printing education in fashion design coursework. *Fashion and Textiles*, 4(1), 26. Advance online publication. DOI: 10.1186/s40691-017-0111-3

Lee, J., Kim, D., Ryoo, H., & Shin, B. (2016). Sustainable wearables: Wearable technology for enhancing the quality of human life. *Sustainability (Basel)*, 8(5), 466. DOI: 10.3390/su8050466

Lindberg, R., Seo, J., & Laine, T. H. (2016). Enhancing Physical Education with Exergames and Wearable Technology. *IEEE Transactions on Learning Technologies*, 9(4), 328–341. DOI: 10.1109/TLT.2016.2556671

Lindeque, B. G., Ponce, B. A., Menendez, M. E., Oladeji, L. O., Fryberger, C. T., & Dantuluri, P. K. (2014). Emerging technology in surgical education: Combining Real-Time augmented reality and wearable computing devices. *Orthopedics*, 37(11), 751–757. DOI: 10.3928/01477447-20141023-05 PMID: 25361359

Liu, S., Hu, S., Weng, J., Zhu, S., & Chen, Z. (2016). A novel asymmetric three-party based authentication scheme in wearable devices environment. *Journal of Network and Computer Applications*, 60, 144–154. DOI: 10.1016/j.jnca.2015.10.001

Ma, S., Wang, Z., Zhu, Y., Tang, Y., Fan, G., Ma, B., Ye, T., & Wei, L. (2022). Micro/nanofiber fabrication technologies for wearable sensors: A review. *Journal of Micromechanics and Microengineering*, 32(6), 064002. DOI: 10.1088/1361-6439/ac6b7e

Malwade, S., Abdul, S. S., Uddin, M., Nursetyo, A. A., Fernandez-Luque, L., Zhu, X., Cilliers, L., Wong, C., Bamidis, P., & Li, Y. (2018). Mobile and wearable technologies in healthcare for the ageing population. *Computer Methods and Programs in Biomedicine*, 161, 233–237. DOI: 10.1016/j.cmpb.2018.04.026 PMID: 29852964

Mardonova, M., & Choi, Y. (2018). Review of wearable device technology and its applications to the mining industry. *Energies*, 11(3), 547. DOI: 10.3390/en11030547

Marelle, C., Thomas, E. V., & Paul, C. D. (2022). A Survey of Wireless Technology Supporting Individuals with Intellectual and Developmental Disabilities in the Workplace. *Journal of Special Education Technology*, 38(4), 434–444. DOI: 10.1177/01626434221131463

Marttinen, R., Landi, D., Fredrick, R. N., & Silverman, S. (2020). Wearable digital technology in PE: Advantages, barriers, and teachers' ideologies. *Journal of Teaching in Physical Education*, 39(2), 227–235. DOI: 10.1123/jtpe.2018-0240

Masters, K., Ellaway, R. H., Topps, D., Archibald, D., & Hogue, R. J. (2016). Mobile technologies in medical education: AMEE Guide No. 105. *Medical Teacher*, 38(6), 537–549. DOI: 10.3109/0142159X.2016.1141190 PMID: 27010681

Mehmood, R., & Lee, H. (2017). Towards building a computer aided education system for special students using wearable sensor technologies. *Sensors (Basel)*, 17(2), 317. DOI: 10.3390/s17020317 PMID: 28208734

Musa, U., Shah, S. M., Majid, H. A., Abidin, Z. Z., Yahya, M. S., Babani, S., & Yunusa, Z. (2022). Recent advancement of wearable Reconfigurable Antenna Technologies: A review. *IEEE Access : Practical Innovations, Open Solutions*, 10, 121831–121863. DOI: 10.1109/ACCESS.2022.3222782

Nakazawa, E., Yamamoto, K., London, A. J., & Akabayashi, A. (2021). Solitary death and new lifestyles during and after COVID-19: Wearable devices and public health ethics. *BMC Medical Ethics*, 22(1), 89. Advance online publication. DOI: 10.1186/s12910-021-00657-9 PMID: 34246258

Ngai, G., Chan, S. C. F., Cheung, J. C. Y., & Lau, W. W. Y. (2010). Deploying a wearable computing platform for computing education. *IEEE Transactions on Learning Technologies*, 3(1), 45–55. DOI: 10.1109/TLT.2009.49

Nogueira, T., De Morais, V. L., De Moraes, C. P., & Lellis-Santos, C. (2020). The wearable bone glossary: Wearable technology at the service of inclusive education. *Medical Education*, 54(5), 447–448. DOI: 10.1111/medu.14102 PMID: 32249961

Nugent, G., Barker, B., Lester, H., Grandgenett, N., & Valentine, D. (2019). Wearable textiles to support student STEM learning and attitudes. *Journal of Science Education and Technology*, 28(5), 470–479. DOI: 10.1007/s10956-019-09779-7

Panigrahi, R. R., Dash, M., Shaikh, Z. H., & Irfan, M. (2023). Review of Machine Learning Techniques in the Supply Chain Management of Indian Industry: A Future Research Agenda. In Irfan, M., Elhoseny, M., Kassim, S., & Metawa, N. (Eds.), *Advanced Machine Learning Algorithms for Complex Financial Applications* (pp. 199–219). IGI Global., DOI: 10.4018/978-1-6684-4483-2.ch013

Patil, V., Singhal, D., Naik, N., Hameed, B., Shah, M., Ibrahim, S., Smriti, K., Chatterjee, G., Kale, A., Sharma, A., Paul, R., Chłosta, P., & Somani, B. (2022). Factors Affecting the Usage of Wearable Device Technology for Healthcare among Indian Adults: A Cross-Sectional Study. *Journal of Clinical Medicine*, 11(23), 7019. DOI: 10.3390/jcm11237019 PMID: 36498594

Pennefather, P., & Krebs, C. (2019). Exploring the role of XR in visualisations for use in medical education. *Advances in Experimental Medicine and Biology*, 1171, 15–23. DOI: 10.1007/978-3-030-24281-7_2 PMID: 31823236

Pillai, K. R., Upadhyaya, P., Prakash, A. V., Ramaprasad, B. S., Mukesh, H. V., & Pai, Y. (2021). End-user satisfaction of technology-enabled assessment in higher education: A coping theory perspective. *Education and Information Technologies*, 26(4), 3677–3698. Advance online publication. DOI: 10.1007/s10639-020-10401-2

Ramírez-Moreno, M. A., De, J., & Lozoya-Santos, J. (2024). Wearable Biosensor Technology in Education: A Systematic review. *Sensors (Basel)*, 24(8), 2437. DOI: 10.3390/s24082437 PMID: 38676053

Sarea, A., Subramanian, S., Alareeni, B., Shaikh, Z. H., Hawaldar, I. T., & Elshaker, A. H. (2021). Financial Disclosures by Using Machine Learning Analysis. In Alareeni, B., Hamdan, A., & Elgedawy, I. (Eds.), *The Importance of New Technologies and Entrepreneurship in Business Development: In The Context of Economic Diversity in Developing Countries. ICBT 2020. Lecture Notes in Networks and Systems* (Vol. 194). Springer., DOI: 10.1007/978-3-030-69221-6_5

Selwin, S., Saseendran, H., Vemavarapu, K. R., Shaikh, I., Shaikh, Z. H., & Harris, S. B. (2024). AI-Driven Employee Engagement: Transforming Banking Workforce Dynamics Using the BEULA Model. In Awwad, B. (Ed.), *The AI Revolution: Driving Business Innovation and Research. Studies in Systems, Decision and Control* (Vol. 525). Springer., DOI: 10.1007/978-3-031-54383-8_32

Shaikh, Z., Irfan, M., Sarea, A., & Panigrahi, R. R. (2024). *The Emergence of Islamic Fintech and Bahrain: Prospect for Global Financial Sectors* (Vol. 503). Springer., DOI: 10.1007/978-3-031-43490-7_52

Shaikh, Z. H., Irfan, M., Nomran, N. M., & Lawin, M. L. (2024). The Effect of Strategic Partnership on Innovation and Business: Performance of the Fintech Industry in Bahrain. In Irfan, M., Kadry, S., Sharif, M., & Khan, H. (Eds.), *Fintech Applications in Islamic Finance: AI, Machine Learning, and Blockchain Techniques* (pp. 236–254). IGI Global., DOI: 10.4018/979-8-3693-1038-0.ch016

Shaikh, Z. H., Irfan, M., Nomran, N. M., & Ratnakaram, S. P. (2024). Unlocking the Potential of AI for Efficient Governance: Innovative Approaches of Bahrain. In Irfan, M., Elmogy, M., Gupta, S., Khalifa, F., & Dias, R. (Eds.), *AI-Driven Decentralized Finance and the Future of Finance* (pp. 115–142). IGI Global., DOI: 10.4018/979-8-3693-6321-8.ch005

Shaikh, Z. H., Sarea, A., & Irfan, M. (2022). *Islamic Banking Strategies in the World of Fintech: Success Story of Bahrain* (Vol. 423). Springer., DOI: 10.1007/978-3-030-93464-4_10

Sheela, P., Kusuma, K., Panigrahi, R. R., & Shaikh, Z. H. (2024). Fintech and Islamic Banking: A Systematic View and Future Research Agenda. In Irfan, M., Kadry, S., Sharif, M., & Khan, H. (Eds.), *Fintech Applications in Islamic Finance: AI, Machine Learning, and Blockchain Techniques* (pp. 1–23). IGI Global., DOI: 10.4018/979-8-3693-1038-0.ch001

Sultan, N. (2015). Reflective thoughts on the potential and challenges of wearable technology for healthcare provision and medical education. *International Journal of Information Management*, 35(5), 521–526. DOI: 10.1016/j.ijinfomgt.2015.04.010

Wei, L., Li, Z., Dai, Z., Ding, L., Wei, H., Li, J., Zhu, H., & Zhu, Y. (2024). Wearable sweat management Technologies. *Advanced Materials Technologies*, 9(7), 2301812. Advance online publication. DOI: 10.1002/admt.202301812

Wen, J., Xu, B., Gao, Y., Li, M., & Fu, H. (2021). Wearable technologies enable high-performance textile supercapacitors with flexible, breathable and wearable characteristics for future energy storage. *Energy Storage Materials*, 37, 94–122. DOI: 10.1016/j.ensm.2021.02.002

Yoshida, Y., & Yuda, E. (2023). Workout detection by wearable device data using machine learning. *Applied Sciences (Basel, Switzerland)*, 13(7), 4280. DOI: 10.3390/app13074280

Zhao, Y., & You, Y. (2021). Design and data analysis of wearable sports posture measurement system based on Internet of Things. *Alexandria Engineering Journal*, 60(1), 691–701. DOI: 10.1016/j.aej.2020.10.001

Chapter 13
Beyond the Classroom: A Comprehensive Guide to Wearable Device Security in Education

Abhishek Bhargava
Wizons Digital Academy LLP, India

Shafiq Ul Rehman
https://orcid.org/0000-0003-2266-218X
College of Information Technology, Kingdom University, Bahrain

ABSTRACT

Academic circles have grown to embrace wearable technology since everyone wears something on their body and their use in teaching-learning context has benefits and limitations. Thus, these devices can facilitate teaching and learning experiences still, they add new risks to cybersecurity. Hence, this chapter presents a detailed cybersecurity framework that can be adopted for education institutions with an integration of wearable technology. Potential concerns are considered, threats are contained, and meaningful instructions for learners and teachers are given to reduce threats and ensure that learning takes place in a safe environment. The chapter delves into the unique characteristics and vulnerabilities of wearable devices, analyzes existing cybersecurity frameworks, and presents a comprehensive framework with key components: Wearable Device Security Assessment, Data Privacy and Protection, Cyberspace Security Awareness and Education as well as Management of Security Incidents and Breaches.

DOI: 10.4018/979-8-3693-7817-5.ch013

1. INTRODUCTION

Wearable technology in education promises to change the way that students learn, and teachers teach (Aldowah H. R., 2017). From fitness trackers and their ability to monitor students' activity levels, to smartwatches that enable real-time testing and assessment, wearable devices are opening exciting opportunities that can engage learners, provide greater personalization and hence help to inform more effective approaches to both teaching and learning. However, along with the promise of wearable technology brings new information, physical and psychological security risks that must be tackled for its application to be safe from attackers (Sadek I. C., 2022).

In this chapter, we present our developed comprehensive wearable technology cybersecurity framework to address the educational risks/challenges with specific reference to wearable technology. These platforms may vary in their user features, but they share a critical infrastructure and threat landscape. Identifying and presenting these shared elements within a framework provides an apt foundation for a campus design overlay. In Microsoft's "Hacking the Nation" report, they sought to answer the question: "For more than a decade, hackers have mostly aimed to discredit and disrupt adversaries. Yet the hacks of 2016 were different as they are said to be affecting the state elections." By using a high-altitude balloon to create a dense cellular network, state-sponsored actors successfully breached security and compromised the smartphones of targets to steal emails, pictures and other data, discredit a company, and access classified documents. Once they achieve access, hackers refuse to relinquish it.

The chapter specifically address the cybersecurity challenges related to the use of wearable devices in the public education and investigate how existing cybersecurity frameworks address security and privacy concerns on wearable devices with their unique features and risks (Aldowah H. U., 2019). The chapter further propose a well-rounded cybersecurity framework amended with a range of essential components, including wearable device security assessment, cybersecurity awareness and training, data privacy and protection, and incident response and recovery (Aldowah H. U., 2021). By implementing the proposed framework, educational institutions can:

- Safeguard sensitive student data collected from wearable devices.
- Mitigate the risk of data breaches, unauthorized access, and other cyberattacks.
- Foster a culture of cybersecurity awareness among educators, students, and staff.
- Enhance the overall security posture of the educational institution.

There are several types of wearable devices which are popularly used in the smart classrooms and they have different applications as well as features as mentioned in Table 1.

This chapter will be helpful for administrators, teachers, and IT workers in schools who wish to be well-informed about the issues as they consider how best to manage these new devices for educational purposes. However, before getting into security issues, and mitigation strategies, we need to get familiar with the wearable devices used for educational purposes as depicted in Table 1.

Table 1. List of popular wearable devices with their educational applications and features

Type of Devices	Educational Applications	Features
Smart Glasses	Interactive learning experiences, virtual field trips, language learning assistance	Augmented reality capabilities, video and audio recording, hands-free interaction
Smartwatches	Attendance tracking, classroom engagement monitoring, personalized learning recommendations	Real-time notifications, fitness tracking, calendar, educational app compatibility
Fitness Trackers	Physical education assessment, health and wellness programs, student activity level tracking	Step counting, heart rate monitoring, sleep tracking, activity tracking
Brainwave Monitoring Devices	Personalized learning adjustments, early detection of learning difficulties	Measure brain activity (EEG- Electroencephalogram), assess cognitive workload and engagement
Smart Clothing	Smart uniforms for tracking attendance and location, interactive learning experiences integrated into clothing	Embedded sensors for data collection, customizable features. The IEC TC 124 classifies smart wearables into four categories i.e. patchable, implantable, near-body, and on-body wearables, each with distinct applications and integration requirements. (Fraga-Lamas, 2018)[1]

1.1. Overview of Wearable Technology in Education and its History

Talking about history of wearables, the first wearable was invented in 1955 and was a timing device which had a small size approximated to the size of a cigarette pack. It was hidden in shoes and made to predict roulette wheels in casinos also it was presented publicly in 1966. After the evolution of calculator watches in mid 1970s, market began to grow. However, the boom in this industry took off in 2010. According to Statista National Purchase Diary Panel wearables were used mostly by those who are aged between 18 and 39. Hence resulting in introduction of wearables in education sector (Sara Khosravi, 2022). Smart technologies have become one of the most dynamic fields in recent years, especially for the education system. A study in Taiwan shows that integrating smart watches and smartphone apps into the education system can enhance the teaching and learning process for educators and school students. They also predict student's learning scores in very less time (by Jia-Ming Liang 1, 2019). Those small, portable devices, which are worn on the body, are provided with functionalities such as sensors, processors as well as wireless connectivity to collect, process and transmit information. The integration of wearable technology into education offers a wide range of potential benefits, including:

1. Personalized Learning- Tailored Instruction: Smart watches have the capability of recording performance, level of interest as well as learning modality of students. The information can be utilized to identify areas that require intervention as well as to adapt instruction.
2. Adaptive Learning Education: By using wearable technology, educational content, and activities can be personalized as the device can detect the knowledge level of students.
3. Enhanced Engagement- Gamification: Several ideas and strategies can be implemented by using wearable devices to enhance learning and make them fun and exciting at the same time for students. Integrating wearable devices enhance engagement by providing immediate feedback, which is essential for maintaining attention and developing interest in learning activities (María A. Hernández-Mustieles, 2024). Wearables may be useful for a few specific learning interventions such as field trips, simulations, and augmented reality.
4. Improved Classroom Management- Attendance Tracking: Smart devices can monitor the attendance of the students thus significantly cutting the workload and enhancing order in classrooms. Behavior Monitoring: There are wearables that allow the tracking of student conduct in class and hence can aid the teacher in noting any concerning incidents and addressing them.
5. Health and Wellness: It concerns such uses of wearable devices as attaining the level of students' physical activity, encouraging them to do proper physical activity, and backing up physical and education classes. Some of them may have features that help in tracking the students' heart rate or how well they slept, information that is useful in determining the students' state of health. (Wei Gao, 2022)
6. Accessibility- Wearable devices are a form of assistive technology that can be adopted in the classroom, to help students with disabilities learn and join other classmates in different tasks. Devices such as sensor gloves can help to achieve interactive learning for those who are suffering from mental disabilities. A study

developed a mobile application that uses flex sensor glove, allowing students to answer questions by bending their fingers (Athif Lanang Untoro), 2021.

There are different types of wearable technology which can be used in smart classrooms for educational purposes and the examples of them with pictures are shown in Figure 1.

Figure 1. Types of wearable technology

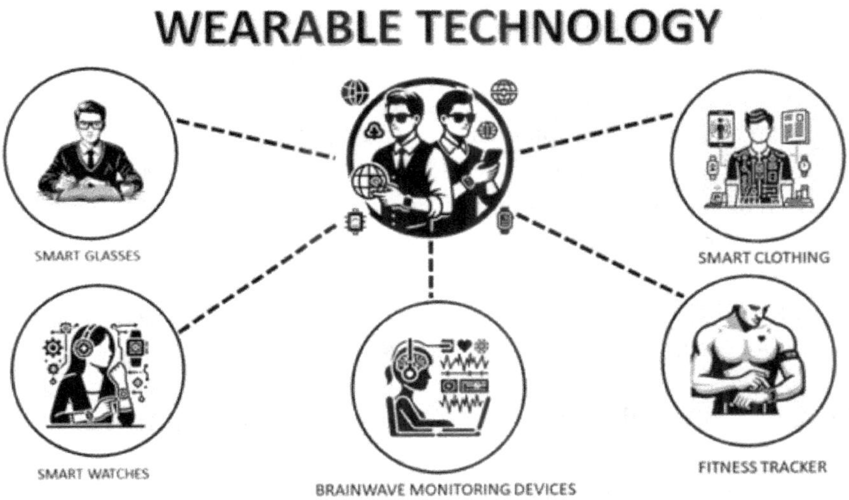

1.2. Cybersecurity Challenges Associated with Wearable Devices

From the standpoint of cybersecurity, introducing wearable technology into educational settings poses several concerns. Most of these devices relay information, accommodate sensors, interface with networks, and contain very sensitive data about students, all of which make them vulnerable to malicious parties. Wearable devices risk data integrity and threaten data integrity impairment. (Abdullah, 2024)

Key Cybersecurity Concerns

1. Data Privacy: A solicitation of personal or sensitive information and data from wearable devices is an issue regarding the privacy of an individual. Ensuring that students' information does not appear in the wrong hands or get hacked

and misused is very important. 1782 internet users were surveyed, and their behavior was studied regarding possible risks associated with wearable devices. It was found that people were concerned about things like video capturing and financial data being collected by these devices (Linda N. Lee1, 2015).

2. Unauthorized Access: Unauthorized access may be possible with wearables, which means that the attacker controls the device as well as the internal data stored in it. As a result, inexperience in using this technology might lead to data leakage, privacy infringement, and eventual interruption of education delivery.
3. Malware and Phishing Attacks: Wearables are not immune to hackers and infected malware that puts the device security at risk thus exposing student information.
4. Supply Chain Vulnerabilities: The supply chain for wearable devices may encounter several potential problems (Rehman S. U., 2023). One issue may stem from the requirement to implement software updates or make modifications to the proposed hardware of the devices. This could affect the smooth flow of the supply chain and impact the production and delivery of wearable devices.
5. Limited Security Features: A good number of wearable devices possess few or minimal security than those commonly found in other computing devices.
6. Appearance Issues: One more problem with the wearable devices is that they are hard to distinguish from similar devices. (Prerit Datta, 2018)

1.3. Purpose and Objectives of Cybersecurity Framework

The main goal of the cybersecurity framework for wearable technology in education is the safety of the student's information, the preservation of the learning system's trustworthiness, and the provision of a secure environment for learning processes. And hence advanced security measures like blockchain and fog computing are necessary to protect sensitive data. (Mohammed Saleh, 22). By addressing the unique cybersecurity challenges associated with wearable devices, the framework aims to, securing health information, performance data which may be collected by wearable devices used by students from such invasions or misuse.

Second is preventing disruptions to educational services, avert risks of cyber threats that may hinder the normal functioning of learning institutions which includes network downtimes, loss of data, or compromised student accounts (Ul Rehman, 2016).

Third is foster a culture of cybersecurity awareness among instructors, learners, and workers and sensitize them on the correct utilization of wearable gadgets, and security measures.

Fourth is to ensure that all legal requirements are met. This includes complying with laws on data privacy effectively as it will help in the protection of student's data and also avoid lawsuits and fines like GDPR, FERPA, etc.

Fifth is enhancing the reputation of educational institutions by ensuring that consumers of students' data are protective of the data, which will foster trust in the community, students, and parents.

Sixth is support the effective use of wearable technology in education. Suggest the ways through which educational institutions can benefit from wearable technology in learning, motivation, and evaluation while avoiding vices. Like one interesting use of wearables in education is their capability to measure physiological signals to understand emotions in educational contexts, revealing insights into student engagement and also anxiety levels (Shen ba, 2023)

Thus, attaining these objectives will support the advancement of the cybersecurity framework to create a safe and reliable learning ecosystem for wearable technology adoption in teaching and learning processes. In addition to that, since wearable devices and phone often works together so it can put one's personal data at risk, to avoid this one should keep the operating system and apps up to date and avoid putting personal data on such devices. (Manal Alshammari, 2022)

2. UNIQUE CHARACTERISTICS AND VULNERABILITIES OF WEARABLE TECHNOLOGY

Several benefits are provided by wrist-worn devices; however, they present specific cybersecurity risks because of specific features. Such gadgets especially are built with lower processing capabilities, and thus are more vulnerable to malware and other attacks. Most wearable devices available only allow for a few basic features hence limiting control by the users over the security settings which in return increases the vulnerability of such gadgets to hacking. Wearable devices as a connected concept with other devices like smartphones or laptops open new possibilities for attacks. Any of the connected devices can be hacked therefore posing a danger to the wearable device and the data collected. (Ayushi hora), 2024.

Solving these cybersecurity threats is possible only with a requisite focus on the kind of wearable devices and educational institutions. Let's have a look at a few important points related to this topic.

2.1 Limited Computational Resources

Unlike computers, wearable devices may have less CPU and memory limit as compared to conventional computers. This makes them less equipped to handle programs of a higher sophistication of security software" and hence makes them more vulnerable to malware, among other dangers. These limitations however can be used by attackers against wearable devices with attack plans that would be difficult on a more powerful device.

Implications: 1) Vulnerability to malware. 2) Reduced security features. 3) Slower response times.

Mitigation Strategies: 1) Regular software updates2) Lightweight security solutions: Authenticate tools that are compact with wearable devices and developed especially for such constrained environments Edge computing can alleviate the limitation of wearable devices by offloading task and hence can provide access to greater resources without compromising size and weight (Xinqi jin, 2022). 3) Cloud-based security

2.2 Battery Constraints

Smart devices must draw their power from batteries; thus, they are likely to be exploited by a hacker for purposes of controlling their power consumption. It is possible that an attacker may try to pull out the battery from a wearable device to cause it to stop working or to get access to some information that he or she is not supposed to get. The wearable system converts mechanical energy by human activities into electrical energy (Abdullah Sayam, 2024).

Implications: 1) Limited usage time. 2) Vulnerability during charging. 3) Data loss

Mitigation Strategies: 1) Power management. 2) Secure charging practices. 3) Regular battery maintenance. 4) For energy harvesting, methods like thermoelectric generators and piezoelectric transducers are used to enhance batter life and make us less independent on traditional batteries (Roberto De Fazio).

2.3 Wireless Connectivity

Wearable devices use a wireless connection (e. g. Bluetooth, Wi-Fi) for transmitting data and thus, introduce new points of attack in wireless technologies and links. It shows that numerical systems such as unsecured wireless network connections or hacked access points can let the attackers decode the data transmitted or inject hazardous code. Wearable devices, like fitness bands and health monitors, collect sensitive personal data. However, there is no standard for how this data is collected and shared that makes it vulnerable to unauthorized access (ARUP BARUA, 2022). Bluetooth Low Energy (BLE) implementations with weak security measures allow attackers to intercept and extract health data from encrypted packets. (Qi Liu, 2024)

Implications: 1) Data interception. 2) Man-in-the-middle attacks. 3) Denial of service attacks

Mitigation Strategies: 1) Secure wireless networks. 2) Avoid public Wi-Fi. 3) Update wireless firmware.

2.4 Limited User Interfaces

Most of the wearables offer minimal interfaces through which the user can control the security features and hence exposes the wearables to others access. These limitations can make it difficult for users to set strong passwords.

Implications: 1) Weak passwords. 2) Lack of 2FA. 3) Limited security options: Users may not be able to apply appropriate measures of protection on their devices and data, thus resulting invulnerability. Wrist-worn devices have small screens and limited input methods, which make it hard to adapt mobile phone interactions to these devices (Byron Lowens, 2015)

Mitigation Strategies: 1) Educate users. 2) Encourage device updates. 3) Consider alternative devices.

2.5 Integration with Other Devices

Wearable devices make communications with other devices such as smartphones, tablets, PCs, and laptops to improve the overall performance of the device in question. But this integration brings extra new cybersecurity threats.

Implications: 1) Expanded attack surface. 2) Data sharing. 3) Dependency on other devices.

Mitigation Strategies: 1) Secure device pairing. 2) Review data sharing settings. 3) Update all devices

3. COMMON VULNERABILITIES

Some of the most common vulnerabilities include:

1. Malware: Wearable devices are also capable of contracting malicious applications including viruses, worms, or spy ware (Sadek I. C., 2019). This malware can cause the theft of information, a reduction in the performance of the device, or its security breach. Malware designed for wearable devices can be created by attackers since those devices have certain features different from other devices.
2. Data Leaks: It is also important to note that students' data collected by wearables can be compromised by various factors such as: A) Unencrypted data transmission: Plain messages transmitted through a network can be easily acquired by unintended recipient besides being transmitted through supported infrastructure. B) Weak access controls: Lack of proper access control exposes student data to be accessed by those people it was not intended for them. C) Data breaches in connected devices or cloud storage platforms: Wearable devices are also likely

to be vulnerable if the connected devices or cloud services are hacked since they contain user data including that of students. Wearable pose unique risks not only in term of data security but also to physical safety, as malicious hacking could lead to harmful consequences for the user. (Adam J. Mills a, 2016)

3. Unauthorized Access: Often, attackers are able to achieve unauthorized access to wearable and other type of devices in the following ways: A) Brute-force attacks: Attempting to unlock the device by inputting random passwords even when they are forbidden by their employers. B) Phishing attacks: Just like the Trojan Horse, it involves deceiving the users either to enter their login details or to click on certain links. C) Exploiting vulnerabilities: Exploiting the vulnerabilities present in the device either in the latter's software or physical layers.
4. Supply Chain Vulnerabilities: Wearable devices are not immune to attackers and other types of intrusion: malicious software may be injected; during manufacturing or transportation phases the supply chain may be compromised. The devices under attack may be tampered with during the manufacturing of the actual devices with additional backdoors or even viruses. Generally, these devices lack robust security and hence are prime targets for such attacks also, by demonstration it was found that smart toys and other IOT devices are one of them (Diane Gan).
5. Social Engineering Attacks: Phishing and other kinds of social engineering can be used to deliver users into furnishing rather delicate data or providing access to their Wearables. Phishing scams involve the attacker mimicking familiar entities such as the user's colleagues or reputable organizations to cheat him out of vital information or make him agree to actions that are unbeneficial to his security (Rehman S. U.-C., 2023). These attacks take advantage of human behavior weaknesses rather than just relying on technical flows (Oveh R.O., 2022).
6. Physical Security Threats: Some of the risks of wearables include physical loss or damage of the device whereby data is stolen, or the real device can as well be stolen. This can be in the form of loss, theft or any form of alteration to the device in question.

Addressing these common vulnerabilities requires a comprehensive cybersecurity approach that includes, 1. Regular software updates. 2. Strong password policies. 3. Two-factor authentication. 4. Data encryption. 5. Regular security audits. 6. User education and training.

4. LIMITATIONS OF EXISTING CYBERSECURITY FRAMEWORKS

Present cybersecurity guidelines and models though useful for conventional IT systems may not be adequate for wearable technology in learning institutions. These frameworks do not usually provide detailed information about protecting wearable devices and the peculiarities of their nature as well as the challenges that relate to data privacy (Rehman S. &., 2023) (Rehman S. U., 2020).

1. Prospective of IT Operations in Conventional IT settings-A lot of the current cybersecurity frameworks are just oriented to the protection of PCs, servers, and network devices. They might not fully address the precise concerns and threats pertaining to wearable devices that differ from other gadgets in several ways including in aspects like the computational capabilities, battery power, or the fact that most of them are wirelessly connected. These frameworks may not offer specific insight into approaching the security challenges that affect wearable gadgets such as malware, data breaches, as well as unauthorized access.

    2.Relatively Little Emphasis of Educational Context- The current framework can be sort of insufficient to capture the precise needs and specifications of learning institutions. They may not cover the peculiarities of students' data protection, compliance with educational legislation, and the creation of a secure learning environment. For instance, the frameworks that have been there previously may not be so effective in the case of the integrated model's requirements such as the need to ensure the security of the devices while at the same time encouraging their use for educative purposes. They may not think about the inherent factors, risks, or challenges that are linked with the kind of data gathered by wearable devices in learning environments including health data, performance data and other personal data.

In addition, there is a lack of information on how to share the security policies and best practices with educators, students and parents on how they can protect their information. To sum up, it is necessary to state that the existing cybersecurity frameworks could be helpful, but currently they do not provide comprehensive coverage of all the factors that can contribute to the protection of wearable technologies in educational environment. For this purpose, it will require a special framework to ensure the safety and privacy of student information in wearable technologies and use of the wearable gadgets in learning environment. Here are some additional points to consider: a) Emerging Threats: Some previous models have limited applicability to novel risks and susceptibilities associated with wearable tech, for instance, conflicting malware forms or fresh invasions of present wearable devices. b) Regulatory Compliance: Schools are bound to operate under many data protection laws for example GDPR as well as FERPA. While there might be certain models designed to help companies figure out how to abide by particular rules, few of these models apply specifically to wearable technology. c) Integration with Existing Infrastructure: Incorporation of wearable technology in learning institutions may require blending with other IT requirements, which may be a subject of vulnerability. d) Cost-Benefit Analysis: The overall establishment of a cybersecurity framework for wearable technology needs dedicated effort and time and can be quite resource-intensive at times. Management therefore must consider the costs and benefits of such an investment when planning to open such institutions for education. In light of, challenging these limitations and identifying the requirements required to fit the education sector, a specialized cyberspace security direction shall be the right pointer and ally in guaranteeing the safety of the student details and appropriateness of the wearing accessories in learning institutions.

5. PROPOSED CYBERSECURITY FRAMEWORK

Figure 2. Proposed cybersecurity framework

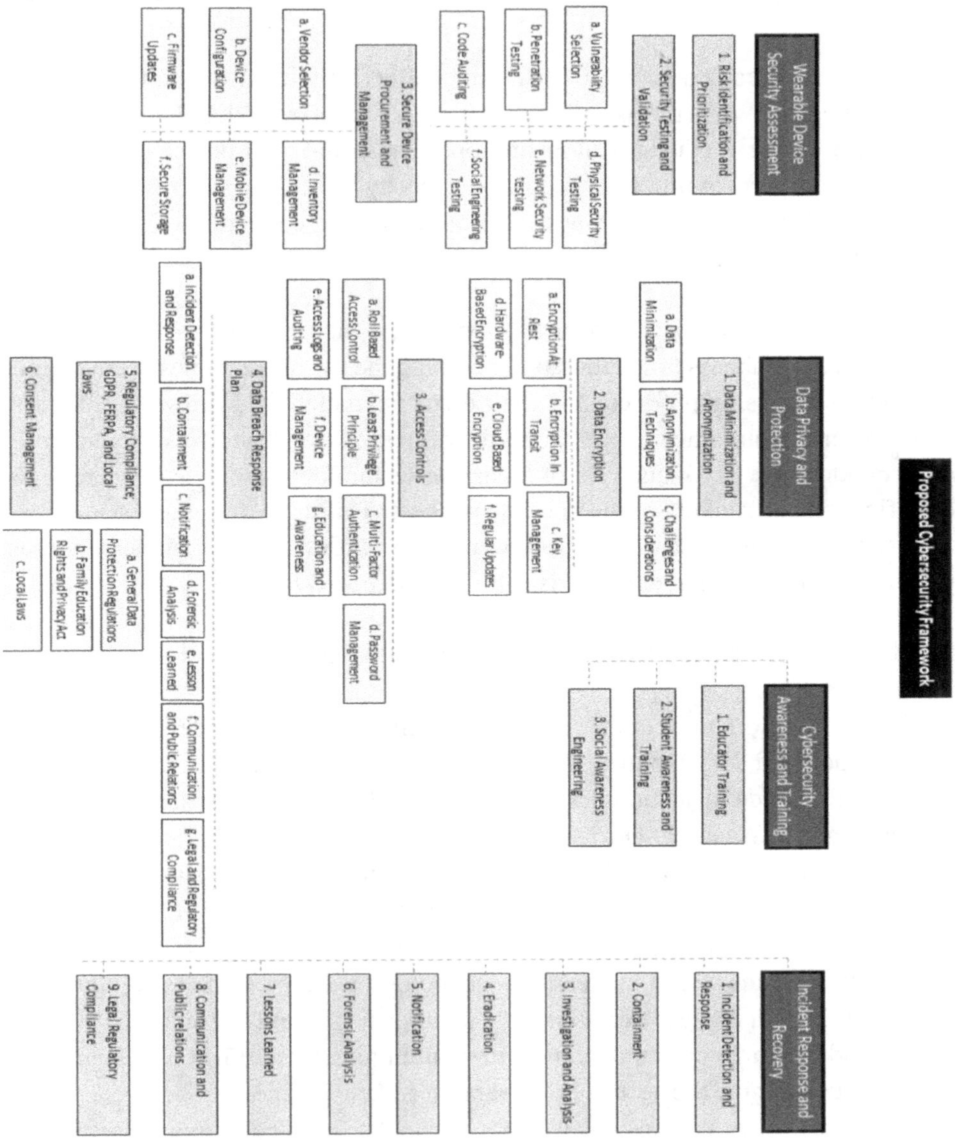

A comprehensive cybersecurity framework for wearable technology in education should encompass all the points covered in Figure 2, elaborated as follows:

5.1. Wearable Device Security Assessment

To evaluate the security measures of wearable devices, we need to address the following questions.

RQ1: How can educational institutions effectively identify and focus on the most critical security risks that comes with wearable devices?
RQ2: How can educational institutions ensure that their device procurement and management processes prioritize security?
RQ3: What are the best ways for minimizing the collection of student data from such devices?

A) Risk Identification and Prioritization: Carry out a risk analysis of what may be considered weaknesses and threats concerning the use of wearables in education. This entails assessing products and their elements including Hardware, software, firmware, and communication protocols for vulnerabilities. The first step, therefore, when conducting a risk analysis of wearable technology into schools entails an identification and ranking of the possible dangers. This concerns a methodological approach to examining the various peculiarities of such devices, their implementation into the learning process, and the risks they are exposed to.

Key Steps in Risk Identification and Prioritization

First is 'Asset Identification'; Teaching institution wearable device – type of, make, model, and data captured by any wearables used throughout the institution. Second is threat identification; Examine possible threats that may seize the opportunities to attack Wearable Devices.

This includes considering: Malware and Phishing Attacks: viruses and Trojan horses and unscrupulous programs and tricks that undertake the theft of data or the violation of device protection. Third is data breaches; Hacking of private information of students which is stored in wearable technology gadgets. Forth is supply chain vulnerabilities; Challenges that arise from manufacturing and distributing wearable devices and the challenges that come with maintaining them. Fifth is vulnerability assessment; Examine the security of wearable gadgets, namely:

Hardware Vulnerabilities: Flaws in a particular subtype of components that form the device, or the manufacturing procedures utilized in producing the device. Sixth is software vulnerabilities: Defects in the Operating system, application software, or firmware. Seventh is communication vulnerabilities: Flaws in base station standards or data forwarding techniques. Eighth is threat modeling; Explain how possible threats can leverage the exposed weaknesses to threaten the security of wearable devices and student information. Next is risk analysis, while prioritizing the threats it is important to evaluate the possibility and the level of their impact on the company. Some of the important attributes to consider include, whether the data collected is confidential or not, the number of students that may have been affected, and the possible implications that may arise in terms of losses or damage to company or institution's reputation in the event of a successful attack. And last is risk prioritization: Evaluate the threats according to their potential and their potential risk that they pose to the system.

Table 2. Types of threats with their impact and risk score

Threat	Impact	Risk Score
Malware infection	High	High
Data breach	High	High
Phishing attacks	Medium	High
Loss or theft of devices	Medium	Medium
Supply chain vulnerabilities	High	Medium

Therefore, by timely risk identification and risk assessment, the educational institution can evaluate the risks that face the wearable devices and properly allocate the available resources to the most important risks. The list of different types of threats are mentioned in Table 2.

B) Security Testing and Validation: A number of testing approaches like penetration testing, vulnerability scanning, and code auditing should be used to uncover security weaknesses of wearable devices. Ensuring security and validating this security can be said to be the rather crucial elements of the general protective model of using wearable devices in education context. These processes make it possible to detect loopholes and gaps in the devices' hardware, software, and firmware for remedial actions to be taken. Vulnerability scanning, it is important to conduct vulnerability scanning using automated tools to identify vulnerability in Wearable device Software and firmware. These tools can search for most crucial misconfigurations, applications which are not updated, poor passwords, among others. It's crucial to making sure that vulnerability scanning tools allow vulnerability databases to be updated to reflect the current threats. Organize attacks against wearable devices to analyze the current security status and determine possible threats that can be threatening for the devices' owners. Hire wearable device security specialists that possess sound ethical hacking knowledge to perform penetration testing and discover vulnerabilities that may not be easily detected via automated testing. Enable the testing of realistic attack scenarios to discover the weaknesses in security measures and find out how these can be addressed.

To ensure that wearable devices are safe, it is important to test their physical Security as well as proper anti-tampering provisions, hardware tamper evidence, port and interface security, and protection against physical attacks. Additionally, it is important to know how these devices behave to environmental conditions for instance high/low temperature, humidity, and shock. Next is network security testing: Evaluate the wireless networks to which wearable devices connect; encryption, access control, and susceptibility to attacks such as Wi-Fi cracking. Data Transmission Security: If applicable consider the security of data transfer through wearable data communication protocols to guarantee that the data is encrypted and thus hard to intercept. 6. Social Engineering Testing: Phishing Simulations: Test email of social engineering by sending security faux-phishing emails with the purpose of checking how the users respond. This can assist to detect gaps in exercising the level of user education and training. Thus, through the strong security test and subsequent verification, it would be possible to reveal potential threats of wearable devices in educational institutions and minimize the likelihood of data leaks and other security threats.

C). Secure Device Procurement and Management: Create specific rules for purchasing and handling wearable devices such as security policies of the supply chain as well as management of devices and firmware. Ensuring the physical security of the devices as well as their acquisition is critical to the protection of these wearable devices and the information they bring. Educational institutions should implement the following strategies. First is vendor selection with good track record. Select vendors with whom you can foster good business relations since they are reputable in the market in terms of security and privacy. A vendor is a business entity that supplies goods and services to others; to minimize risks there is a need to look at the vendor's past records on how they have been handling their clients, the security measures they use, and Customers' feedback.

Next is device configuration, check and update the modifying default settings on the wearable gadgets in compliance with the security policies of the institution. Also, it is recommended that users need to select difficult and specific passwords when they use wearable devices. Use two-factor authentication (2FA) wherever allowed in order to increase the level of protection. Next is disable unnecessary features. This involves switching off options which might be a security menace like Bluetooth or Wi-Fi in case they are not relevant. Take care of regular updates, Wearable devices must upgrade their firmware frequently to patch vulnerabilities in firmware and enhance the devices' security. Setting devices to download or install firmware updates will also help to reduce chances of having wrong firmware versions. For inventory management keep accurate records. Keeping records of all wearable devices that are used in the course of the activities in an educational institution. Track devices used, with the help of automated systems which allow the location and other usage by workers to be monitored. Develop strict protocols on how to deal with computers that are old or damaged to avoid giving out sensitive information to the wrong people. Next is mobile device management (MDM)solutions in that it helps in the management of the wearable devices by configuring them, applying security measures and monitoring of usage.

5.2. Data Privacy and Protection

> **RQ4.** How can educational institute balance the need for data collection as well as data privacy concerns?
> **RQ5.**What are data privacy regulations that apply to these institutions while using wearable devices?

Privacy of data is an important issue when it comes to the use of wearable technology in teaching and learning. Wearable devices collect a variety of sensitive student data; *a) Health data*: For instance, it could be the heart rate, activity levels, sleep patterns, and any other health-related markers. b) *Performance data*: This may include data on the students, and their performance in various aspects of learning activities like the results, dropout rates, and interaction rates at classes. *c)Personal data*: This may include names of the students, identification numbers, and any other information that can easily identify the students. This data must be secured to prevent violation of student's privacy as well as the encouraged EU data protection laws.

A. Data Minimization and Anonymization

Limitation and anonymity of data are effective measures of reducing student's privacy in the use of wearable technology. This is why educational institutions can secure their data centers and limit the amount of personal data stored by excluding all non-relevant data and anonymizing the data that remains.

1. Data Minimization: Firstly, collect only essential data and find out subjective parameters that require values in order to fulfill the goals and objectives of the education process. Among the recommendations given to improve the limitations faced during the data collection process, was to avoid getting too much information that would compromise the privacy of the clients. That calls for appropriate consideration of the scope of collecting data to avoid the accumulation of possibly unnecessary information. For defining data retention policies, implementation of policies on when data should be erased or anonymized should also be clearly

defined as to the number of years it will be retained. It also assists in reducing the time that the sensitive data will have been stored hence minimizing cases of unauthorized access or data leaks.

2. Anonymization Techniques: Mask the identities and replace the values with junk or dummy values in order to provide anonymity as seen by replacing names with ID numbers. This makes such data lacking in attributes that may be used to pinpoint the identity of a person. Aggregate information in such a way that the results cannot be traced back to the specific student to make statistics that encompass different students and make an analysis of results without revealing the names of the students that were involved. For example, instead of storing student-specific heart rates, store overall statistics of the class in order to calculate the overall average pulse rate. Substitute an individual value with a more general and regular type of value. For example, instead of spreading the exact position of the students, the data can be generalized to a city or a region. And lastly pseudonymization, this should be common by replacing direct identifiers, for instance, the names, and social security numbers with pseudonyms or alphanumeric numbers. While it is still possible to find out more information about such users and reconnect the pseudonym with the person's real identity, it is a lot more challenging to do so.
3. Challenges and Considerations: a. Balancing Privacy and Utility: Although there have been attempts to minimize the amount of data collected as well as anonymize data to protect the users' privacy, efforts were made to ensure that the results are still relevant and should help in the education process. Some features should be public and easily accessible and at the same time some should be private b. Technical Feasibility: In realizing anonymization, it may be possible in some cases but not on other cases depending on the type of data being collected or the tools that are available, and techniques c. Emerging Technologies: Continuously assess various technologies and strategies for data anonymization and data privacy preservation.

Therefore, by applying data minimization and anonymization techniques to the collected data such educational institutions can minimize the risk of data leakage and more importantly protect students' data when using wearable devices in learning.

B. Data Encryption

The privacy of students is ensured through security measures such as data encryption in wearable devices for the collection of data. Special techniques involved in converting the data into a form that cannot be understood by anyone even if the data is tapped or retrieved by the wrong hands, is referred to as encryption.

1. **Encryption at Rest:** To prevent loss of data in the event that a wearable device is lost this information should be encrypted to prevent unauthorized access to data that may be stored online or in cloud storage. Secondly, employ reliable cryptographic methods of coding that are difficult to decipher such as AES or RSA. Also, ensure proper management of keys with adequate security measures due to the importance of the keys in decrypting of data.
2. **Encryption in Transit:** Use a secure communication channel that has been established as secure to transmit data from wearable devices to the servers (for instance using TLS, HTTPS). As a result, it avoids the possibility of interception and eavesdropping by unauthorized people during the transfer of data. For, end-to-end encryption there are methods like full or end-to-end encryption which provides for the encryption of data in the watch and the decryption in the recipient end only can be used.
3. **Key Management:** The encryption key should be stored in an environment that is very safe such as the HSM or should be part of the key management system. One should change encryption keys after some time so that there is no possibility of exposure of the encryption keys. If we talk about access controls, this can best be achieved by having measures such as strict access controls to ensure that only authorized people have access to encryption keys for decryption.
4. **Hardware-Based Encryption:** Some wearable devices may contain briefly integrated hardware encryption to supplement the functionality prescribed to the device's central processor.
5. **Cloud-Based Encryption:** If storing data collected by wearable devices uses a cloud, then one should take advantage of the encryption support offered by the cloud provider. Be careful and make sure that the provider is using a high level of encryption and follows the standards.
6. **Regular Updates:** Ensure that wearable devices and the software that comes with them have the latest updates that would counter all the security flaws that could lead to compromised encryption.

It is advisable to implement strong data encryption measures so that educational institutions can minimize the threat of data leakage incidents and secure personal student data obtained from wearable devices.

C. Access Controls

Choice controls are important to reduce the probability of unauthorized access to the student data gathered from wearables. Therefore, through effective adoption of access controls, educational institutions will be in a position to prevent any unauthorized persons from accessing and exposing, and individuals from accessing the information they need in order to perform their duties.

1. **Role-Based Access Control (RBAC):** Define Roles, Specify the power relations of various users of the educational institution including administrators, faculty, employees, and students. Secondly assign permissions, specify the roles so that you can grant suitable privileges to positions of the organization based on tasks they perform and information they are to process. This makes certain that users can only retrieve the information required by them to perform their tasks. And lastly It is necessary to monitor role-based access controls on a frequent and ongoing basis with the aim to revoke or grant access rights in the context of organizational changes and shifts in job responsibilities.
2. **Least Privilege Principle:** Grant the user the least level of permission that will enable him or her to complete a specific task or perform a given function. Refrain from providing permissions than are necessary in running your programs or using your device since this exposes your device to unauthorized people. Make the data available in a 'need-to-use' basis; this entails allowing the user to access certain data only when he/she needs it and denying him/her access to it when he/she does not need it.
3. **Multi-Factor Authentication (MFA):** Employ MFA so to enhance the security of login procedures by providing an additional layer. This involves the user having to prove his identity up to three times using a username, password, TOTP, or a biometric factor.
4. **Password Management:** Ensure good password practices, that is, ensure users set intricate passwords that cannot be easily cracked. *Password Restrictions*: Ensure that even though the user's accounts appear to be in different organizations they do not use the same password for all their accounts. Put or set up password expiry policies so that user has to change the existing password after a specific number of days.
5. **Access Logs and Auditing:** Control the access of user to the student through logs and audit trail. This can assist in identifying would-be unauthorized access as well as detecting security breaches. In addition to that daily, weekly or monthly depending on the traffic, analyze the access logs to find out if there is any strange behavior.
6. **Device Management:** The adoption of remote wiping for wearable devices should be done to ensure that losing the device does not lead to the compromise of the data they hold. Devices should also be put on standby locks after some time of inactivity so that nobody can access the devices without authorization.

7. **Education and Awareness:** Watch users and teach them about the proper password, don't open unknown emails, and if they receive such message, then it is better to immediately inform about it. Guide users towards relevant measures of device protection like the overall protection that means not connecting to unknown Wi-Fi networks, protection from viruses, constant updates, and others.

Through strengthened access protocols, the institutions minimize the susceptibility to access of students' information and/or invasion of personal privacy.

D. Data Breach Response Plan

This makes it imperative for an educational institution to have a data breach response plan that will act as a measure of minimizing the damages of a security breach of wearable devices in the same institution. The plan should identify measures to be undertaken if there is leakage, this will help in having a proper and efficient measure.

1. **Incident Detection and Response:** Ensure that the user management systems have strong auditing, logging and monitoring mechanisms to alert an organization of any activity which is out of the ordinary in the event of a breach. Identify procedures for notifying suspected data breach to other staff like IT security personnel, administrative staff and legal advisors. Work out a way through which the extent of the incident can be determined and therefore the response effort.
2. **Containment:** Identify affected devices or networks to reduce the spread of the breach in the network within the shortest time possible. One should consider denying users, particularly the unauthorized ones, access to the compromised systems or data for some time in a bid to minimize the impact created by the vice.
3. **Notification:** Establish guidelines for communicating with the parties involved like the students, parents, and the staff on the data breach incident. And obey legal guidelines pertaining to data privileges' violation like GDPR and FERPA with regards to reports of data violations. GDPR emphasizes the need for accountability and transparency in data managements practices (Ziccardi, 2020)
4. **Forensic Analysis:** Investigate the situation to find out details about the breach including the culprit behind the act.
5. **Lessons Learned:** Perform a formal review of the process that was used in responding to the incident in order to be able to pinpoint the problems as well as develop new strategies for change.

6. **Communication and Public Relations:** Develop a communication strategy on how to forward media inquiries or is to alert the general public about the incident. And for Crisis Management, establish a crisis management team to intervene for public relations and to contact the media any time there is a leakage of data.
7. **Legal and Regulatory Compliance:** Consult with a legal officer, especially for matters involving data privacy laws and regulations that may have been put in place: Inform the security incident to the relevant regulatory authorities like the data protection authorities or police as provided by the regulations on security incidents.

Having an effective data breach response plan is effective for educational institutions to reduces the vulnerability access by hackers, and the loss of student information and shows commitment to the students regarding their privacy.

E. Regulatory Compliance: GDPR, FERPA, and Local Laws

There are a number of data privacy laws that have to be adhered to by educational institutions while using wearable technology to capture and analyze students' data. The present regulations are designed to safeguard confidentiality and people's rights alongside guaranteeing proper data processing techniques.

1. **General Data Protection Regulation (GDPR):** According to the GDPR, any organization involved in the processing of the personal data of residents of the European Union is bound by the regulation regardless of the region, in which the organization is located. This means that those institutions of education located outside the EU but having data on students from the EU may still be required to apply the GDPR. Another factor that GDPR entails on the organizations includes; Lawful, fair, and transparent processing: The four principles of data processing of data include: the data processing must be Lawful, fair and be done in a transparent manner, that is, the individuals must be informed on how their data is collected, used and processed. It is recommended that organizations gather as little data as possible but data that are crucial in the decision-making process. Personal data should not be stored for more than is necessary to fulfill the reason for which the data is collected Also, Data subjects have access and the right to data portability which allows them to obtain and reuse their personal data in an easy understandable format for products and services by another organization. Under the GDPR regulations it is mandatory for organizations that experience a data breach must inform the authorities and the respective affected individuals. Organizations are accountable for providing evidence of compliance with GDPR putting adequate measures in place to ensure the protection of personal data.

2. **Family Educational Rights and Privacy Act (FERPA):** FERPA is effective for all the institutions that participate in any federal education program. It safeguards the rights of the student with regard to education records. The following are specific rights given to students under FERPA; Student records; 1) The rights of students to access their education records. 2) The right to request records corrections if they are erroneous. 3) The right to consent as to the disclosure of student education records. 4) The right to exercise complaint powers in the event FERPA provisions are infringed by the U. S. Department of Education.
3. **Local Laws:** However, besides GDPR and FERPA, educational institutions may require other laws in their country that protect data. These laws may have measures that have to be followed when it comes to the collection, storage, and processing of data. When it comes to compliance considerations it's important to carry out a risk analysis to establish the various risks that are inherent in the use of wearable technology to your institution's data privacy. Creating effective data privacy policies and procedures with existing laws is crucial. In addition to that educate the staff and students on various data privacy rights as well as their responsibilities. Be prepared to handle instances of data breach and report the same to the affected individuals and authorities is also essential.

F. Consent Management

This means that gaining the well-informed consent of the students (or their parents/guardians) is an essential precondition for data collection and data use based on wearable devices in school contexts. It means that consent management allows people to know the consequences of data collection, make decisions or at least give approvals themselves, and fulfil their rights to their own personal data.

- **Key Elements of Consent Management:** A) *Clear and Meaningful Information.* B) *Informed Decision.* C) *Opt-In or Opt-Out.* D) *Granular Consent.* E) *Documentation.* F) *Withdrawal of Consent.* G) *Age-Appropriate Consent*
- **Best Practices for Consent Management**: A) *Use Plain Language*: Do not use hard to comprehend terminologies and rather use simple and easy to comprehend language when defining data collection practices. when proposing data collection methods in a particular population then the research should use the host language. B) *Provide Options:* Create for students or for parents, multiple ways to give consent including consent to specific data collection or no consent at all. C) *Educate Users:* Develop information sheets that will be useful to the students and parents when it comes to gathering the information that is being collected as well as when seeking consent for the

same. D)*Review and Update*: It is necessary to have consistency in reviewing and updating the consent procedures with the advancement of technology, rules, regulations, and standards. Ensuring sound and feasible procedures in the management of consent shall showcase commitment to data privacy of learners and in turn enhance trust with the students as well as parents.

5.3. Cybersecurity Awareness and Training

RQ6. What are most effective methods for connecting students and educators in cybersecurity education?
RQ7. How can educational institute measure the effectiveness of their cybersecurity programs?

It is argued that teaching all the members of the institution about cybersecurity is critical in ensuring students' data and overall security in institutions that incorporate wearable technologies. Among the changes for educational institutions, there is a necessity to establish effective informative and methodological work to ensure the identification of the actions and behavior that promote appropriate and responsible behavior on the Internet and the protection of personal information (Joshi, Raising Awareness of Social Engineering among Adolescents: Psychological and Cybersecurity Perspective., 2023).

A. Educator Training

It is, therefore, the responsibility of educators to ensure that students embrace cybersecurity and have the best practice on contemporary technology. Ensuring that the educators who have direct access to students' information are well trained in cybersecurity empowers them to handle student's issues proactively by having protective measures on their own devices and training the students on proper conduct and proper use of the internet.

Key areas of focus for educator training:

- **Wearable Device Security Fundamentals:** 1) Key distinctions and risks of wearables of the case. 2) Also, the threats associated with security issues in wearables. 3) Gaining a knowledge of the recommendation on protection of wearable gadgets.
- **Data Privacy and Protection:** This article aims to help the reader to identify various data privacy regulations and their potential consequences for educational institutions.1) Discussing data minimization and anonymization and

getting to know about such methods. 2) Possible ways of safeguarding students' data produced by wearable technology.

- **Phishing and Social Engineering Awareness:** Firstly, the most notable type of attack common among employees on the social web is phishing and other social engineering. Approximately 90% of cyber-attacks involve social engineering which highlights the critical need for awareness and preventions to be taken (Lilit Manukyan, 2024). Secondly, Training teachers on how they can possibly recognize scam e-mails, messages or even a website. Thirdly, to educate educators on how to defend themselves against phishing scams and how to report the cases to the relevant security authorities.
- **Password Management and Authentication:** 1) Ensuring that people adopt proper passwords such as implementing passwords with a minimum of eight characters. 2) Easy adoption of using 2FA security measures to improve on safety of accounts 3) Offering proper recommendations concerning passwords policy.
- **Incident Response:** 1) Educating our educators about ways and means of recognizing and reporting security breach associated with wearables. 2) Guiding the cluster on how to handle and report incidents or in dealing with incidents such as how to contain it, how to investigate as well as the process of notifying the superiors.
- **Best Practices for Device Usage:** Making educators aware of the appropriate ways to use wearable devices including not to use free Wi-Fi, update the device frequently and do not install third party applications.
- **Curriculum Integration:** Introducing educators to different resources and lesson presentation options to foster understanding of cybersecurity topics and to teach students decision making and problem-solving skills in relation to cybersecurity.

Based on these findings, it is possible to train educators to acquire awareness and skills regarding cybersecurity thus enabling institutions of learning to cultivate positive cybersecurity culture among learners.

B. Student Awareness

Inculcating knowledge to the students regarding cybersecurity is vital so as to secure their identity along with avoiding a security compromise. When educational institutions establish the culture of cyber security, this will help students to make right decisions when it comes to cybersecurity and take necessary precautions online.

Key areas of focus for student awareness:

- **Phishing and Social Engineering:** 1) Educate students on the dangers of phishing procedures like emails or messages which may contain request for student's identity or a link they are supposed to click. 2) Enumerate some of the techniques employed by social engineers to get the targeted victims to disclose the sensitive information that they are seeking. 3) Teach students to be wary of any such messages and do not supply any information until one is sure it is not a hoax.
- **Strong Password Practices:** 1) Encourage the use of hard-to-guess passwords, that are not easily identifiable by other people. 2) Some of these tips include avoiding passwords such as dates of birth, children's or pets' names and dates. 3) People should be encouraged to use password managers when it comes to the use of passwords.
- **Data Privacy:** 1) That is why the protection of personal data is critical and sharing the information that can be potentially attractive to hackers would be rather unwise. 2) Explain students about data privacy laws and how they can invoke those laws. 3) Advise students to be careful about the kind of information they post on the social networks and other forums online. Research shows that only 54.3% of users demonstrating adequate awareness of privacy issues (Ghada Alrabia, 2022).
- **Safe Online Behavior:** 1) The mining companies should offer advice on how to stay protected online and this should include but not limited to avoiding opening any attachment from an unknown source, and when accessing the site, avoid unmetered Wi-Fi services among others, frequent update of the applications. 2) Explain to the students all the dangers that are associated with internet usage such as sharing location details, photos, and more financial details. 3) Teach the students to always question the content on social media and always cross-check on the information that they want to post.
- **Incident Reporting:**1) Explain to students what safe and unsafe behavior is, including receiving phishing emails, and having a virus on a computer or a stranger logged into a classroom account. 2) Some of them should be encouraged to report such cases to the school administrators or even the IT department.
- **Age-Appropriate Content:** Cater education programs in cybersecurity to the age of the students and the level of indoctrination that is expected from students. It is good to know that younger students may need more basic information and directions whereas old students can handle more complicated information.
- **Real-World Examples:** Share several examples and situations that happened, as well as case studies to demonstrate the effects of cyber threats and the need to stay safe online.

Thus, by ensuring students obtain the adequate knowledge and skills to be able to make these choices, educational institutions can contribute to the fight against cybercrimes by guaranteeing the students' safety from scammers.

C. Social Engineering Awareness

Some of the most popular and successful attacks carried out by hackers involve social engineering where they are able to trick people into releasing certain information or giving them [(Joshi, Raising Awareness of Social Engineering among Adolescents: Psychological and Cybersecurity Perspective., 2023)]. Social engineering attacks use human psychology to trick people into providing sensitive information or doing things that can harm security. It's important to educate and make people aware to fight these threats. Different methods and motivations are highlighted, showing the need for user education and awareness (Hussain Aldawood, 2019). Schools have the responsibility of availing information to students and teaching staff on how social engineering tricks are implemented so that one can avoid such tricks.

Key areas of focus for social engineering awareness:

- **Phishing Attacks:**1) List down practices of phishes, for instance fake emails or messages from known sources, containing a request for personal details or a prompting for or better yet clicks on dangerous links.2) The students should be made aware that the communications that they receive are from strangers and they should avoid getting such messages, especially those that are demanding very crucial issues or threatening ones.3) Be keen on the sender's credentials and the links sent to you by the sender before you accept and respond to them.
- **Pretexting:** 1) Discuss how pretexting can be used by attackers to pose as a trusted entity to get legitimize information access. 2) Educate the child to be wary of any messages from unknown people or people he has not communicated with before asking them questions like, 'What is your mother's/male guardian's/father's name?
- **Tailgating:** Inform students regarding the consequences of tailgating which is a case where unauthorized people access restricted areas by following the authorized persons. Ensure students carry out their activities in a very careful manner but make them aware to report anyone behaving in a suspicious manner to security.
- **Shoulder Surfing:** Describe how the attackers may be able to notice people as they type in passwords or any other secrets.

So, by teaching students and staff about social engineering techniques, institutions can make them aware of these attacks and how to avoid them.

5.4. Incident Response and Recovery

RQ8. What are best ways to communicating with stakeholders during data breach?
RQ9. How can institutions check their incident response plans to ensure their effectiveness?

That is why an effective action plan for a particular case in educational institutions is a mandatory factor when dealing with such cybercrimes as those related to wearable technology. The plan should outline procedures for:

1. **Incident Detection and Response:**
    a. *Monitoring and Logging*: Such systems should have efficient monitoring and logging systems that shall help track activities that look out of the norm and may depict a security breach.
    b. *Alert Systems*: Define specific communication processes regarding sending signals to specific people or groups (IT security personnel, system administrators etc.) about a security breach. c)*Incident Triage*: Determine how severe the incident is in order to plan the response.
2. **Containment:**
    a. *Isolate Affected Systems*: It should also be able to disconnect someone who has been infected quickly so that the malware does not spread any further.
    b. *Disable Access*: Suspend as many systems or databases as possible in order to avoid a greater loss.
3. **Investigation and Analysis:**
    a. *Gather Evidence*: Gather set of relevant information, records, and documents in order to identify the extent and nature of the occurrence.
    b. *Identify Root Cause*: Thus, one should take the matter a step further and find out why such an occurrence is possible in the first place.
4. **Eradication**:
    a. *Remove Malware or Threats*: Assist in eradicating the source of the incident, for instance; Query malware bugs or other vulnerabilities.
    b. *Restore Systems:* The elimination of effects that have taken place in the systems and data that were breached must also be done and this can be done using backups of the system or retrieved in other ways.

5. **Notification**: Notify Affected Individuals: Establish protocols for informing the people who may be involved like students, parents, teachers or school administrators of the occurrence and the subsequent consequences. Be mindful with the data privacy laws specifically in relation to the data breach notification provisions.
6. **Forensic Analysis**: Perform a forensics analysis to compile information on the breach so as to establish the culprit.
7. **Lessons Learned**: *Review and Improve:* it is necessary to produce a detailed assessment of the actions that were undertaken after the occurrence of the scenario and to adjust the plan impacting the field adequately.
8. **Communication and Public Relations**:
    a. *Develop a Communication Strategy*: Develop an action plan that would deal with media people or journalists and the public once the incident is reported.
    b. *Crisis Management*: There should be appointed crisis management team who would be responsible for answering the public and media in case of leak.
9. **Legal and Regulatory Compliance**:
    a. *Consult Legal Counsel*: To avoid violating data privacy regulations and laws if any, it is recommended to consult a lawyer.
    b. *Report the Breach*: Inform the concerned authorities including data protection authorities or law enforcement agencies depending on the regulations in force because data breaches are increasingly becoming common and students' data is sensitive and private, it is crucial to have a strategic plan in case of security threats, and having a strategic plan helps the institutions to reduce the consequences of security threats in the future.

6. SUMMARY

The use of wearable devices in education contains the promise and the issue. In this chapter, various vulnerabilities of wearable devices such as integrated computational capabilities, power supply, wireless channel, and data confidentiality have been discussed. In order to tackle these issues, a good cybersecurity framework is required. These comprise wearables security sounding, data privacy/protection policies, cybersecurity sensitivity/resources, and managing security breaches. The components highlighted in this framework will help educational institutions ward off risks while upholding the rights and integrity of students, something that holds learning institutions in high regard. Main recommendations refer to security auditing, protection of data and information, security awareness, and security incident contingency. Thus, addressing the issue of cybersecurity will let the educational institutions take advantages of the wearable technologies and, at the same time, protect the student's information and have a safer environment.

REFERENCES

Abdullah, B. (2024, April). The effects of inadequate cybersecurity controls chapter 1. *The effects of inadequate cybersecurity controls*.

Aldowah, H. R. (2017). Internet of Things in higher education: a study on future learning. *Journal of Physics: Conference Series (Vol. 892, No. 1, p. 012017). IOP Publishing., 892*.

Aldowah, H. U. (2019). Security in internet of things: issues, challenges and solutions. *In Recent Trends in Data Science and Soft Computing:Proceedings of the 3rd International Conference of Reliable Information and Communication.* Springer International Publishing.

Aldowah, H. U. (2021). Trust in iot systems: a vision on the current issues, challenges, and recommended solutions. 329-339.

AlZohbi, G. (2022, July). Green Hydrogen Generation: Recent Advances and Challenges. [). IOP Publishing.]. *IOP Conference Series. Earth and Environmental Science*, 1050(1), 012003.

Athif Lanang Untoro, D. A. (n.d.). *Design of an Interactive Learning System Using a Wearable Sensor Glove to Help Special Needs Student*. IEEE.

Ba, S., & Hu, X. (2023). Measuring emotions in education using wearable devices: A systematic review. *Computers & Education*, 200, 104797.

Byron Lowens, V. M. (2015). Design Recommendations to Improve the User Interaction with Wrist Worn Devices. *Workshop on Sensing Systems and Applications Using Wrist Worn Smart Devices,* 2015. Clemson, SC, USA: IEEE.

Diane Gan, R. H. (n.d.). Social engineering in the internet of everything. *Cutter IT Journal*, 20–29.

Fraga-Lamas, T. M.-C. (2018, December 7). owards The Internet of Smart Clothing: A Review on IoT Wearables and Garments for Creating Intelligent Connected E-Textiles. *Electronics (Basel)*, 36.

Ghada Alrabia, B. A. (2022). User Awareness of Wearable IoT Devices Privacy Issues. *2022 International Conference on Innovation and Intelligence for Informatics, Computing, and Technologies (3ICT).* IEEE Xplore.

Grøseth, L. K., & Sammerud, V. K. (2023). Solid Metal Fuels for Hydrogen Production for the Norwegian Heritage Fjords Maritime Sector (Bachelor's thesis, NTNU).

Hora, A., & Kulkarni, P. (2024, July). Wearables and Cybersecurity: Navigating the Threat Landscape. In 2024 2nd International Conference on Sustainable Computing and Smart Systems (ICSCSS) (pp. 563-567). IEEE.

Hussain Aldawood, G. S. (2019). A Taxonomy for Social Engineering Attacks via Personal Devices. *International Journal of Computer Applications (0975 – 8887).*

Jin, X., Li, L., Dang, F., Chen, X., & Liu, Y. (2022). A survey on edge computing for wearable technology. *Digital Signal Processing*, 125, 103146.

Joshi, R. (2023). Raising Awareness of Social Engineering among Adolescents: Psychological and Cybersecurity Perspective. In *Cybersecurity for Decision Makers* (pp. 99-109 (11)). CRC Press.

Joshi, R. (2023). Raising Awareness of Social Engineering among Adolescents: Psychological and Cybersecurity Perspective. In *Cybersecurity for decision makers* (pp. 99–109). CRC Press. DOI: 10.1201/9781003319887-7

Lee, L., Egelman, S., Lee, J. H., & Wagner, D. (2015). Risk perceptions for wearable devices. arXiv preprint arXiv:1504.05694.

Liang, J. M., Su, W. C., Chen, Y. L., Wu, S. L., & Chen, J. J. (2019). Smart interactive education system based on wearable devices. *Sensors (Basel)*, 19(15), 3260.

Lilit Manukyan, M. G. (2024). Social Engineering Attacks: How to Prevent. In *Journal of Digital Science* (Vol. 6).

Liu, Q., Zhang, Y., & Sun, Y. (2024, May). Intercepting Bluetooth Traffic from Wearable Health Devices. In 2024 IEEE Security and Privacy Workshops (SPW) (pp. 267-273). IEEE.

Manal Alshammari, M. A. (2022). Cyber Security for Connected wearable Devices. *International Conference on Business Analytics for Technology and Security* (p. 4). Dubai: IEEE Xplore.

María, A. Hernández-Mustieles, Y. E.-C.-R. (2024, march 14). Wearable Biosensor Technology in Education: A Systematic Review. preprints.*org*, 1-39.

Mills, A. J., Watson, R. T., Pitt, L., & Kietzmann, J. (2016). Wearing safe: Physical and informational security in the age of the wearable device. *Business Horizons*, 59(6), 615–622.

Mohammed, S. (2022). Health Wearable IoT (WIoT) Technology Devices Security and Privacy Vulnerability Analysis. 8th *International Conference on Information Technology Trends (ITT)*. (p. 5).

Ochaya, W. B. (2018). The cyberthreat in IoT, mobile learning, and wearable devices in education. *International Journal of Online Graduate Education*, 1(1).

Oveh, R. O. A. G. (2022). Mitigating Social Engineering Attack: A Focus on the Weak Human Link. *2022 5th Information Technology for Education and Development (ITED)* (p. 4). IEEE.

Rehman, S. &. (2023). Towards a Cybersecurity Awareness Program for Non-Technical Audiences in Malaysia. 85-97.

Rehman, S. U. (2020). Towards sustainable IoT ecosystem. *In 2020 2nd International conference on industrial electrical and electronics (ICIEE) (pp. 135-138). IEEE.* (pp. 135-138). IEEE.

Rehman, S. U. (2023). Impact of DoS/DDoS attacks in IoT environment: A study. *AIP Conference Proceedings (Vol. 2760, No. 1).* AIP Publishing.

Rehman, S. U.-C. (2023). *Privacy calculus model for online social networks: a study of Facebook users in a Malaysian university.* Springer.

Roberto De Fazio, A.-R. A.-H. (n.d.). An Energy-Autonomous Smart Shirt Employing Wearable Sensors for Users' Safety and Protection in Hazardous Workplaces. *MDPI*, 2022.

Sadek, I. C. (2019). *Memory snapshot dataset of a compromised host with malware using obfuscation evasion techniques.* Elsevier. DOI: 10.1016/j.dib.2019.104437

Sadek, I. C., Codjo, J., Rehman, S. U., & Abdulrazak, B. (2022). Security and privacy in the Internet of Things healthcare systems: Toward a robust solution in real-life deployment. *Computer Methods and Programs in Biomedicine Update*, 2, 100071. DOI: 10.1016/j.cmpbup.2022.100071

Sara Khosravi, S. G. (2022). Wearable Sensors for Learning Enhancement in Higher Education. *Sensors (Basel)*, 22(19), 17. DOI: 10.3390/s22197633

Sayam, A., Rahman, M. M., Sayem, A. S. M., Ahmed, A. F., & Alimuzzaman, S. (2024). Carbon-Based Textile-Structured Triboelectric Nanogenerators for Smart Wearables. *Advanced Energy and Sustainability Research*, 5(11), 2400127.

Ul Rehman, S., & Manickam, S. (2016). A study of smart home environment and its security threats. *Ul Rehman, S., & Manickam. International Journal of Reliability Quality and Safety Engineering*, 23(03), 1640005. DOI: 10.1142/S0218539316400052

Wei Gao, T. W. (2022, August 1). Toward a Systematic Survey on Wearable Computing for Education Applications. *IEEE Xplore*, 9, 15.

Ziccardi, G. (2020, may 29). Wearable Technologies and Smart Clothes in the Fashion Business: Some Issues Concerning Cybersecurity and Data Protection. *MDPI*.

Compilation of References

A brief history of wearable computing. (n.d.). Retrieved October 2, 2024, from https://www.media.mit.edu/wearables/lizzy/timeline.html#1268

Abd-alrazaq, A., Alajlani, M., Ahmad, R., AlSaad, R., Aziz, S., Ahmed, A., Alsahli, M., Damseh, R., & Sheikh, J. (2024). The Performance of Wearable AI in Detecting Stress Among Students: Systematic Review and Meta-Analysis. *Journal of Medical Internet Research*, 26, e52622. DOI: 10.2196/52622 PMID: 38294846

Abdullah, B. (2024, April). The effects of inadequate cybersecurity controls chapter 1. *The effects of inadequate cybersecurity controls*.

About Physical Activity | Physical Activity | CDC. (n.d.). Retrieved September 11, 2024, from https://www.cdc.gov/physicalactivity/php/about/index.html?CDC_AAref_Val=https://www.cdc.gov/physicalactivity/basics/pdfs/FrameworkGraphicV9.pdf

Adeghe, E. P., Okolo, C. A., & Ojeyinka, O. T. (2024). A review of wearable technology in healthcare: Monitoring patient health and enhancing outcomes. *OARJ of Multidisciplinary Studies*, 7(01), 142–148.

Adel, A. (2024). The Convergence of Intelligent Tutoring, Robotics, and IoT in Smart Education for the Transition from Industry 4.0 to 5.0. *Smart Cities, 7*(1), 325–369. https://www.mdpi.com/2624-6511/7/1/14

Aderinto, N., Olatunji, G., Abdulbasit, M. O., Edun, M., Aboderin, G., & Egbunu, E. (2023). Exploring the efficacy of virtual reality-based rehabilitation in stroke: a narrative review of current evidence. In *Annals of Medicine* (Vol. 55, Issue 2). Taylor and Francis Ltd. DOI: 10.1080/07853890.2023.2285907

Aeffner, F., Wilson, K., Martin, N. T., Black, J. C., Hendriks, C. L. L., Bolon, B., Rudmann, D. G., Gianani, R., Koegler, S. R., Krueger, J., & Young, G. D. (2017). The gold standard paradox in digital image analysis: Manual versus automated scoring as ground truth. In *Archives of Pathology and Laboratory Medicine* (Vol. 141, Issue 9, pp. 1267–1275). College of American Pathologists. DOI: 10.5858/arpa.2016-0386-RA

Afshari, M. (2014). *Wearable technology in education*. Cloud Tweaks–Connecting Our Planet.

Agarwal, M., Narcisse, D., Khouzam, N., & Khouzam, R. N. (2018). Wearable Cardioverter Defibrillator "The Lifevest": Device Design, Limitations, and Areas of Improvement. *Current Problems in Cardiology*, 43(2), 45–55. DOI: 10.1016/j.cpcardiol.2017.04.002 PMID: 29277221

Aggarwal, D., Sharma, D., & Saxena, A. (2023). Adoption of Artificial Intelligence (AI) For Development of Smart Education as the Future of a Sustainable Education System Journal of Artificial Intelligence, Machine Learning and Neural Network Adoption of Artificial Intelligence (AI) For Development of Smart Education as the Future of a Sustainable Education System. Journal of Artificial Intelligence. *Machine Learning and Neural Network*, 03(06), 23–28. Advance online publication. DOI: 10.55529/jaimlnn.36.23.28

Ahajjam, T., Moutaib, M., Aissa, H., Azrour, M., Farhaoui, Y., & Fattah, M. (2022). Predicting Students' Final Performance Using Artificial Neural Networks. *Big Data Mining and Analytics*, 5(4), 294–301. DOI: 10.26599/BDMA.2021.9020030

Ahmed, A., Aziz, S., Alzubaidi, M., Schneider, J., Irshaidat, S., Serhan, H. A., & Househ, M. (2023). Wearable devices for anxiety & depression: A scoping review. *Computer Methods and Programs in Biomedicine Update*, 3, 100095. DOI: 10.1016/j.cmpbup.2023.100095 PMID: 36743720

Al Ghawail, E. A., & Ben Yahia, S. (2022). Using the E-Learning Gamification Tool Kahoot! to Learn Chemistry Principles in the Classroom. *Procedia Computer Science*, 207, 2667–2676. https://doi.org/https://doi.org/10.1016/j.procs.2022.09.325. DOI: 10.1016/j.procs.2022.09.325

Al-Alusi, M. A., Ding, E., McManus, D. D., & Lubitz, S. A. (2019). Wearing Your Heart on Your Sleeve: the Future of Cardiac Rhythm Monitorin. In *Current Cardiology Reports* (Vol. 21, Issue 12). Springer. DOI: 10.1007/s11886-019-1223-8

Alam, A. (2022). Employing adaptive learning and intelligent tutoring robots for virtual classrooms and smart campuses: Reforming education in the age of artificial intelligence. In *Advanced computing and intelligent technologies* [Singapore: Springer Nature Singapore.]. *Proceedings of ICACIT*, 2022, 395–406.

Alam, A. (2023). Harnessing the Power of AI to Create Intelligent Tutoring Systems for Enhanced Classroom Experience and Improved Learning Outcomes. In *Intelligent Communication Technologies and Virtual Mobile Networks* (pp. 571–591). Springer Nature Singapore. DOI: 10.1007/978-981-99-1767-9_42

Al-Azawi, R., Joe, S. A., Al-Obaidy, M., & Westlake, J. (2019). The Use of Gamification Technique in Agile Development Methodology. In Sanchez, G., Rodríguez-González, S., Liberona, D., & Uden, L. (Eds.), *Communications in Computer and Information Science* (Vol. 1011, pp. 3–13). Springer Verlag., DOI: 10.1007/978-3-030-20798-4_1

Alban, A. Q., Alhaddad, A. Y., Al-Ali, A., So, W. C., Connor, O., Ayesh, M., Ahmed Qidwai, U., & Cabibihan, J. J. (2023). Heart Rate as a Predictor of Challenging Behaviours among Children with Autism from Wearable Sensors in Social Robot Interactions. *Robotics (Basel, Switzerland)*, 12(2), 55. Advance online publication. DOI: 10.3390/robotics12020055

Albogami, Y., Alfakhri, A., Alaqil, A., Alkoraishi, A., Alshammari, H., Elsharawy, Y., Alhammad, A., & Alhossan, A. (2024). Safety and quality of AI chatbots for drug-related inquiries: A real-world comparison with licensed pharmacists. *Digital Health*, 10, 20552076241253523. Advance online publication. DOI: 10.1177/20552076241253523 PMID: 38757086

Aldiabat, K. M., Matani, N. A., & Navenec, C. L. (2014). Mental health among undergraduate university students: A background paper for administrators, educators and healthcare providers. *Universal Journal of Public Health*, 2(8), 209–214.

Aldowah, H., Ul Rehman, S., & Umar, I. (2019). Security in internet of things: issues, challenges and solutions. In *Recent Trends in Data Science and Soft Computing:Proceedings of the 3rd International Conference of Reliable Information and Communication Technology (IRICT 2018)* (pp. 396-405). Springer International Publishing. DOI: 10.1007/978-3-319-99007-1_38

Aldowah, H. R. (2017). Internet of Things in higher education: a study on future learning. *Journal of Physics: Conference Series (Vol. 892, No. 1, p. 012017). IOP Publishing., 892.*

Aldowah, H. U. (2019). Security in internet of things: issues, challenges and solutions. *In Recent Trends in Data Science and Soft Computing:Proceedings of the 3rd International Conference of Reliable Information and Communication.* Springer International Publishing.

Aldowah, H. U. (2021). Trust in iot systems: a vision on the current issues, challenges, and recommended solutions. 329-339.

Aldowah, H., Rehman, S. U., Ghazal, S., & Umar, I. N. (2017, September). Internet of Things in higher education: A study on future learning. [). IOP Publishing.]. *Journal of Physics: Conference Series*, 892(1), 012017. DOI: 10.1088/1742-6596/892/1/012017

Aldowah, H., Ul Rehman, S., & Umar, I. (2021). Trust in iot systems: a vision on the current issues, challenges, and recommended solutions. *Advances on Smart and Soft Computing:Proceedings of ICACIn 2020*, 329-339.

Al-Emran, M., Al-Nuaimi, M. N., Arpaci, I., Al-Sharafi, M. A., & Jnr, B. A. (2022). Towards a wearable education: Understanding the determinants affecting students' adoption of wearable technologies using machine learning algorithms. *Education and Information Technologies*, 28(3), 2727–2746. DOI: 10.1007/s10639-022-11294-z

Al-Emran, M., Granić, A., Al-Sharafi, M. A., Ameen, N., & Sarrab, M. (2020). Examining the roles of students' beliefs and security concerns for using smartwatches in higher education. *Journal of Enterprise Information Management*, 34(4), 1229–1251. DOI: 10.1108/JEIM-02-2020-0052

Alhejaili, R., & Alomainy, A. (2023). The use of wearable technology in providing assistive solutions for mental well-being. *Sensors (Basel)*, 23(17), 7378. DOI: 10.3390/s23177378 PMID: 37687834

Alim, A., & Imtiaz, M. H. (2023). Wearable Sensors for the Monitoring of Maternal Health—A Systematic Review. In *Sensors* (Vol. 23, Issue 5). MDPI. DOI: 10.3390/s23052411

Aliverti, A. (2017). Wearable technology: Role in respiratory health and disease. *Breathe (Sheffield, England)*, 13(2), e27–e36. DOI: 10.1183/20734735.008417 PMID: 28966692

Alkhawaldeh, M. A., & Khasawneh, M. A. S. (2023). Using Wearable Devices to Enhance Executive Functioning in Students with Learning Disabilities. *Migration Letters : An International Journal of Migration Studies*, 20(S3), 1076–1089.

Allegra, A., Tonacci, A., Sciaccotta, R., Genovese, S., Musolino, C., Pioggia, G., & Gangemi, S. (2022). Machine Learning and Deep Learning Applications in Multiple Myeloma Diagnosis, Prognosis, and Treatment Selection. In *Cancers* (Vol. 14, Issue 3). MDPI. DOI: 10.3390/cancers14030606

Allen, L. M., Hay, M., & Palermo, C. (2022). Evaluation in health professions education—Is measuring outcomes enough? *Medical Education*, 56(1), 127–136. DOI: 10.1111/medu.14654 PMID: 34463357

Almusawi, H. A., & Durugbo, C. M. (2024). Determinants of teacher attitudes and innovative use of wearable technology. *IEEE Transactions on Learning Technologies*, 17, 1428–1439. DOI: 10.1109/TLT.2024.3386464

Almusawi, H. A., Durugbo, C. M., & Bugawa, A. M. (2021). Wearable technology in education: A systematic review. *IEEE Transactions on Learning Technologies*, 14(4), 540–554. DOI: 10.1109/TLT.2021.3107459

Almusawi, H. A., Durugbo, C. M., & Bugawa, A. M. (2021a). Innovation in physical education: Teachers' perspectives on readiness for wearable technology integration. *Computers & Education*, 167, 104185. DOI: 10.1016/j.compedu.2021.104185

Alowais, S. A., Alghamdi, S. S., Alsuhebany, N., Alqahtani, T., Alshaya, A. I., Almohareb, S. N., Aldairem, A., Alrashed, M., Bin Saleh, K., Badreldin, H. A., Al Yami, M. S., Al Harbi, S., & Albekairy, A. M. (2023). Revolutionizing healthcare: the role of artificial intelligence in clinical practice. In *BMC Medical Education* (Vol. 23, Issue 1). BioMed Central Ltd. DOI: 10.1186/s12909-023-04698-z

Alrayes, A., Henari, T. F., & Ahmed, D. A. (2024). ChatGPT in Education – Understanding the Bahraini Academics Perspective. *Electronic Journal of e-Learning*, 22(2), 112–134. DOI: 10.34190/ejel.22.2.3250

Al-Tamimi, A., & Al-Smadi, O. (2022). The Impact of Internet of Things (IoT) on Education: A Review. *International Journal of Advanced Research in Computer Science and Engineering*, 11(12), 106–111.

Altmann, A., Ebersberger, B., Mössenlechner, C., & Wieser, D. (2018). The Disruptive Power of Online Education: Challenges, Opportunities, Responses. In *The Disruptive Power of Online Education: Challenges, Opportunities, Responses.* Emerald Group Publishing Ltd. https://www.scopus.com/inward/record.uri?eid=2-s2.0-85115278629&partnerID=40&md5=26565d3224daa5c211b62e480f0ad9ca

Alvarez, V., Bower, M., de Freitas, S., Gregory, S., & de Wit, B. (2016). *The Use of Wearable Technologies in Australian Universities: Examples from Environmental Science*. Cognitive and Brain Sciences and Teacher Training.

AlZohbi, G. (2022, July). Green Hydrogen Generation: Recent Advances and Challenges. []. IOP Publishing.]. *IOP Conference Series. Earth and Environmental Science*, 1050(1), 012003.

Amano, I., & Murakami, A. (2013). Use of non-invasive total hemoglobin measurement as a screening tool for anemia in children. *Pediatrics International*, 55(6), 803–805. DOI: 10.1111/ped.12236 PMID: 24330295

Amer-Yahia, S. (2022). Towards AI-powered data-driven education. *Proceedings of the VLDB Endowment International Conference on Very Large Data Bases*, 15(12), 3798–3806. DOI: 10.14778/3554821.3554900

Amft, O., & Lukowicz, P. (2009). From backpacks to smartphones: Past, present, and future of wearable computers. *IEEE Pervasive Computing*, 8(3), 8–13. DOI: 10.1109/MPRV.2009.44

Amos, Z. (2024, October 25). The Futuristic Role of Wearables in Education. IoT for All. https://www.iotforall.com/the-futuristic-role-of-wearables-in-education

Amrutha, A.M., & Sidenur, B., P.S, B. *et al.* (. (2024). Estimation of haemoglobin using non-invasive portable device with spectroscopic signal application. *Scientific Reports*, 14, 8697. DOI: 10.1038/s41598-024-58990-z PMID: 38622231

Annansingh, F. (2017). An investigation into the gamification of e-learning in higher education. In *Disability and Equity in Higher Education Accessibility* (pp. 268–284). IGI Global., DOI: 10.4018/978-1-5225-2665-0.ch012

Apple Support (BH). (n.d.). *How to use the Blood Oxygen app on Apple Watch* Retrieved October 1, 2024, from https://support.apple.com/en-bh/120358

Apple. (n.d.). *Healthcare - Apple Watch* Retrieved September 30, 2024, from https://www.apple.com/healthcare/apple-watch/

Araújo, A. A., Gardim, L., Salma, J., Stephen, T., dos Santos, S. S., Silva, Í. R., de Godoy, S., & Mendes, I. A. C. (2024). Advancing Nursing Education through Wearable Electronic Devices: A Scoping. *Nurse Education in Practice*, 79, 104032. DOI: 10.1016/j.nepr.2024.104032 PMID: 38964082

Arroyo, I., Micciollo, M., Casano, J., Ottmar, E., Hulse, T., & Rodrigo, M. M. (2017). *"Wearable learning: multiplayer embodied games for math," in Proceedings of the Annual Symposium on Computer-Human Interaction in Play*. ACM., https://dl.acm.org/doi/pdf/10.1145/3116595.3116637 DOI: 10.1145/3116595.3116637

Askin, S., Burkhalter, D., Calado, G., & El Dakrouni, S. (2023). Artificial Intelligence Applied to clinical trials: opportunities and challenges. In *Health and Technology* (Vol. 13, Issue 2, pp. 203–213). Springer Science and Business Media Deutschland GmbH. DOI: 10.1007/s12553-023-00738-2

Athif Lanang Untoro, D. A. (n.d.). *Design of an Interactive Learning System Using a Wearable Sensor Glove to Help Special Needs Student*. IEEE.

Attallah, B., & Il-agure, Z. (2018). Wearable Technology: Facilitating or Complexing Education? *International Journal of Information and Education Technology (IJIET)*, 8(6), 433–436. DOI: 10.18178/ijiet.2018.8.6.1077

Attention-Deficit/Hyperactivity Disorder (ADHD). (n.d.). Retrieved September 15, 2024, from https://my.clevelandclinic.org/health/diseases/4784-attention-deficithyperactivity-disorder-adhd

Ayearst, L. E., Brancaccio, R., & Weiss, M. D. (2023). An Open-Label Study of a Wearable Device Targeting ADHD, Executive Function, and Academic Performance. *Brain Sciences*, 13(12), 1728. Advance online publication. DOI: 10.3390/brainsci13121728 PMID: 38137176

Baber, C. (2001). Wearable Computers: A Human Factors review. *International Journal of Human-Computer Interaction*, 13(2), 123–145. DOI: 10.1207/S15327590IJHC1302_3

Bahrain Business Laws | Personal Data Protection Law. (n.d.). Bahrainbusinesslaws.com. https://bahrainbusinesslaws.com/laws/Personal-Data-Protection-Law

Bahrain Pioneers AI Regulation with New Standalone Law - HyScaler. (2023). HyScaler. https://hyscaler.com/insights/bahrain-pioneers-ai-regulation

Bahrain.bh. (2023). Main Template. Bahrain.bh. https://bahrain.bh/wps/portal/en/BNP/HomeNationalPortal/ContentDetailsPage/

Baker, S. B., Xiang, W., & Atkinson, I. (2017). Internet of Things for Smart Healthcare: Technologies, Challenges, and Opportunities. In *IEEE Access* (Vol. 5, pp. 26521–26544). Institute of Electrical and Electronics Engineers Inc., DOI: 10.1109/ACCESS.2017.2775180

Bano, K. S., Bhuyan, P., & Ray, A. (2022, December). EEG-based brain computer interface for emotion recognition. In *2022 5th International Conference on Computational Intelligence and Networks (CINE)* (pp. 1-6). IEEE. DOI: 10.1109/CINE56307.2022.10037255

Ba, S., & Hu, X. (2023). Measuring emotions in education using wearable devices: A systematic review. *Computers & Education*, 200, 104797. Advance online publication. DOI: 10.1016/j.compedu.2023.104797

Basnet, S. (2024). Artificial Intelligence and machine learning in human resource management: Prospect and future trends. *International Journal of Research Publication and Reviews*, 5(1), 281–287. DOI: 10.55248/gengpi.5.0124.0107

Beard, H. R., Marquez-Lara, A. J., & Hamid, K. S. (2016). Using wearable video technology to build a Point-of-View surgical education library. *JAMA Surgery*, 151(8), 771. DOI: 10.1001/jamasurg.2016.0976 PMID: 27276102

Begg, E. J., Barclay, M. L., & Kirkpatrick, C. J. M. (1999). The therapeutic monitoring of antimicrobial agents. *British Journal of Clinical Pharmacology*, 47(1), 23–30. DOI: 10.1046/j.1365-2125.1999.00850.x PMID: 10073735

Bellotti, F., Berta, R., De Gloria, A., Lavagnino, E., Antonaci, A., Dagnino, F. M., & Ott, M. (2013). A gamified short course for promoting entrepreneurship among ICT engineering students. *Proceedings - 2013 IEEE 13th International Conference on Advanced Learning Technologies, ICALT 2013*, 31–32. DOI: 10.1109/ICALT.2013.14

Bezzina, S., & Dingli, A. (2023, July). Rethinking gamification through artificial intelligence. In *International Conference on Human-Computer Interaction* (pp. 252-263). Cham: Springer Nature Switzerland.

Bhaltadak, V., Ghewade, B., & Yelne, S. (2024). A Comprehensive Review on Advancements in Wearable Technologies: Revolutionizing Cardiovascular Medicine. *Cureus*. Advance online publication. DOI: 10.7759/cureus.61312 PMID: 38947726

Bhawra, J., Khadilkar, A., Krishnaveni, G. V., Kumaran, K., & Katapally, T. R. (2023). The 2022 India Report Card on physical activity for children and adolescents. *Journal of Exercise Science and Fitness*, 21(1), 74–82. DOI: 10.1016/j.jesf.2022.10.013 PMID: 36408207

Blanco-González, A., Cabezón, A., Seco-González, A., Conde-Torres, D., Antelo-Riveiro, P., Piñeiro, Á., & Garcia-Fandino, R. (2023). The Role of AI in Drug Discovery: Challenges, Opportunities, and Strategies. In *Pharmaceuticals* (Vol. 16, Issue 6). Multidisciplinary Digital Publishing Institute (MDPI). DOI: 10.3390/ph16060891

Bloomfield, L. S. P., Fudolig, M. I., Kim, J., Llorin, J., Lovato, J. L., McGinnis, E. W., McGinnis, R. S., Price, M., Ricketts, T. H., Dodds, P. S., Stanton, K., & Danforth, C. M. (2024). Predicting stress in first-year college students using sleep data from wearable devices. *PLOS Digital Health*, 3(4), e0000473. DOI: 10.1371/journal.pdig.0000473 PMID: 38602898

Borthwick, A. C., Anderson, C. L., Finsness, E. L., & Foulger, T. S. (2015). Special article personal wearable technologies in education: Value or villain? *Journal of Digital Learning in Teacher Education*, 31(3), 85–92. DOI: 10.1080/21532974.2015.1021982

Bose, P. (2024). *Augmented Reality in Anatomy Education and Training*. News Medical Life Sciences. https://www.news-medical.net/health/Augmented-Reality-in-Anatomy-Education-and-Training.aspx

Boskic, N., & Hu, S. (2015). Gamification in higher education: How we changed roles. In L. Kolas & R. Munkvold (Eds.), *Proceedings of the European Conference on Games-based Learning* (Vols. 2015-January, pp. 741–748). Dechema e.V. https://www.scopus.com/inward/record.uri?eid=2-s2.0-84955154619&partnerID=40&md5=b0fcf9bcb3a235f668a0fff8aa1a224a

Bower, M., & Sturman, D. (2020). *Wearable technology and immersive learning*. Journal of Educational Technology.

Bower, M., Sturman, D., & Alvarez, V. (2016), "Perceived utility and feasibility of wearable technologies in higher education," in Proc. *15th World Conference on Mobile and Contextual Learning, mLearn*, Sydney.

Bradley, P. (2006). The history of simulation in medical education and possible future directions. *Medical Education*, 40(3), 254–262. DOI: 10.1111/j.1365-2929.2006.02394.x PMID: 16483328

Brandt, J. D., DuBiner, H. B., Benza, R., Sall, K. N., Walker, G. A., Semba, C. P., Budenz, D., Day, D., Flowers, B., Lee, S., Nguyen, Q., & Wirta, D. (2017). Long-term Safety and Efficacy of a Sustained-Release Bimatoprost Ocular Ring. *Ophthalmology*, 124(10), 1565–1566. DOI: 10.1016/j.ophtha.2017.04.022 PMID: 28528010

Breed, R., Kay, A., Spittle, M., & Orth, D. (2024). The effect of pedagogical approach on physical activity of girls during physical education. *Research Quarterly for Exercise and Sport*, 95(4), 1–10. DOI: 10.1080/02701367.2024.2329165 PMID: 38771866

Bridging the Digital Divide in Education | LapSafe News & Articles. (2024, February 22). Www.lapsafe.com. https://www.lapsafe.com/blog/bridging-the-digital-divide-in-education

Brown, W. J., & Reid, C. (2022). Implementing a cost effective and configurable hybrid simulation platform in healthcare education, using wearable and web-based technologies. *Smart Learning Environments*, 9(1), 20. DOI: 10.1186/s40561-022-00201-1

Bruijns, B. A., Truelove, S., Johnson, A. M., Gilliland, J., & Tucker, P. (2020). Infants' and toddlers' physical activity and sedentary time as measured by accelerometry: A systematic review and meta-analysis. *The International Journal of Behavioral Nutrition and Physical Activity*, 17(1), 14. DOI: 10.1186/s12966-020-0912-4 PMID: 32028975

BrundhaDrM. (2019, July 1). Priyadharshini S. Comparisonof haemoglobin estimationby Sahli's two—time average, Sahli's three—time average methods and automatedanalyzermethod:A different approach in clinical pathology. *Int J Clin Diagn Pathol.*, 2(2), 291–295. DOI: 10.33545/pathol.2019.v2.i2e.118

Budnick, C. J., Stults-Kolehmainen, M., Dadina, C., Bartholomew, J. B., Boullosa, D., Ash, G. I., Sinha, R., Blacutt, M., Haughton, A., & Lu, T. (2023). Motivation states to move, be physically active and sedentary vary like circadian rhythms and are associated with affect and arousal. *Frontiers in Sports and Active Living*, 5, 1094288. DOI: 10.3389/fspor.2023.1094288 PMID: 37143586

Burnham, J. P., Lu, C., Yaeger, L. H., Bailey, T. C., & Kollef, M. H. (2018). Using wearable technology to predict health outcomes: A literature review. *Journal of the American Medical Informatics Association : JAMIA*, 25(9), 1221–1227. DOI: 10.1093/jamia/ocy082 PMID: 29982520

Byron Lowens, V. M. (2015). Design Recommendations to Improve the User Interaction with Wrist Worn Devices. *Workshop on Sensing Systems and Applications Using Wrist Worn Smart Devices,* 2015. Clemson, SC, USA: IEEE.

Campbell, L. O., & Frawley, C. (2023). An exploration of factors that predict higher education faculty members' intentions to utilize emerging technologies. *Educational Technology Research and Development*, 72(2), 643–659. DOI: 10.1007/s11423-023-10321-1

Campbell, L.. (2020). Ethical considerations in AI for education. *Educational Research Review*.

Cao, L. (2023). Trans-AI/DS: Transformative, transdisciplinary and translational artificial intelligence and data science. *International Journal of Data Science and Analytics*, •••, 1–13.

Caponetto, I., Earp, J., & Ott, M. (2014). Gamification and education: A literature review. In C. Busch (Ed.), *Proceedings of the European Conference on Games-based Learning* (Vol. 1, pp. 50–57). Dechema e.V. https://www.scopus.com/inward/record.uri?eid=2-s2.0-84923559781&partnerID=40&md5=84bb4279ab429280fa63ad588e2e6552

Cappon, G., Acciaroli, G., Vettoretti, M., Facchinetti, A., & Sparacino, G. (2017). Wearable continuous glucose monitoring sensors: A revolution in diabetes treatment. *Electronics (Basel)*, 6(3), 65. Advance online publication. DOI: 10.3390/electronics6030065

Cardona, M. A., Rodríguez, R. J., & Ishmael, K. (2023). Artificial Intelligence and the Future of Teaching and Learning Insights and Recommendations. In Office of Educational Technology (pp. 18–34). U.S. Department of Education, Office of Educational Technology, https://tech.ed.gov/

Carlozzi, N. E., Gade, V., Rizzo, A. S., & Tulsky, D. S. (2013). Using virtual reality driving simulators in persons with spinal cord injury: Three screen display versus head mounted display. *Disability and Rehabilitation. Assistive Technology*, 8(2), 176–180. DOI: 10.3109/17483107.2012.699990 PMID: 22775982

Chamola, P., & Mittal, P. (2023). A non-invasive optical method for anaemia detection. *Physica Scripta*, 98(6), 065023. DOI: 10.1088/1402-4896/acd5b8

Chang, Y., Lee, S., Wong, S. F., & Jeong, S. P. (2022). AI-powered learning application use and gratification: An integrative model. *Information Technology & People*, 35(7), 2115–2139. DOI: 10.1108/ITP-09-2020-0632

Chau, K. Y., Lam, M. H. S., Cheung, M. L., Tso, E. K. H., Flint, S. W., Broom, D. R., Tse, G., & Lee, K. Y. (2019). Smart technology for healthcare: Exploring the antecedents of adoption intention of healthcare wearable technology. *Health Psychology Research*, 7(1). Advance online publication. DOI: 10.4081/hpr.2019.8099 PMID: 31583292

Chen, L., Chen, P., & Lin, Z. (2020). Artificial Intelligence in Education: A Review. *IEEE Access : Practical Innovations, Open Solutions*, 8, 75264–75278. DOI: 10.1109/ACCESS.2020.2988510

Chen, X., Xie, H., Zou, D., & Hwang, G. (2020). Application and theory gaps during the rise of Artificial Intelligence in Education, *Computers and Education. Artificial Intelligence*, 1, 100002. Advance online publication. DOI: 10.1016/j.caeai.2020.100002

Chih, H., Ahmed, T., Chiu, A. P., Liu, Y., Kuo, H., Yang, A. C., & Lien, D. (2023). Multitask learning for automated sleep staging and wearable technology integration. *Advanced Intelligent Systems*. Advance online publication. DOI: 10.1002/aisy.202300270

Chiu, W., Oh, G. E., & Cho, H. (2021). Factors Influencing Consumers' adoption of Wearable Technology: A Systematic Review and Meta-Analysis. *International Journal of Information Technology & Decision Making*, 20(03), 933–958. DOI: 10.1142/S0219622021500206

Choi, Y., Jeon, Y., Wang, L., Sensors, K. K.-, & 2017, undefined. (n.d.). A biological signal-based stress monitoring framework for children using wearable devices. *Mdpi.ComY Choi, YM Jeon, L Wang, K KimSensors, 2017•mdpi.Com*. DOI: 10.3390/s17091936

Choo, D., Dettman, S., & Care, R. D.-… and C., & 2017, undefined. (2017). Talking to toddlers: Drawing on mothers' perceptions of using wearable and mobile technology in the home. *Ebooks.Iospress.NlD Choo, S Dettman, R Dowell, R CowanIntegrating and Connecting Care, 2017•ebooks.Iospress.Nl*, 239, 21–27. DOI: 10.3233/978-1-61499-783-2-21

Christie, M., Carey, M., Robertson, A., & Grainger, P. (2015). Putting transformative learning theory into practice. *Australian Journal of Adult Learning*, 55(1).

Christmann, C. A., Hoffmann, A., Zolynski, G., & Bleser, G. (2018). Stress-mentor: Linking gamification and behavior change theory in a stress management application. In, S. C. (Ed.), *20th International Conference on HCI, HCI International 2018* (Vol. 851, pp. 387–393). Springer Verlag. DOI: 10.1007/978-3-319-92279-9_52

Chu, S. L., Garcia, B. M., & Rani, N. (2023). Research on wearable technologies for learning: a systematic review. In Frontiers in Education (Vol. 8). https://doi.org/ DOI: 10.3389/feduc.2023.1270389

Clements, A. J., Ahmed, S., & Henderson, B. (2017). Student experience of gamified learning: A qualitative approach. In Pivec, M., & Grundler, J. (Eds.), *Proceedings of the 11th European Conference on Games Based Learning, ECGBL 2017* (pp. 88–94). Academic Conferences and Publishing International Limited., https://www.scopus.com/inward/record.uri?eid=2-s2.0-85036452359&partnerID=40&md5=9f24fc6f67145e6969bd3c6974a90ea1

Cliffe Schreuders, Z., & Butterfield, E. (2016). Gamification for teaching and learning computer security in higher education. *2016 USENIX Workshop on Advances in Security Education, ASE 2016, Co-Located with the 25th USENIX Security Symposium*. https://www.scopus.com/inward/record.uri?eid=2-s2.0-85081683632&partnerID=40&md5=3d08087cb4e82536a51f8c9c113e1963

Corbo, I., Forte, G., Favieri, F., & Casagrande, M. (2023). Poor sleep quality in aging: The association with mental health. *International Journal of Environmental Research and Public Health*, 20(3), 1661. DOI: 10.3390/ijerph20031661 PMID: 36767029

Cornejo-Plaza, I., & Cippitani, R. (2023). Ethical and Legal Considerations of Artificial Intelligence in Higher Education: Challenges and Prospects. *Revista de Educacion y Derecho*, 28. Advance online publication. DOI: 10.1344/REYD2023.28.43935

Costello, R. (2017). Gaming innovations in higher education: Emerging research and opportunities. In *Gaming Innovations in Higher Education: Emerging Research and Opportunities*. IGI Global., DOI: 10.4018/978-1-5225-2981-1

Cowperthwait, A. L., Campagnola, N., Doll, E. J., Downs, R. G., Hott, N. E., Kelly, S. C., Montoya, A., Bucha, A. C., Wang, L., & Buckley, J. M. (2015). Tracheostomy overlay system: An effective learning device using standardized patients. *Clinical Simulation in Nursing*, 11(5), 253–258. DOI: 10.1016/j.ecns.2015.03.001

Dalledonne, S. (2020). *Remote sensing data access policy, data products regulatory framework and intellectual property rights challenges in an era of environmental protection urgency. 2020-Octob*. https://www.scopus.com/inward/record.uri?eid=2-s2.0-85100923787&partnerID=40&md5=200d351cdc4f11e90f366bf237a40ff7

Darejeh, A., & Salim, S. S. (2016). Gamification Solutions to Enhance Software User Engagement—A Systematic Review. *International Journal of Human-Computer Interaction*, 32(8), 613–642. DOI: 10.1080/10447318.2016.1183330

Daru, J. (2022, May). Sustainable Development Goals for anaemia: 20 years later, where are we now? *The Lancet. Global Health*, 10(5), e586–e587. DOI: 10.1016/S2214-109X(22)00127-9 PMID: 35427502

Das Mahapatra, P. P., Roy, C., Agarwal, K., Banerjee, J., & Sharma, S. (2024). Performance of the non-invasive point-of-care device, EzeCheck, for haemoglobin assessment in adults and children in community and institutional care settings. *PLOS Digital Health*, 3(5), e0000500. DOI: 10.1371/journal.pdig.0000500 PMID: 38717987

Das Mahapatra, P. P., Roy, C., Agarwal, K., Sharma, S. D., Chowdhury, S. R., Sharma, S., & Meher, D. (2024). Non-invasive hemoglobin screening device: A promising digital method for reducing anemia prevalence through routine screening and timely intervention. *Hematology*, 29(1), 2365078.

Dash, S., Shakyawar, S. K., Sharma, M., & Kaushik, S. (2019). Big data in healthcare: Management, analysis and future prospects. *Journal of Big Data*, 6(1), 54. Advance online publication. DOI: 10.1186/s40537-019-0217-0

Data protection in schools. (2024, August 7). Unicef.org. https://www.unicef.org/eca/reports/data-protection-schools

Datta, N., MacQueen, I. T., Schroeder, A. D., Wilson, J. J., Espinoza, J. C., Wagner, J. P., Filipi, C. J., & Chen, D. C. (2015). Wearable technology for global surgical teleproctoring. *Journal of Surgical Education*, 72(6), 1290–1295. DOI: 10.1016/j.jsurg.2015.07.004 PMID: 26276303

Daud, R., Salam, S., Mohamad, S. N. M., & Yusoff, A. M. (2017). Modeling a mobile gamification model to increase student engagement: An analysis using analytic hierarchy process. *Advanced Science Letters*, 23(9), 8707–8712. DOI: 10.1166/asl.2017.9955

Davidesco, I., Matuk, C., Bevilacqua, D., Poeppel, D., & Dikker, S. (2021). Neuroscience Research in the Classroom: Portable Brain Technologies in Education Research. *Educational Researcher*, 50(9), 649–656. DOI: 10.3102/0013189X211031563

Dayal, S., & Pratibha, . (2023). Roadblocks in education amidst global crisis—A study based in India. *PLoS One*, 18(10), e0292465. DOI: 10.1371/journal.pone.0292465 PMID: 37847726

de Arriba-Pérez, F., Caeiro-Rodríguez, M., & Santos-Gago, J. M. (2017). "Towards the use of commercial wrist wearables in education," 4th Experiment@International Conference (exp.at'17), Faro, Portugal, 2017, pp. 323-328, Doi: DOI: 10.1109/EXPAT.2017.7984354

De Freitas, S., Irving, L., Verbert, K., Gibson, D., Star, K., Alvarez, V., & Charleer, S. (2017). How to use gamified dashboards and learning analytics for providing immediate student feedback and performance tracking in higher education. *26th International World Wide Web Conference 2017, WWW 2017 Companion*, 429–434. DOI: 10.1145/3041021.3054175

De Zambotti, M., Cellini, N., Goldstone, A., Colrain, I. M., & Baker, F. C. (2019). Wearable sleep technology in clinical and research settings. *Medicine and Science in Sports and Exercise*, 51(7), 1538–1557. DOI: 10.1249/MSS.0000000000001947 PMID: 30789439

Debard, G., De Witte, N., Sels, R., Mertens, M., Van Daele, T., & Bonroy, B. (2020). Making wearable technology available for mental healthcare through an online platform with stress detection algorithms: The CareWear project. *Journal of Sensors*, 2020, 1–15. Advance online publication. DOI: 10.1155/2020/8846077

Deci, E. L., & Ryan, R. M. (1985). The general causality orientations scale: Self-determination in personality. *Journal of Research in Personality*, 19(2), 109–134. DOI: 10.1016/0092-6566(85)90023-6

De-La-Fuente-Robles, Y., Ricoy-Cano, A., Albín-Rodríguez, A., López-Ruiz, J. L., & Espinilla-Estévez, M. (2022). Past, Present and Future of Research on Wearable Technologies for Healthcare: A bibliometric analysis using Scopus. *Sensors (Basel)*, 22(22), 8599. DOI: 10.3390/s22228599 PMID: 36433195

Della Ventura, M. (2017). Creating inspiring learning environments by means of digital technologies: A case study of the effectiveness of WhatsApp in music education. *E-Learning, E-Education (Eds.), Online training*, Springer (2017), pp. 36-45. http://dx.doi.org/DOI: 10.4108/eai.26-7-2017.152906

Department of Health. T. (n.d.). *Start Active, Stay Active publications/physical-activity-guidelines-uk-chief-medical-officers-report.* Retrieved September 11, 2024, from https://www.gov.uk/government/

Dexcom. (n.d.). *Dexcom Continuous Glucose Monitoring* Retrieved September 30, 2024, from https://www.dexcom.com/

Dhar, P., Rocks, T., Samarasinghe, R. M., Stephenson, G., & Smith, C. (2021). Augmented reality in medical education: Students' experiences and learning outcomes. *Medical Education Online*, 26(1), 1953953. DOI: 10.1080/10872981.2021.1953953 PMID: 34259122

Dhar, R. (2023). Smart wearable devices for real-time health monitoring. *Asian Journal of Medical Sciences*, 14(12), 1–3. DOI: 10.3126/ajms.v14i12.58664

Dhiman, V. K. (2024). The emergence of AI in mental health: A transformative journey. *World Journal of Advanced Research and Reviews*, 22(1), 1867–1871. DOI: 10.30574/wjarr.2024.22.1.1298

Diane Gan, R. H. (n.d.). Social engineering in the internet of everything. *Cutter IT Journal*, 20–29.

Díaz-Quesada, G., Bahamonde-Pérez, C., Giménez-Egido, J. M., & Torres-Luque, G. (2021). Use of wearable devices to study physical activity in early childhood education. *Sustainability (Basel)*, 13(24), 13998. DOI: 10.3390/su132413998

Diaz-Ramos, R. E., Noriega, I., Trejo, L. A., Stroulia, E., & Cao, B. (2023). Using wearable devices and speech data for personalized machine learning in early detection of mental disorders: Protocol for a participatory research study. *JMIR Research Protocols*, 12(1), e48210. DOI: 10.2196/48210 PMID: 37955959

Dimauro, G., De Ruvo, S., Di Terlizzi, F., Ruggieri, A., Volpe, V., Colizzi, L., & Girardi, F. (2020). Estimate of anemia with new non-invasive systems—A moment of reflection. *Electronics (Basel)*, 9(5), 780. DOI: 10.3390/electronics9050780

Dimitrov, D. V. (2016). Medical internet of things and big data in healthcare. In *Healthcare Informatics Research* (Vol. 22, Issue 3, pp. 156–163). Korean Society of Medical Informatics. DOI: 10.4258/hir.2016.22.3.156

Ding, L., Hua, H., Zhu, H., Zhu, S., Lu, J., Zhao, K., & Xu, Q. (2020). Effects of virtual reality on relieving postoperative pain in surgical patients: A systematic review and meta-analysis. In *International Journal of Surgery* (Vol. 82, pp. 87–94). Elsevier Ltd., DOI: 10.1016/j.ijsu.2020.08.033

Dong, P., & A, L. (2020). RETRACTED: Simulation of physical education teaching based on FPGA and wearable VR equipment. *Microprocessors and Microsystems*, 81, 103773. DOI: 10.1016/j.micpro.2020.103773

dos Reis Lívero, F. A., da Silva, G. R., Amaral, E. C., de Souza, A. N. V., Baretta, I. P., Diegues, M. E. M., Arpini, E., & Lovato, E. C. W. (2021). Playfulness in the classroom: Gamification favor the learning of pharmacology. *Education and Information Technologies*, 26(2), 2125–2141. DOI: 10.1007/s10639-020-10350-w

Duggal, K., & Gupta, L. R. (2020). Hope Enabler: A Novel Gamification-Based Approach to Enhance Classroom Engagement. In *Lecture Notes in Networks and Systems* (Vol. 121, pp. 501–519). Springer. DOI: 10.1007/978-981-15-3369-3_38

Du, J., & Nan, Z. (2021). RETRACTED: Research on the intelligent model of progress in physical education training based on motion sensor. *Microprocessors and Microsystems*, 82, 103903. DOI: 10.1016/j.micpro.2021.103903

ECRI. (n.d.). *Pulse Oximetry: Choosing the Right Sensor and Applying It Correctly* Retrieved September 30, 2024, from https://www.ecri.org/components/HDJournal/Pages/Pulse-Oximetry-Choosing-the-Right-Sensor.aspx

El Asmar, W. (2022). *The Effectiveness of AI-Powered Digital Educational Platforms: Students' Attainment and Teachers' Teaching strategies in a private high school in Dubai* (Master's thesis, The British University in Dubai).

Emilio Standoli, C., Renata Guarneri, M., Perego, P., Mazzola, M., Mazzola, A., Andreoni, G., & Sp, N. (2016). *A Smart Wearable Sensor System for Counter-Fighting Overweight in Teenagers*. DOI: 10.3390/s16081220

Erenli, K. (2013). The impact of gamification-recommending education scenarios. [IJET]. *International Journal of Emerging Technologies in Learning*, 8(S1), 15. Advance online publication. DOI: 10.3991/ijet.v8iS1.2320

Esangbedo, C., Zhang, J., Esangbedo, M., Kone, S., & Xu, L. (2023). The role of industry-academia collaboration in enhancing educational opportunities and outcomes under the digital driven Industry 4.0. *Journal of Infrastructure. Policy and Development*, 8(1). Advance online publication. DOI: 10.24294/jipd.v8i1.2569

Escobar-Linero, E., Muñoz-Saavedra, L., Luna-Perejón, F., Sevillano, J. L., & Domínguez-Morales, M. (2023). Wearable Health Devices for Diagnosis Support: Evolution and Future Tendencies. In *Sensors* (Vol. 23, Issue 3). MDPI. DOI: 10.3390/s23031678

eSight Eyewear. (n.d.). *How it Works* Retrieved September 30, 2024, from https://www.esighteyewear.com/how-it-works/

Evans, S., & Willis, C. (2024). The use of digital technologies in the sport and physical education lesson: Fostering need-supportive behaviours in physical education teachers. *Wearable Technologies*, 5(1), 2527. DOI: 10.54517/wt.v5i1.2527

Evenson, K. R., Goto, M. M., & Furberg, R. D. (2015). Systematic review of the validity and reliability of consumer-wearable activity trackers. In *International Journal of Behavioral Nutrition and Physical Activity* (Vol. 12, Issue 1). BioMed Central Ltd. DOI: 10.1186/s12966-015-0314-1

Evmenova, A. S., Graff, H. J., Motti, V. G., Giwa-Lawal, K., & Zheng, H. (2018). Designing a wearable technology intervention to support young adults with intellectual and developmental disabilities in inclusive postsecondary academic environments. *Journal of Special Education Technology*, 34(2), 92–105. DOI: 10.1177/0162643418795833

Faisal, S. (2020). Tamkeen, Bahrain Polytechnic & Microsoft collaborate to launch the 1st batch of the Artificial Intelligence Academy – StartUp Bahrain. Startup-bahrain.com. https://startupbahrain.com/news/tamkeen-collaborates-with-bahrain-polytechnic-microsoft-to-kickstart-the-1st-batch-of-the-artificial-intelligence-academy/#:~:text=Bahrain%20Polytechnic%20has%20officially%20launched%20a

Fedele, A., Punzi, C., & Tramacere, S. (2024). The ALTAI checklist as a tool to assess ethical and legal implications for a trustworthy AI development in education. *Computer Law & Security Report*, 53, 105986. DOI: 10.1016/j.clsr.2024.105986

Feng, S., Chong, Y., Yu, H., Ye, X., & Li, G. (2022). Digital financial development and ecological footprint: Evidence from green-biased technology innovation and environmental inclusion. *Journal of Cleaner Production*, 380, 135069. https://doi.org/https://doi.org/10.1016/j.jclepro.2022.135069. DOI: 10.1016/j.jclepro.2022.135069

Ferreira, J. J., Fernandes, C. I., Rammal, H. G., & Veiga, P. M. (2021). Wearable technology and consumer interaction: A systematic review and research agenda. *Computers in Human Behavior*, 118, 106710. DOI: 10.1016/j.chb.2021.106710

Ferrier, B., Lee, J., Mbuli, A., & James, D. A. (2022). Translational Applications of Wearable Sensors in Education: Implementation and Efficacy. [CrossRef] [PubMed]. *Sensors (Basel)*, 22(4), 1675. DOI: 10.3390/s22041675 PMID: 35214578

Figg, C., & Jaipal-Jamani, K. (2018a). Developing teacher knowledge about gamification as an instructional strategy. In *Ophthalmology: Breakthroughs in Research and Practice* (pp. 227–255). IGI Global. DOI: 10.4018/978-1-5225-5198-0.ch013

Figg, C., & Jaipal-Jamani, K. (2018b). Developing teacher knowledge about gamification as an instructional strategy. In *Teacher Training and Professional Development: Concepts, Methodologies, Tools, and Applications* (Vol. 3, pp. 1215–1243). IGI Global. DOI: 10.4018/978-1-5225-5631-2.ch056

Fitbit Sense. (n.d.). *Advanced Health Smartwatch* Retrieved September 30, 2024, from https://www.fitbit.com/global/us/products/smartwatches/sense

Flack, K. D., Stults-Kolehmainen, M. A., Creasy, S. A., Khullar, S., Boullosa, D., Catenacci, V. A., & King, N. (2023). Altered motivation states for physical activity and 'appetite'for movement as compensatory mechanisms limiting the efficacy of exercise training for weight loss. *Frontiers in Psychology*, 14, 1098394. DOI: 10.3389/fpsyg.2023.1098394 PMID: 37187558

Fonseca, D., Navarro, I., De Renteria, I., Moreira, F., Ferrer, Á., & De Reina, O. (2017). Assessment of wearable virtual reality Technology for visiting world Heritage Buildings: An Educational approach. *Journal of Educational Computing Research*, 56(6), 940–973. DOI: 10.1177/0735633117733995

Forero-Corba, W. 1, & Bennasar, F. N. 1 1 U. de les I. B. (. (2024). Techniques and applications of Machine Learning and Artificial Intelligence in education: A systematic review. *ProQuest*, 27(1), 209–238. DOI: 10.5944/ried.27.1.37491

Fornell, C., & Larcker, D. F. (1981). Evaluating structural equation models with unobservable variables and measurement error. *JMR, Journal of Marketing Research*, 18(1), 39–50. DOI: 10.1177/002224378101800104

Fraga-Lamas, T. M.-C. (2018, December 7). owards The Internet of Smart Clothing: A Review on IoT Wearables and Garments for Creating Intelligent Connected E-Textiles. *Electronics (Basel)*, 36.

Frajhof, L., Borges, J., Hoffmann, E., Lopes, J., & Haddad, R. (2018). Virtual reality, mixed reality and augmented reality in surgical planning for video or robotically assisted thoracoscopic anatomic resections for treatment of lung cancer. *Journal of Visualized Surgery*, 4, 143–143. DOI: 10.21037/jovs.2018.06.02

Frederick, J., & Van Gelderen, S. (2021). Revolutionizing simulation education with smart glass technology. *Clinical Simulation in Nursing*, 52, 43–49. DOI: 10.1016/j.ecns.2020.12.007

Frenk, J., Chen, L. C., Chandran, L., Groff, E. O. H., King, R., Meleis, A., & Fineberg, H. V. (2022). Challenges and opportunities for educating health professionals after the COVID-19 pandemic. *Lancet*, 400(10362), 1539–1556. DOI: 10.1016/S0140-6736(22)02092-X PMID: 36522209

Fu, D., Chen, L., & Cheng, Z. (2021). Integration of Wearable Smart Devices and Internet of Things Technology into Public Physical Education. DOI: 10.1155/2021/6740987

Fucarino, A., Fabbrizio, A., Garrido, N. D., Iuliano, E., Reis, V. M., Sausa, M., & Cantoia, M. (2024, July). Emerging Technologies and Open-Source Platforms for Remote Physical Exercise: Innovations and Opportunities for Healthy Population—A Narrative Review. [). MDPI.]. *Health Care*, 12(15), 1466. PMID: 39120170

Gao, W., Wei, T., Huang, H., Chen, X., & Li, Q. (2022). Toward a systematic survey on wearable computing for education applications. *IEEE Internet of Things Journal*, 9(15), 12901–12915. DOI: 10.1109/JIOT.2022.3168324

Gerber, K., Flores, I. G., Ruiz, A. C., Ali, I., Ginsberg, N. L., & Schenberg, E. E. (2021). Ethical Concerns about Psilocybin Intellectual Property. *ACS Pharmacology & Translational Science*, 4(2), 573–577. DOI: 10.1021/acsptsci.0c00171 PMID: 33860186

Ghada Alrabia, B. A. (2022). User Awareness of Wearable IoT Devices Privacy Issues. *2022 International Conference on Innovation and Intelligence for Informatics, Computing, and Technologies (3ICT)*. IEEE Xplore.

Gibson, W., Wand, B. M., & O'Connell, N. E. (2017). Transcutaneous electrical nerve stimulation (TENS) for neuropathic pain in adults. In *Cochrane Database of Systematic Reviews* (Vol. 2017, Issue 9). John Wiley and Sons Ltd. DOI: 10.1002/14651858.CD011976.pub2

Glez Almaguer, C. A., Aguirre Acosta, A. C., & Roman Jimenez, O. R. (2021). "Wow" Experience with Immersive Reality: Gamification in the Tec Virtual Campus. *Future of Educational Innovation Workshop Series - Machine Learning-Driven Digital Technologies for Educational Innovation Workshop 2021*. DOI: 10.1109/IEEECONF53024.2021.9733757

Gligorea, I., Cioca, M., Oancea, R., Gorski, A. T., Gorski, H., & Tudorache, P. (2023). Adaptive learning using artificial intelligence in e-learning: A literature review. *Education Sciences*, 13(12), 1216. DOI: 10.3390/educsci13121216

Godfrey, A., Hetherington, V., Shum, H., Bonato, P., Lovell, N. H., & Stuart, S. (2018). From A to Z: Wearable technology explained. *Maturitas*, 113, 40–47. DOI: 10.1016/j.maturitas.2018.04.012 PMID: 29903647

Gomes, N., Pato, M., Lourenco, A. R., & Datia, N. (2023). A survey on wearable sensors for mental health monitoring. *Sensors (Basel)*, 23(3), 1330. DOI: 10.3390/s23031330 PMID: 36772370

González Crespo, R., & Burgos, D. (2019). Advanced Sensors Technology in Education. *Sensors (Basel)*, 19(19), 4155. DOI: 10.3390/s19194155 PMID: 31557927

Gonzo, F. (2023). The Impact of Digital Learning Technology on Higher Education Students' Mental Health. *Advances in Hospitality, Tourism, and the Services Industry*.

Goodfellow, I., Bengio, Y., & Courville, A. (2016). *Deep Learning*. MIT Press.

Goodwin, M. S., Intille, S. S., Albinali, F., & Velicer, W. F. (2011). Automated detection of stereotypical motor movements. *Journal of Autism and Developmental Disorders*, 41(6), 770–782. DOI: 10.1007/s10803-010-1102-z PMID: 20839042

Gordon, M. S., Forker, A. D., Gessner, I., McGuire, C., Mayer, J. W., Sajid, D. P. A., Ewy, G. A., Felner, J. M., & Waugh, R. A. (1980). Teaching bedside cardiologic examination skills using "Harvey", the cardiology patient simulator. *The Medical Clinics of North America*, 64(2), 305–313. DOI: 10.1016/S0025-7125(16)31620-0 PMID: 6155573

Gordon, N., Brayshaw, M., & Grey, S. (2013). Maximising gain for minimal pain: Utilising natural game mechanics. *ITALICS Innovations in Teaching and Learning in Information and Computer Sciences*, 12(1), 27–38. DOI: 10.11120/ital.2013.00004

Gore, T. (n.d.). *How Wearable Simulators Can Improve Clinical Simulation Realism*. https://www.healthysimulation.com/how-wearable-simulators-can-improve-clinical-simulation-realism/#:~:text=Wearable%20simulators%20allow%20clinical%20professional,available%20to%20educate%20healthcare%20learners

Goshevski, D., Veljanoska, J., & Hatziapostolou, H. (2017). A review of gamification platforms for higher education. *ACM International Conference Proceeding Series, Part F130953*. DOI: 10.1145/3136273.3136299

Government of Bahrain. (2023). Home. Bahrain.bh. https://www.bahrain.bh/wps/portal/en/

Greiwe, J., & Nyenhuis, S. M. (2020). Wearable technology and how this can be implemented into clinical practice. *Current Allergy and Asthma Reports*, 20(8), 1–10. DOI: 10.1007/s11882-020-00927-3 PMID: 32506184

Grøseth, L. K., & Sammerud, V. K. (2023). Solid Metal Fuels for Hydrogen Production for the Norwegian Heritage Fjords Maritime Sector (Bachelor's thesis, NTNU).

Group, W. H. O. M. G. R. S., & de Onis, M. (2006). WHO Motor Development Study: Windows of achievement for six gross motor development milestones. *Acta Paediatrica (Oslo, Norway)*, 95(S450), 86–95. DOI: 10.1111/j.1651-2227.2006.tb02379.x PMID: 16817682

Gullett, N., Zajkowska, Z., Walsh, A., Harper, R., & Mondelli, V. (2023). Heart rate variability (HRV) as a way to understand associations between the autonomic nervous system (ANS) and affective states: A critical review of the literature. *International Journal of Psychophysiology*, 192, 35–42. DOI: 10.1016/j.ijpsycho.2023.08.001 PMID: 37543289

Guo, J., Wan, B., Zheng, S., Song, A., & Huang, W. (2022). A Teenager Physical Fitness Evaluation Model Based on 1D-CNN with LSTM and Wearable Running PPG Recordings. *Biosensors (Basel)*, 12(4), 202. Advance online publication. DOI: 10.3390/bios12040202 PMID: 35448262

Gupta, A., Scott, K., & Dukewich, M. (2018). Innovative technology using virtual reality in the treatment of pain: Does it reduce pain via distraction, or is there more to it? *Pain Medicine*, 19(1), 151–159. DOI: 10.1093/pm/pnx109 PMID: 29025113

Gupta, S., Srivastava, R., Shaikh, Z. H., & Irfan, M. (2024). Revolutionizing Rural Finance: Exploring the Impact of FinTech on Financial Inclusion in India. In Irfan, M., Muhammad, K., Naifar, N., & Khan, M. A. (Eds.), *Applications of Block Chain technology and Artificial Intelligence. Financial Mathematics and Fintech*. Springer., DOI: 10.1007/978-3-031-47324-1_3

Gurcan, M. N., Boucheron, L. E., Can, A., Madabhushi, A., Rajpoot, N. M., & Yener, B. (2009). Histopathological Image Analysis: A Review. *IEEE Reviews in Biomedical Engineering*, 2, 147–171. DOI: 10.1109/RBME.2009.2034865 PMID: 20671804

Guze, P. A. (2015). Using Technology to Meet the Challenges of Medical Education. *Transactions of the American Clinical and Climatological Association*, 126, 260–270. PMID: 26330687

Hagadone-Bedir, M., Voithofer, R., & Kulp, J. T. (2023). Contextualizing privacy with wearable data in higher education. *British Journal of Educational Technology*, 54(6), 1619–1635. DOI: 10.1111/bjet.13378

Haghi, M., Thurow, K., & Stoll, R. (2017). Wearable devices in medical internet of things: Scientific research and commercially available devices. *Healthcare Informatics Research*, 23(1), 4–15. DOI: 10.4258/hir.2017.23.1.4 PMID: 28261526

Hamza-Lup, F. G., Popovici, D. M., & Bogdan, C. M. (2013). Haptic feedback systems in medical education. *Journal of Advanced Distributed Learning Technology*, 1(2), 7.

Hamza, M. A., Abdalla Hashim, A. H., Alsolai, H., Gaddah, A., Othman, M., Yaseen, I., Rizwanullah, M., & Zamani, A. S. (2023). Wearables-Assisted Smart Health Monitoring for Sleep Quality Prediction Using Optimal Deep Learning. *Sustainability (Basel)*, 15(2), 1084. Advance online publication. DOI: 10.3390/su15021084

Harry, A., & Sayudin. (2023). Role of AI in Education. Injuruty: Interdiciplinary Journal and Humanity. https://injurity.pusatpublikasi.id/index.php/in

Havard, B., & Podsiad, M. (2020). A meta-analysis of wearables research in educational settings published 2016–2019. *Educational Technology Research and Development*, 68(4), 1829–1854. Advance online publication. DOI: 10.1007/s11423-020-09789-y

Hee Keum, D., Kim, S.-K., Koo, J., Lee, G.-H., Jeon, C., Won Mok, J., Ho Mun, B., Jae Lee, K., Kamrani, E., Joo, C.-K., Shin, S., Sim, J.-Y., Myung, D., Hyun Yun, S., Bao, Z., & Kwang Hahn, S. (2020). *Wireless smart contact lens for diabetic diagnosis and therapy*. https://www.science.org

HemoCueHb301Operating-Manual_US[Internet]. [cited2023Jan 14].Available from: https://www. hemocue.us/wp-content/uploads/2020/08/HB-301_Operating-Manual_US.pdf_https://biovoicenews.com/ezerx-registers-a-167-growth-in-year-on-year-net-revenue/

Hendry, D., Rohl, A. L., Rasmussen, C. L., Zabatiero, J., Cliff, D. P., Smith, S. S., Mackenzie, J., Pattinson, C. L., Straker, L., & Campbell, A. (2023). Objective Measurement of Posture and Movement in Young Children Using Wearable Sensors and Customised Mathematical Approaches: A Systematic Review. *Sensors (Basel)*, 23(24), 9661. Advance online publication. DOI: 10.3390/s23249661 PMID: 38139507

Herbert, B. J. P., Charles, D., McNeill, M., Moore, A., & Charles, M. (2011). Dynamic virtual learning landscapes to enhance student reflective processes. In M. Meimaris & D. Gouscos (Eds.), *Proceedings of the European Conference on Games-based Learning* (Vols. 2011-January, pp. 691–702). Dechema e.V. https://www.scopus.com/inward/record.uri?eid=2-s2.0-84938566496&partnerID=40&md5=d0854015789791774ab5505c0f680960

Hercog, D., Lerher, T., Truntič, M., & Težak, O. (2023). Design and Implementation of ESP32-Based IoT Devices. *Sensors (Basel)*, 23(15), 6739. Advance online publication. DOI: 10.3390/s23156739 PMID: 37571523

Hernandez, R., & Patel, S. (2021). *Inclusive design in educational wearables*. Journal of Inclusive Education.

Hernández-Mustieles, M. A., Lima-Carmona, Y. E., Pacheco-Ramírez, M. A., Mendoza-Armenta, A. A., Romero-Gómez, J. E., Cruz-Gómez, C. F., & Lozoya-Santos, J. D. J. (2024). Wearable Biosensor Technology in Education: A Systematic Review. *Sensors (Basel)*, 24(8), 2437.

Hernández-Mustieles, M. A., Lima-Carmona, Y. E., Pacheco-Ramírez, M. A., Mendoza-Armenta, A. A., Romero-Gómez, J. E., Cruz-Gómez, C. F., Rodríguez-Alvarado, D. C., Cruz-Garza, J. G., Arceo, A., Ramírez-Moreno, M. A., & de Lozoya-Santos, J. (2024). *Wearable Biosensor Technology in Education: A Systematic Review*. https://doi.org/DOI: 10.20944/preprints202403.0831.v1

Hew, K. F., Huang, B., Chu, K. W. S., & Chiu, D. K. W. (2016). Engaging Asian students through game mechanics: Findings from two experiment studies. *Computers & Education*, 92–93, 221–236. DOI: 10.1016/j.compedu.2015.10.010

HHS. (n.d.). *Physical Activity Guidelines for Americans 2 nd edition.*

Hickey, B. A., Chalmers, T., Newton, P., Lin, C. T., Sibbritt, D., McLachlan, C. S., Clifton-Bligh, R., Morley, J., & Lal, S. (2021a). Smart devices and wearable technologies to detect and monitor mental health conditions and stress: A systematic review. In *Sensors* (Vol. 21, Issue 10). MDPI AG. DOI: 10.3390/s21103461

Holland, J. (Ed.). (2016). *Wearable technology and mobile innovations for next-generation education*. IGI Global. DOI: 10.4018/978-1-5225-0069-8

Holmes, B. (2021). *Personalized learning through AI wearables*. Technology in Education.

Hong, L. (2023). Study on the influence of social network for college students' mental health through big data technology. *ACM International Conference Proceeding Series*, 147–153. DOI: 10.1145/3584871.3584892

Hooda, M., Rana, C., Dahiya, O., Rizwan, A., & Hossain, M. S. (2022). Artificial intelligence for assessment and feedback to enhance student success in higher education. *Mathematical Problems in Engineering*, 2022(1), 5215722. DOI: 10.1155/2022/5215722

Hopcan, S., Türkmen, G., & Polat, E. (2024). Exploring the artificial intelligence anxiety and machine learning attitudes of teacher candidates. *Education and Information Technologies*, 29(6), 7281–7301. DOI: 10.1007/s10639-023-12086-9

Hora, A., & Kulkarni, P. (2024, July). Wearables and Cybersecurity: Navigating the Threat Landscape. In 2024 2nd International Conference on Sustainable Computing and Smart Systems (ICSCSS) (pp. 563-567). IEEE.

Horvers, A., Tombeng, N., Bosse, T., Lazonder, A. W., & Molenaar, I. (2021). Detecting emotions through electrodermal activity in learning contexts: A systematic review. *Sensors (Basel)*, 21(23), 7869. Advance online publication. DOI: 10.3390/s21237869 PMID: 34883870

Hosny, A., Parmar, C., Quackenbush, J., Schwartz, L. H., & Aerts, H. J. W. L. (2018). Artificial intelligence in radiology. In *Nature Reviews Cancer* (Vol. 18, Issue 8, pp. 500–510). Nature Publishing Group. DOI: 10.1038/s41568-018-0016-5

Hossain, M. I., Zahid, M. S., Chowdhury, M. A., Hossain, M. M. M., Hossain, N., Islam, M. A., & Mobarak, M. H. (2024). Smart bandage: A device for wound monitoring and targeted treatment. In *Results in Chemistry* (Vol. 7). Elsevier B.V., DOI: 10.1016/j.rechem.2023.101292

Hou, Y., Li, Z., & Li, H. (2025). Sensor based interactive digital entertainment and gamified training to alleviate basketball player fatigue. *Entertainment Computing*, 52, 100838. https://doi.org/https://doi.org/10.1016/j.entcom.2024.100838. DOI: 10.1016/j.entcom.2024.100838

Hovnanyan, A., Mastromatteo, L. Y., Rubaltelli, E., & Scrimin, S. (2022). Stress and Emotional Intelligence Shape Giving Behavior: Are There Different Effects of Social, Cognitive, and Emotional Stress? *Frontiers in Psychology*, 13, 800742. Advance online publication. DOI: 10.3389/fpsyg.2022.800742 PMID: 35282189

Hovsepian, K., Al'absi, M., Ertin, E., Kamarck, T., Nakajima, M., & Kumar, S. (2015). CStress: Towards a gold standard for continuous stress assessment in the mobile environment. *UbiComp 2015 - Proceedings of the 2015 ACM International Joint Conference on Pervasive and Ubiquitous Computing*, 493–504. DOI: 10.1145/2750858.2807526

How Wearable Technology is Revolutionizing the Education Sector? (2024). VLink. https://www.vlinkinfo.com/blog/how-wearable-technology-is-revolutionizing-the-education-sector/

Hsiung Chang, I., Keh, H. C., Dande, B., & Roy, D. S. (2020). Smart hat: Design and implementation of a wearable learning device for kids using AI and IoTS techniques. *Journal of Internet Technology*, 21(2), 593–604. DOI: 10.3966/160792642020032102026

Hsu, C., Chen, Y., Lin, C., & Lien, W. (2021). Cognitive development, self-efficacy, and wearable technology use in a virtual reality language learning environment: A structural equation modeling analysis. *Current Psychology (New Brunswick, N.J.)*, 41(3), 1618–1632. DOI: 10.1007/s12144-021-02252-y

Hu, J. R., Martin, G., Iyengar, S., Kovell, L. C., Plante, T. B., van Helmond, N., Dart, R. A., Brady, T. M., Turkson-Ocran, R. A. N., & Juraschek, S. P. (2023). Validating cuffless continuous blood pressure monitoring devices. In *Cardiovascular Digital Health Journal* (Vol. 4, Issue 1, pp. 9–20). Elsevier Inc. DOI: 10.1016/j.cvdhj.2023.01.001

Hu, Y. (2020). Improve the mental health of college students by means of computer network technology. *Proceedings - 2020 International Conference on Computer Engineering and Intelligent Control, ICCEIC 2020*, 5–9. DOI: 10.1109/ICCEIC51584.2020.00008

Huang, Y., Xiong, H., Leach, K., Zhang, Y., Chow, P., Fua, K., Teachman, B. A., & Barnes, L. E. (2016). Assessing social anxiety using GPS trajectories and point-of-interest data. *UbiComp 2016 - Proceedings of the 2016 ACM International Joint Conference on Pervasive and Ubiquitous Computing*, 898–903. DOI: 10.1145/2971648.2971761

Huang, B., & Hew, K. F. (2018). Implementing a theory-driven gamification model in higher education flipped courses: Effects on out-of-class activity completion and quality of artifacts. *Computers & Education*, 125, 254–272. DOI: 10.1016/j.compedu.2018.06.018

Huang, L., & Wang, F. (2019). The impact of artificial intelligence on education: A systematic review. *International Journal of Educational Technology in Higher Education*, 16(1), 1–23.

Huang, Y., Chen, Y., & Liu, Y. (2021). Wearable technology in education: A systematic review. *Journal of Educational Technology*, 48(1), 1–22.

Hudson-Thomas, M., Bingham, K. C., & Simmons, W. K. (1994). An evaluation of the HemoCue for measuring haemoglobin in field studies in Jamaica. *Bulletin of the World Health Organization*, 72(3), 423.

Huifeng, W., Kadry, S. N., & Raj, E. D. (2020). Continuous health monitoring of sportsperson using IoT devices based wearable technology. *Computer Communications*, 160, 588–595. DOI: 10.1016/j.comcom.2020.04.025

Hu, J. R., Park, D. Y., Agarwal, N., Herzig, M., Ormseth, G., Kaushik, M., Giao, D. M., Turkson-Ocran, R. A. N., & Juraschek, S. P. (2023). The Promise and Illusion of Continuous, Cuffless Blood Pressure Monitoring. *Current Cardiology Reports*, 25(10), 1139–1149. DOI: 10.1007/s11886-023-01932-4 PMID: 37688763

Hung, A. C. Y. (2017). A critique and defense of gamification. *Journal of Interactive Online Learning*, 15(1), 57–72. https://www.scopus.com/inward/record.uri?eid=2-s2.0-85020689516&partnerID=40&md5=8556a4eb9a9170e270389dc58fbe48b8

Hussain Aldawood, G. S. (2019). A Taxonomy for Social Engineering Attacks via Personal Devices. *International Journal of Computer Applications (0975 – 8887).*

Hussein, M., & Nätterdal, C. (2015). *The benefits of virtual reality in education.* Department of Computer Science and Engineering, Chalmers University of Technology, University of Gothenburg.

Hyndman, B. P. (2017). A simulation pedagogical approach to engaging generalist pre-service teachers in physical education online: The GoPro trial 1.0. *The Australian Journal of Teacher Education*, 42(1), 84–102. DOI: 10.14221/ajte.2017v42n1.6

Ijeh, S., Okolo, C. A., Arowoogun, J. O., Adeniyi, A. O., & Omotayo, O. (2024). Predictive modeling for disease outbreaks: a review of data sources and accuracy. *International Medical Science Research Journal*, 4(4), 406–419. DOI: 10.51594/imsrj.v4i4.999

Ikegami, H., Hiromine, Y., & Noso, S. (2022). Insulin-dependent diabetes mellitus in older adults: Current status and future prospects. In *Geriatrics and Gerontology International* (Vol. 22, Issue 8, pp. 549–553). John Wiley and Sons Inc. DOI: 10.1111/ggi.14414

Immanuel, S., Teferra, M. N., Baumert, M., & Bidargaddi, N. (2023). Heart rate variability for evaluating psychological stress changes in healthy adults: A scoping review. *Neuropsychobiology*, 82(4), 187–202. DOI: 10.1159/000530376 PMID: 37290411

Iqbal, S. M. A., Mahgoub, I., Du, E., Leavitt, M. A., & Asghar, W. (2021). Advances in healthcare wearable devices. In *npj Flexible Electronics* (Vol. 5, Issue 1). Nature Research. DOI: 10.1038/s41528-021-00107-x

Iqbal, M. H., Aydin, A., Brunckhorst, O., Dasgupta, P., & Ahmed, K. (2016). A review of wearable technology in medicine. *Journal of the Royal Society of Medicine*, 109(10), 372–380. DOI: 10.1177/0141076816663560 PMID: 27729595

Irvine, U. C. (n.d.). *UCI School of Medicine first to integrate Google Glass into curriculum – wearable computing technology will transform training of future doctors. UCIrvine News.*https://news.uci.edu/press-releases/uci-school-of-medicine-first-to-integrate-google-glass-into-curriculum/

Jafarlou, S., Lai, J., Azimi, I., Mousavi, Z., Labbaf, S., Jain, R. C., Dutt, N., Borelli, J. L., & Rahmani, A. (2023). Objective Prediction of Next-Day's Affect Using Multimodal Physiological and Behavioral Data: Algorithm Development and Validation Study. *JMIR Formative Research*, 7, e39425. Advance online publication. DOI: 10.2196/39425 PMID: 36920456

Jamison, R. N., Wan, L., Edwards, R. R., Mei, A., & Ross, E. L. (2019). Outcome of a High-Frequency Transcutaneous Electrical Nerve Stimulator (hfTENS) Device for Low Back Pain: A Randomized Controlled Trial. *Pain Practice*, 19(5), 466–475. DOI: 10.1111/papr.12764 PMID: 30636101

Jarraya, S. K., Masmoudi, M., & Hammami, M. (2021). A comparative study of Autistic Children Emotion recognition based on Spatio-Temporal and Deep analysis of facial expressions features during a Meltdown Crisis. *Multimedia Tools and Applications*, 80(1), 83–125. DOI: 10.1007/s11042-020-09451-y

Järveläinen, J. (2021). Collective Mindful and Mindless Behavior in Game-Based Learning – Comparing Physical and Digital Educational Escape Rooms. *42nd International Conference on Information Systems, ICIS 2021 TREOs: "Building Sustainability and Resilience with IS: A Call for Action."*https://www.scopus.com/inward/record.uri?eid=2-s2.0-85159101926&partnerID=40&md5=999bdac99f198f4b41a344f39d831652

Jeng, M., Pai, F., & Yeh, T. (2022). Antecedents for older adults' intention to use smart Health Wearable Devices-Technology Anxiety as a moderator. *Behavioral Sciences (Basel, Switzerland)*, 12(4), 114. DOI: 10.3390/bs12040114 PMID: 35447686

Jiang, L. (2020). Research on the Construction of New Mode of Mental Health Education for Higher Vocational College Students Based on Computer Big Data Technology. *Journal of Physics: Conference Series*, 1578(1), 012113. Advance online publication. DOI: 10.1088/1742-6596/1578/1/012113

Jiangxin, H., & Zhiguo, X. (2021). Research on Mental Health Assessment and Crisis Intervention Mechanism of University Students with Data Technology. *Proceedings - 2021 2nd International Conference on Education, Knowledge and Information Management, ICEKIM 2021*, 451–454. DOI: 10.1109/ICEKIM52309.2021.00105

Ji, D. (2024). Research on Intelligent Multimedia Technology Path of College Students' Mental Health Education in Higher Vocational Colleges and Universities. *Applied Mathematics and Nonlinear Sciences*, 9(1), 20243379. Advance online publication. DOI: 10.2478/amns-2024-3379

Jin, X., Li, L., Dang, F., Chen, X., & Liu, Y. (2022). A survey on edge computing for wearable technology. *Digital Signal Processing*, 125, 103146.

Johnson, R.. (2021). Gamification in the AI-powered classroom. *Interactive Learning Environments*.

Jose, B., Strawderman, L., Burch, R. F., Saucier, D., & Kikuta, R. (2024). Passive vs. active wearable technology monitoring trunk flexion in elementary teachers. *International Journal of Human Factors and Ergonomics*, 11(2), 172–187. DOI: 10.1504/IJHFE.2024.139190

Joshi, R. (2023). Raising Awareness of Social Engineering among Adolescents: Psychological and Cybersecurity Perspective. In *Cybersecurity for Decision Makers* (pp. 99-109 (11)). CRC Press.

Joshi, R. (2023). Raising Awareness of Social Engineering among Adolescents: Psychological and Cybersecurity Perspective. In *Cybersecurity for decision makers* (pp. 99–109). CRC Press. DOI: 10.1201/9781003319887-7

Jovanovic, P., & Kay, R. (2023). Exploring Ethical Issues with Using Wearable Technology in K-12 Classrooms: A Review of the Literature. *Journal of Digital Life and Learning*, 2(2), 3–11. DOI: 10.51357/jdll.v2i2.203

K201525-Luke Olson. (n.d.). https://www.fda.gov/medical-

Kabudi, T., Pappas, I., & Olsen, D. H. (2021). AI-enabled adaptive learning systems: A systematic mapping of the literature. *Computers and Education: Artificial Intelligence*, 2, 100017. DOI: 10.1016/j.caeai.2021.100017

Kanschik, D., Bruno, R. R., Wolff, G., Kelm, M., & Jung, C. (2023). Virtual and augmented reality in intensive care medicine: a systematic review. In *Annals of Intensive Care* (Vol. 13, Issue 1). Springer Science and Business Media Deutschland GmbH. DOI: 10.1186/s13613-023-01176-z

Kapoor, V., Singh, R., Reddy, R., & Churi, P. (n.d.). *International Conference on Innovative Computing and Communication (ICICC 2020) Privacy Issues in Wearable Technology: An Intrinsic Review*. https://ssrn.com/abstract=3566918

Karagianni, E., & Drigas, A. (2023). New Technologies for Inclusive Learning for Students with Special Educational Needs. *International Journal of Online & Biomedical Engineering*, 19(5), 4–21. DOI: 10.3991/ijoe.v19i05.36417

Kayyali, M. (2024). Future possibilities and challenges of AI in education. In *Transforming education with generative AI: Prompt engineering and synthetic content creation* (pp. 118–137). IGI Global. DOI: 10.4018/979-8-3693-1351-0.ch006

Kazanskiy, N. L., Khonina, S. N., & Butt, M. A. (2024). A review on flexible wearables-Recent developments in non-invasive continuous health monitoring. *Sensors and Actuators. A, Physical*, 366, 114993. DOI: 10.1016/j.sna.2023.114993

Kennedy, H. L. (2013). The evolution of ambulatory ECG monitoring. *Progress in Cardiovascular Diseases*, 56(2), 127–132. DOI: 10.1016/j.pcad.2013.08.005 PMID: 24215744

Khalifa, M., Alrasheed, R., Khalifa, M., Alrasheed, R., & Aldada, A. (2023). Education in Light of the Sustainable Development Goals in the Kingdom of Bahrain "An Analytical Study" Education in Light of the Sustainable Development Goals in the Kingdom of Bahrain "An Analytical Study.". *Journal of Statistics Applications & Probability*, 12(S1), 1419–1423. DOI: 10.18576/jsap/12S103

Khosravi, S., Bailey, S. G., Parvizi, H., & Ghannam, R. (2021). Learning enhancement in higher education with wearable technology. arXiv preprint arXiv:2111.07365.

Khosravi, S., Bailey, S. G., Parvizi, H., & Ghannam, R. (2022). Wearable Sensors for Learning Enhancement in Higher Education. *In Sensors* (Vol. 22, Issue 19). https://doi.org/DOI: 10.3390/s22197633

Kim, I., Kang, K., Song, Y., & Kim, T. J. (2022). Application of Artificial Intelligence in Pathology: Trends and Challenges. In *Diagnostics* (Vol. 12, Issue 11). Multidisciplinary Digital Publishing Institute (MDPI). DOI: 10.3390/diagnostics12112794

Kim, H., & Lee, J. (2017). The impact of virtual reality on student learning outcomes in chemistry. *Journal of Educational Technology*, 8(2), 123–142.

Kim, J., Heo, N., & Kang, H. (2023). Validity and reliability of positive attitudes toward and perceived importance of wearable display technology as an effective learning tool among nursing students. *Nurse Education in Practice*, 73, 103812. DOI: 10.1016/j.nepr.2023.103812 PMID: 37922737

Kim, J., & Hwang, S. (2023). Enhancing student engagement through AI-powered real-time feedback in educational wearables. *Journal of Educational Technology Research*, 45(2), 132–147.

Kingdom of Bahrain Portal for Sustainable Development Goals. (n.d.). Sdgs.gov.bh. https://sdgs.gov.bh

Kinny, F., Schlottau, S., Ali Sherazi, B., Obarcanin, E., & Läer, S. (2024). Digital health in pharmacy education: Elective practical course integrating wearable devices and their generated health data. *Exploratory Research in Clinical and Social Pharmacy*, 15, 100465. DOI: 10.1016/j.rcsop.2024.100465 PMID: 38983639

Kirk, M. A., Amiri, M., Pirbaglou, M., & Ritvo, P. (2019). Wearable technology and physical activity behavior change in adults with chronic cardiometabolic disease: A systematic review and meta-analysis. *American Journal of Health Promotion*, 33(5), 778–791. DOI: 10.1177/0890117118816278 PMID: 30586996

Koh, A., Kang, D., Xue, Y., Lee, S., Pielak, R. M., Kim, J., Hwang, T., Min, S., Banks, A., Bastien, P., Manco, M. C., Wang, L., Ammann, K. R., Jang, K.-I., Won, P., Han, S., Ghaffari, R., Paik, U., Slepian, M. J., . . . Rogers, J. A. (2016). *A soft, wearable microfluidic device for the capture, storage, and colorimetric sensing of sweat*. https://stm.sciencemag.org/

Kohnke, L., & Moorhouse, B. L. (2021). Using Kahoot! to Gamify Learning in the Language Classroom. *RELC Journal*, 53(3), 769–775. DOI: 10.1177/00336882211040270

Kolb, D. A., Boyatzis, R. E., & Mainemelis, C. (2014). Experiential learning theory: Previous research and new directions. In *Perspectives on thinking, learning, and cognitive styles* (pp. 227–247). Routledge. DOI: 10.4324/9781410605986-9

Komalasari, R. (2024). AI-Powered Wearables Revolutionizing Health Tracking and Personalized Wellness Management. *Timor Leste Journal of Business and Management*, 6, 42–50.

Korn, E. L., Moscow, J. A., & Freidlin, B. (2023). Dose optimization during drug development: whether and when to optimize. In *Journal of the National Cancer Institute* (Vol. 115, Issue 5, pp. 492–497). Oxford University Press. DOI: 10.1093/jnci/djac232

Koutromanos, G., & Kazakou, G. (2020). The use of smart wearables in primary and secondary education: A systematic review. *Themes in ELearning*, 13, 33–53.

Kraiss, K. N. (2017). *Evaluating the Effects of Wearable Technologies to Improve Evaluating the Effects of Wearable Technologies to Improve Physical Activity Levels for College Students with Intellectual and Physical Activity Levels for College Students with Intellectual and Developmental Disabilities Developmental Disabilities*. https://trace.tennessee.edu/utk_graddiss

Kumar, K., & Al-Besher, A. (2022). IoT enabled e-learning system for higher education. *Measurement. Sensors*, 24, 100480. https://doi.org/https://doi.org/10.1016/j.measen.2022.100480. DOI: 10.1016/j.measen.2022.100480

Kustimah, K., Devy Kumalasari, A., & Meindy, N. (n.d.). *The Open Psychology Journal When do College Students Seek Mental Health Services?* https://doi.org/ DOI: 10.2174/18743501-v16-e230420-2022-112

Kusuma, K., Sheela, P., Panigrahi, R. R., & Shaikh, Z. H. (2024). Blockchain Technology and Islamic Finance: A Decade of Growth Trends. In Irfan, M., Kadry, S., Sharif, M., & Khan, H. (Eds.), *Fintech Applications in Islamic Finance: AI, Machine Learning, and Blockchain Techniques* (pp. 59–79). IGI Global., DOI: 10.4018/979-8-3693-1038-0.ch004

Kuzu Demir, E., & Demir, K. (2017). *Enhancing Learning with Wearable Technologies In and Out of Educational Settings*. DOI: 10.4018/978-1-5225-1692-7.ch006

Kvedar, J., Coye, M. J., & Everett, W. (2014). Connected health: A review of technologies and strategies to improve patient care with telemedicine and telehealth. *Health Affairs*, 33(2), 194–199. DOI: 10.1377/hlthaff.2013.0992 PMID: 24493760

Kwon, Y. M., Lee, Y., & Kim, S. J. (2017). Case study on 3D printing education in fashion design coursework. *Fashion and Textiles*, 4(1), 26. Advance online publication. DOI: 10.1186/s40691-017-0111-3

Laak, K. J., & Aru, J. (2024). AI and personalized learning: bridging the gap with modern educational goals. *arXiv preprint arXiv:2404.02798*.

Lagera, P. G. D., Chan, S. R., & Yellowlees, P. M. (2023). Asynchronous technologies in mental health care and education. *Current Treatment Options in Psychiatry*, 10(2), 59–71. DOI: 10.1007/s40501-023-00286-6 PMID: 37360962

Lambert, N., & Watkins, L. (2013). Meet Mohammed: Using simulation and technology to support learning. *The Journal of Mental Health Training, Education and Practice*, 8(2), 66–75. DOI: 10.1108/JMHTEP-03-2012-0001

Lampropoulos, G., & Kinshuk, . (2024). Virtual reality and gamification in education: A systematic review. *Educational Technology Research and Development*, 72(3), 1691–1785. DOI: 10.1007/s11423-024-10351-3

Lan, L., Sikov, J., Lejeune, J., Ji, C., Brown, H., Bullock, K., & Spencer, A. E. (2023). A Systematic Review of using Virtual and Augmented Reality for the Diagnosis and Treatment of Psychotic Disorders. In *Current Treatment Options in Psychiatry* (Vol. 10, Issue 2, pp. 87–107). Springer Science and Business Media Deutschland GmbH. DOI: 10.1007/s40501-023-00287-5

Lane, J. L., Slavin, S., & Ziv, A. (2001). Simulation in medical education: A review. *Simulation & Gaming*, 32(3), 297–314. DOI: 10.1177/104687810103200302

Lattie, E. G., Lipson, S. K., & Eisenberg, D. (2019). Technology and college student mental health: Challenges and opportunities. *Frontiers in Psychiatry*, 10(APR), 246. Advance online publication. DOI: 10.3389/fpsyt.2019.00246 PMID: 31037061

Laverde, R., Rueda, C., Amado, L., Rojas, D., & Altuve, M. (2018). Artificial neural network for laparoscopic skills classification using motion signals from apple watch. *2018 40th Annual International Conference of the Ieee Engineering in Medicine and Biology Society (Embc)*, 5434–5437.

Lee, C., Yoo, S. K., Park, Y., Kim, N., Jeong, K., & Lee, B. (2006). Using neural network to recognize human emotions from heart rate variability and skin resistance. *2005 IEEE Engineering in Medicine and Biology 27th Annual Conference*, 5523–5525.

Lee, D. A. of, D., & Higginbotham, E. J. (n.d.). Clinical Review Glaucoma and its treatment: A review. In *Am J Health-Syst Pharm*.

Lee, L., Egelman, S., Lee, J. H., & Wagner, D. (2015). Risk perceptions for wearable devices. arXiv preprint arXiv:1504.05694.

Lee, J., & Chen, C. (2019). The impact of microlearning modules on professional development: A transformative learning perspective. *Journal of Educational Technology*, 10(2), 123–142.

Lee, J., Kim, D., Ryoo, H., & Shin, B. (2016). Sustainable wearables: Wearable technology for enhancing the quality of human life. *Sustainability (Basel)*, 8(5), 466. DOI: 10.3390/su8050466

Lee, U., Han, K., Cho, H., Chung, K.-M., Hong, H., Lee, S.-J., Noh, Y., Park, S., & Carroll, J. M. (2019). Intelligent positive computing with mobile, wearable, and IoT devices: Literature review and research directions. *Ad Hoc Networks*, 83, 8–24. https://doi.org/https://doi.org/10.1016/j.adhoc.2018.08.021. DOI: 10.1016/j.adhoc.2018.08.021

Li, L. R., Du, B., Liu, H. Q., & Chen, C. (2021). Artificial Intelligence for Personalized Medicine in Thyroid Cancer: Current Status and Future Perspectives. In *Frontiers in Oncology* (Vol. 10). Frontiers Media S.A. DOI: 10.3389/fonc.2020.604051

Liang, J.-M., Su, W.-C., Chen, Y.-L., Wu, S.-L., & Chen, J.-J. (n.d.). *Smart Interactive Education System Based on Wearable Devices*. DOI: 10.3390/s19153260

Liang, J. M., Su, W. C., Chen, Y. L., Wu, S. L., & Chen, J. J. (2019). Smart interactive education system based on wearable devices. *Sensors (Basel)*, 19(15), 3260.

Li, J., & Washington, P. (2024). A comparison of personalized and generalized approaches to emotion recognition using consumer wearable devices: Machine learning study. *JMIR AI*, 3(1), e52171. DOI: 10.2196/52171 PMID: 38875573

Li, K., Cardoso, C., Moctezuma-Ramirez, A., Elgalad, A., & Perin, E. (2023). Heart Rate Variability Measurement through a Smart Wearable Device: Another Breakthrough for Personal Health Monitoring? *International Journal of Environmental Research and Public Health*, 20(24), 7146. DOI: 10.3390/ijerph20247146 PMID: 38131698

Lilit Manukyan, M. G. (2024). Social Engineering Attacks: How to Prevent. In *Journal of Digital Science* (Vol. 6).

Limna, P., Jakwatanatham, S., Siripipattanakul, S., Kaewpuang, P., & Sriboonruang, P. (2022). A review of artificial intelligence (AI) in education during the digital era. *Advance Knowledge for Executives*, 1(1), 1–9.

Lin, D. T. A., Ganapathy, M., & Kaur, M. (2018). Kahoot! It: Gamification in higher education. *Pertanika Journal of Social Science & Humanities*, 26(1), 565–582. https://www.scopus.com/inward/record.uri?eid=2-s2.0-85045525024&partnerID=40&md5=2f289beb827fad108109309ed6388c45

Lindberg, R., Seo, J., & Laine, T. H. (2016). Enhancing physical education with exergames and wearable technology. *IEEE Transactions on Learning Technologies*, 9(4), 328–341. DOI: 10.1109/TLT.2016.2556671

Lindeque, B. G., Ponce, B. A., Menendez, M. E., Oladeji, L. O., Fryberger, C. T., & Dantuluri, P. K. (2014). Emerging technology in surgical education: Combining Real-Time augmented reality and wearable computing devices. *Orthopedics*, 37(11), 751–757. DOI: 10.3928/01477447-20141023-05 PMID: 25361359

Linskyi, A. (2023). *Virtual Reality for Surgical Training: Theory & Practice*. https://smarttek.solutions/blog/vr-for-surgery/

Lin, Y. R., Hung, C. C., Chiu, H. Y., Chang, B. H., Li, B. R., Cheng, S. J., Yang, J. W., Lin, S. F., & Chen, G. Y. (2018). Noninvasive glucose monitoring with a contact lens and smartphone. *Sensors (Basel)*, 18(10), 3208. Advance online publication. DOI: 10.3390/s18103208 PMID: 30249021

Liu, F., & Mai, Y. (2022). Application of Computer Technology in College Students' Mental Health Education. *Proceedings - 2022 3rd International Conference on Education, Knowledge and Information Management, ICEKIM 2022*, 1019–1024. DOI: 10.1109/ICEKIM55072.2022.00222

Liu, Q., Zhang, Y., & Sun, Y. (2024, May). Intercepting Bluetooth Traffic from Wearable Health Devices. In 2024 IEEE Security and Privacy Workshops (SPW) (pp. 267-273). IEEE.

Liu, Y., & Zhang, W. (2020). *AI in education: Emerging trends*. International Journal of AI in Education.

Liu, S., Hu, S., Weng, J., Zhu, S., & Chen, Z. (2016). A novel asymmetric three-party based authentication scheme in wearable devices environment. *Journal of Network and Computer Applications*, 60, 144–154. DOI: 10.1016/j.jnca.2015.10.001

Liu, Y., Wang, H., & Xu, J. (2020). Smart classroom: A review of its concept, application, and challenges. *International Journal of Educational Technology in Higher Education*, 7(4), 1–16.

Li, X., Zou, L., & Li, H. (2024). Multilayer Perceptron-Based Wearable Exercise-Related Heart Rate Variability Predicts Anxiety and Depression in College Students. *Sensors (Basel)*, 24(13), 4203. DOI: 10.3390/s24134203 PMID: 39000984

Llorente, A. M., Amado, A. J., Voigt, R. G., Berretta, M. C., Fraley, J. K., Jensen, C. L., & Heird, W. C. (2001). Internal consistency, temporal stability, and reproducibility of individual index scores of the Test of Variables of Attention in children with attention-deficit/hyperactivity disorder. *Archives of Clinical Neuropsychology*, 16(6), 535–546. DOI: 10.1093/arclin/16.6.535 PMID: 14590152

Lockley, A., & Boyle, A. (2014). Towards a game-based learning ecosystem: An institutional strategy. In C. Busch (Ed.), *Proceedings of the European Conference on Games-based Learning* (Vol. 1, pp. 312–320). Dechema e.V. https://www.scopus.com/inward/record.uri?eid=2-s2.0-84923587008&partnerID=40&md5=bbaaab7906d3490e3c2c20d4f35389bf

Loncar-Turukalo, T., Zdravevski, E., da Silva, J. M., Chouvarda, I., & Trajkovik, V. (2019). Literature on wearable technology for connected health: Scoping review of research trends, advances, and barriers. *Journal of Medical Internet Research*, 21(9), e14017. DOI: 10.2196/14017 PMID: 31489843

Loos, L. A., & Crosby, M. E. (2017). Gamification methods in higher education. In P. Zaphiris & A. Ioannou (Eds.), *Lecture Notes in Computer Science (including subseries Lecture Notes in Artificial Intelligence and Lecture Notes in Bioinformatics): Vol. 10295 LNCS* (pp. 474–486). Springer Verlag. DOI: 10.1007/978-3-319-58509-3_37

Lopez, A., Garcia, T., & Lee, P. (2022). Adaptive learning in physical education: AI-driven fitness trackers and student engagement. *Physical Education and Technology*, 14(1), 58–73.

Lopez-Jimenez, F., Attia, Z., Arruda-Olson, A. M., Carter, R., Chareonthaitawee, P., Jouni, H., Kapa, S., Lerman, A., Luong, C., Medina-Inojosa, J. R., Noseworthy, P. A., Pellikka, P. A., Redfield, M. M., Roger, V. L., Sandhu, G. S., Senecal, C., & Friedman, P. A. (2020). Artificial Intelligence in Cardiology: Present and Future. In *Mayo Clinic Proceedings* (Vol. 95, Issue 5, pp. 1015–1039). Elsevier Ltd. DOI: 10.1016/j.mayocp.2020.01.038

Luca, J. (2020). Ensuring privacy in wearable tech for education. *Computers & Security.*

Mahapatra, P.P.D., Roy, C., Chowdhury, S.R., Sahoo, P. & Mahapatra, P.D. (2023). A Non-Invasive Approach for Anemia Eradication Through Community Based Mass Screening of Rural Population. International Journal of Social Impact, 8(3), 037-046. DIP: 18.02.005/20230803, DOI: DOI: 10.25215/2455/0803005

Malwade, S., Abdul, S. S., Uddin, M., Nursetyo, A. A., Fernandez-Luque, L., Zhu, X., Cilliers, L., Wong, C., Bamidis, P., & Li, Y. (2018). Mobile and wearable technologies in healthcare for the ageing population. *Computer Methods and Programs in Biomedicine*, 161, 233–237. DOI: 10.1016/j.cmpb.2018.04.026 PMID: 29852964

Manal Alshammari, M. A. (2022). Cyber Security for Connected wearable Devices. *International Conference on Business Analytics for Technology and Security* (p. 4). Dubai: IEEE Xplore.

Manickam, S., & Rehman, S. U. (2023). Secured-KDS: Secret key distribution and authentication scheme for resource-constrained devices. *IET Networks*, 12(1), 1–8. DOI: 10.1049/ntw2.12054

Manikeswari, D., Ratnakaram, S., Harris, S. B., & Saseendran, H. (2023), A literature review on the effectiveness of Problem-based learning (PBL) approach in Higher education with reference to the Education 4.0 Taxonomy, *Proceedings of Global conference on Emerging technologies, Retrieved from*: https://www.researchgate .net/publication/384184489_Conf_Proceedings_Dec'23

Manikeswari, D., Ratnakaram, S., Shaikh, Z. H., Alawi, S. H., & Kumar, A. (2024). PBL in Higher Education: Empowering Students for Industry Dynamics. *European Economic Letters*, 14(3), 1137–1140. DOI: 10.52783/eel.v14i3.1873

Manikeswari, G. D., Kumar, R. S. P., & Sirisetti, S. (2024). Psychometric testing - the game changer: A 360-degree view of psychometric testing and its applicability in HR functions. *International Journal of Business Excellence*, 33(1), 48–75. DOI: 10.1504/IJBEX.2024.139112

Manov, A. E., Chauhan, S., Dhillon, G., Dhaliwal, A., Antonio, S., Donepudi, A., Jalal, Y. N., Nazha, J., Banal, M., & House, J. (2023). The Effectiveness of Continuous Glucose Monitoring Devices in Managing Uncontrolled Diabetes Mellitus: A Retrospective Study. *Cureus*. Advance online publication. DOI: 10.7759/cureus.42545 PMID: 37637581

Maples-Keller, J. L., Bunnell, B. E., Kim, S. J., & Rothbaum, B. O. (2017). The use of virtual reality technology in the treatment of anxiety and other psychiatric disorders. In *Harvard Review of Psychiatry* (Vol. 25, Issue 3, pp. 103–113). Lippincott Williams and Wilkins. DOI: 10.1097/HRP.0000000000000138

Mardonova, M., & Choi, Y. (2018). Review of wearable device technology and its applications to the mining industry. *Energies*, 11(3), 547. DOI: 10.3390/en11030547

Marelle, C., Thomas, E. V., & Paul, C. D. (2022). A Survey of Wireless Technology Supporting Individuals with Intellectual and Developmental Disabilities in the Workplace. *Journal of Special Education Technology*, 38(4), 434–444. DOI: 10.1177/01626434221131463

María, A. Hernández-Mustieles, Y. E.-C.-R. (2024, march 14). Wearable Biosensor Technology in Education: A Systematic Review. preprints.*org*, 1-39.

Martinez, J.. (2021). Future innovations in AI-driven education. *AI & Society*.

Marttinen, R., Landi, D., Fredrick, R. N., & Silverman, S. (2020). Wearable digital technology in PE: Advantages, barriers, and teachers' ideologies. *Journal of Teaching in Physical Education*, 39(2), 227–235. DOI: 10.1123/jtpe.2018-0240

Ma, S., Wang, Z., Zhu, Y., Tang, Y., Fan, G., Ma, B., Ye, T., & Wei, L. (2022). Micro/nanofiber fabrication technologies for wearable sensors: A review. *Journal of Micromechanics and Microengineering*, 32(6), 064002. DOI: 10.1088/1361-6439/ac6b7e

Masri, D., Jaber, L., Mashal, R., Albourini, F., Alsaoud, M. A., & Al-Tarawneh, A. M. (2024, February). The Role of Wearables & Technology in Mental Health. In *2024 2nd International Conference on Cyber Resilience (ICCR)* (pp. 1-5). IEEE. DOI: 10.1109/ICCR61006.2024.10532880

Masters, K., Ellaway, R. H., Topps, D., Archibald, D., & Hogue, R. J. (2016). Mobile technologies in medical education: AMEE Guide No. 105. *Medical Teacher*, 38(6), 537–549. DOI: 10.3109/0142159X.2016.1141190 PMID: 27010681

Mator Design. (2024, March 5). Thomas & Navid invented gloves that can translate sign language into text or speech. Medium. https://matorconsultancy.medium.com/thomas-navid-invented-gloves-that-can-translate-sign-language-into-text-or-speech-601baca9ae91

Medtronic. (n.d.). *Nellcor™ SpO_2 Forehead Sensor with OxiMax™ Technology* Retrieved September 30, 2024, from https://www.medtronic.com/covidien/en-us/products/pulse-oximetry/nellcor-spo2-forehead-sensor.html

Mehmood, R. M., & Lee, H. J. (2017). Towards building a computer aided education system for special students using wearable sensor technologies. *Sensors (Basel)*, 17(2), 317. Advance online publication. DOI: 10.3390/s17020317 PMID: 28208734

Meng, X., Zhang, J., & Ren, G. (2021). The Evaluation Model of College Students' Mental Health in the Environment of Independent Entrepreneurship Using Neural Network Technology. *Journal of Healthcare Engineering*, 2021, 1–9. Advance online publication. DOI: 10.1155/2021/4379623 PMID: 34608410

Mezirow, J. (1978). Transformative dimensions of adult learning. *Adult Education Quarterly*, 28(4), 230–244.

Mezirow, J. (1997). *Transformative dimensions of adult learning*. Jossey-Bass.

Middle East Briefing. (2024, August 7). GCC Countries Prioritize Investing in AI Capabilities. Middle East Briefing. https://www.middleeastbriefing.com/news/gcc-countries-invest-in-ai-capabilities/

Mihai, D., Stefan, D., Stegaru, D., Bernea, G., Vacaroiu, I., Papacocea, T., Lupu o-ru, M., Nica, A., Stiru, O., Dragos, D., & Olaru, O. (2021). Continuous glucose monitoring devices: A brief presentation [Review]. *Experimental and Therapeutic Medicine*, 23(2), 174. Advance online publication. DOI: 10.3892/etm.2021.11097 PMID: 35069855

Mills, A. J., Watson, R. T., Pitt, L., & Kietzmann, J. (2016). Wearing safe: Physical and informational security in the age of the wearable device. *Business Horizons*, 59(6), 615–622.

Minn, S. (2022). AI-assisted knowledge assessment techniques for adaptive learning environments. *Computers and Education: Artificial Intelligence*, 3, 100050.

Mishra, P., & Koehler, M. J. (2006). Technological pedagogical content knowledge: A framework for teacher knowledge. *Journal of Educational Computing Research*, 35(1), 60–75.

Mohammed, S. (2022). Health Wearable IoT (WIoT) Technology Devices Security and Privacy Vulnerability Analysis. *8th International Conference on Information Technology Trends (ITT)*. (p. 5).

Mohanty, A. K., Ahamed, S., Kamra, R., & Junnarkar, A. A. (2023). Challenges and Future Prospects of IoT and AI Integration in Education. *Progress in Language, Literature and Education Research*, 94.

Mora, A., Planas, E., & Arnedo-Moreno, J. (2016). Designing game-like activities to engage adult learners in higher education. In F. J. Garcia-Penalvo & R. I. for E. S. (IUCE) University of Salamanca Department of Computer Science and Automatics Plaza de los Caidos s/n Salamanca (Eds.), *ACM International Conference Proceeding Series* (Vols. 02-04-November-2016, pp. 755–762). Association for Computing Machinery. DOI: 10.1145/3012430.3012603

Mora, A., Tondello, G. F., Nacke, L. E., & Arnedo-Moreno, J. (2018). Effect of personalized gameful design on student engagement. *IEEE Global Engineering Education Conference, EDUCON, 2018-April*, 1925–1933. DOI: 10.1109/EDUCON.2018.8363471

Morash, M., Kwak, D.-H., Hoffman, V., Lee, C. H., Cho, S. H., & Moon, B. (2008). Stressors, coping resources and strategies, and police stress in South Korea. *Journal of Criminal Justice*, 36(3), 231–239. https://doi.org/https://doi.org/10.1016/j.jcrimjus.2008.04.010. DOI: 10.1016/j.jcrimjus.2008.04.010

Motti, V. G. (2019). Wearable technologies in education: a design space. In Learning and Collaboration Technologies. Ubiquitous and Virtual Environments for Learning and Collaboration: 6th International Conference, LCT 2019, Held as Part of the 21st HCI International Conference, HCII 2019, Orlando, FL, USA, July 26–31, 2019 [Springer International Publishing.]. *Proceedings*, 21(Part II), 55–67.

Motti, V. G. (2019). Wearable Technologies in Education: A Design Space. *Lecture Notes in Computer Science*, 11591, 55–67. DOI: 10.1007/978-3-030-21817-1_5

Mubin, O., Stevens, C. J., Shahid, S., Al Mahmud, A., & Dong, J.-J. (2013). A review of the applicability of robots in education. *Journal of Technology in Education and Learning, 1*(209–0015), 13.

Munir, H., Vogel, B., & Jacobsson, A. (2022). Artificial intelligence and machine learning approaches in digital education: A systematic revision. *Information (Basel)*, 13(4), 203. DOI: 10.3390/info13040203

Murphy, R. (2019). *Artificial Intelligence Applications to Support K–12 Teachers and Teaching: A Review of Promising Applications.* Challenges, and Risks., DOI: 10.7249/PE315

Murray, A.. (2020). Adaptive learning through wearable tech. *Educational Technology Research and Development.*

Musa, U., Shah, S. M., Majid, H. A., Abidin, Z. Z., Yahya, M. S., Babani, S., & Yunusa, Z. (2022). Recent advancement of wearable Reconfigurable Antenna Technologies: A review. *IEEE Access : Practical Innovations, Open Solutions*, 10, 121831–121863. DOI: 10.1109/ACCESS.2022.3222782

MyHeart. (n.d.). *Your Guide to Heart Health* Retrieved September 30, 2024, from https://myheart.net/

Nahavandi, D., Alizadehsani, R., Khosravi, A., & Acharya, U. R. (2022). Application of artificial intelligence in wearable devices: Opportunities and challenges. *Computer Methods and Programs in Biomedicine*, 213(1), 106541. DOI: 10.1016/j.cmpb.2021.106541 PMID: 34837860

Nakazawa, E., Yamamoto, K., London, A. J., & Akabayashi, A. (2021). Solitary death and new lifestyles during and after COVID-19: Wearable devices and public health ethics. *BMC Medical Ethics*, 22(1), 89. Advance online publication. DOI: 10.1186/s12910-021-00657-9 PMID: 34246258

Naz, I., & Robertson, R. (2024). Exploring the Feasibility and Efficacy of hatGPT3 for Personalized Feedback in Teaching. *Electronic Journal of e-Learning*, 22(2), 98–111. DOI: 10.34190/ejel.22.2.3345

Negi, R. (2024). Improving women's mental health through AI-powered interventions and diagnoses. In *Artificial Intelligence and Machine Learning for Women's Health Issues* (pp. 173–191). Academic Press. DOI: 10.1016/B978-0-443-21889-7.00017-8

Negoi ă, D. O., & Popescu, M. A. M. (2023, December). The use of artificial intelligence in education. In *International Conference of Management and Industrial Engineering* (Vol. 11, pp. 208-214).

Ngai, G., Chan, S. C. F., Cheung, J. C. Y., & Lau, W. W. Y. (2010). Deploying a wearable computing platform for computing education. *IEEE Transactions on Learning Technologies*, 3(1), 45–55. DOI: 10.1109/TLT.2009.49

Nguyen, D. K., Sermpinis, G., & Stasinakis, C. (2023). Big data, artificial intelligence and machine learning: A transformative symbiosis in favour of financial technology. *European Financial Management*, 29(2), 517–548. DOI: 10.1111/eufm.12365

Nguyen, T., & Lee, K. (2022). AI wearables for accessibility in education: Enhancing learning for students with disabilities. *Journal of Inclusive Educational Technology*, 11(2), 140–155.

Nkereuwem, I. V., Obi, N. C., &Oyovwikoghene, S. (2023). Relationship Between Chatbots and AI-Powered Gamification on Chemistry Students' Motivation.

Nogueira, T., De Morais, V. L., De Moraes, C. P., & Lellis-Santos, C. (2020). The wearable bone glossary: Wearable technology at the service of inclusive education. *Medical Education*, 54(5), 447–448. DOI: 10.1111/medu.14102 PMID: 32249961

Nojima, I., Sugata, H., Takeuchi, H., & Mima, T. (2022). Brain–computer interface training based on brain activity can induce motor recovery in patients with stroke: A meta-analysis. *Neurorehabilitation and Neural Repair*, 36(2), 83–96. DOI: 10.1177/15459683211062895 PMID: 34958261

Noran, O. (2016). On gamification in action learning. *ACM International Conference Proceeding Series, 01-05-February-2016*. DOI: 10.1145/2843043.2843344

Nosta, J. (n.d.). *How Google Glass is changing medical education. Forbes Technology*.https://www.forbes.com/sites/johnnosta/2013/06/27/google-glass-teach-me-medicine-how-glass-is-helping-change-medical-education/

Nugent, G., Barker, B., Lester, H., Grandgenett, N., & Valentine, D. (2019). Wearable textiles to support student STEM learning and attitudes. *Journal of Science Education and Technology*, 28(5), 470–479. DOI: 10.1007/s10956-019-09779-7

Ochaya, W. B. (2018). The cyberthreat in IoT, mobile learning, and wearable devices in education. *International Journal of Online Graduate Education*, 1(1).

OECD. (2024). *The Future of Education and Skills: A Global Perspective.* OECD Publishing. https://www.oecd.org/en/about/projects/future-of-education-and-skills-2030.html

Ometov, A., Shubina, V., Klus, L., Skibińska, J., Saafi, S., Pascacio, P., & Lohan, E. S. (2021). A survey on wearable technology: History, state-of-the-art and current challenges. *Computer Networks*, 193, 108074.

Orban, M., Elsamanty, M., Guo, K., Zhang, S., & Yang, H. (2022). A review of brain activity and EEG-based brain–computer interfaces for rehabilitation application. *Bioengineering (Basel, Switzerland)*, 9(12), 768. DOI: 10.3390/bioengineering9120768 PMID: 36550974

Ouyang, F., & Zhang, L. (2024). AI-driven learning analytics applications and tools in computer-supported collaborative learning: A systematic review, *Educational Research Review*, Volume 44, 2024, 100616, ISSN 1747-938X, https://doi.org/DOI: 10.1016/j.edurev.2024.100616

Oveh, R. O. A. G. (2022). Mitigating Social Engineering Attack: A Focus on the Weak Human Link. *2022 5th Information Technology for Education and Development (ITED)* (p. 4). IEEE.

Oyebode, O., Fowles, J., Steeves, D., & Orji, R. (2023). Machine learning techniques in adaptive and personalized systems for health and wellness. *International Journal of Human-Computer Interaction*, 39(9), 1938–1962. DOI: 10.1080/10447318.2022.2089085

Ozkan-Yildirim, S., & Pancar, T. (2021). Smart wearable technology for health tracking: What are the factors that affect their use?. IoT in healthcare and ambient assisted living, 165-199.

Palanivel, K. (2020). Emerging Technologies to Smart Education. *International Journal of Computer Trends and Technology*, 68(2).

Palumbo, A. (2022). Microsoft HoloLens 2 in Medical and Healthcare Context: State of the Art and Future Prospects. *Sensors (Basel)*, 22(20), 7709. https://www.mdpi.com/1424-8220/22/20/7709. DOI: 10.3390/s22207709 PMID: 36298059

Panakaje, N., Ur Rahiman, H., Parvin, S. M. R., Shareena, P., Madhura, K., Yatheen, M., & Irfana, S. (2024). Revolutionizing pedagogy: Navigating the integration of technology in higher education for teacher learning and performance enhancement. *Cogent Education*, 11(1), 2308430. Advance online publication. DOI: 10.1080/2331186X.2024.2308430

Panigrahi, R. R., Dash, M., Shaikh, Z. H., & Irfan, M. (2023). Review of Machine Learning Techniques in the Supply Chain Management of Indian Industry: A Future Research Agenda. In Irfan, M., Elhoseny, M., Kassim, S., & Metawa, N. (Eds.), *Advanced Machine Learning Algorithms for Complex Financial Applications* (pp. 199–219). IGI Global., DOI: 10.4018/978-1-6684-4483-2.ch013

Panula, T., Sirkia, J. P., Wong, D., & Kaisti, M. (2023). Advances in Non-Invasive Blood Pressure Measurement Techniques. *IEEE Reviews in Biomedical Engineering*, 16, 424–438. DOI: 10.1109/RBME.2022.3141877 PMID: 35015647

Park, S., & Lee, H. (2021). *Wearables and student well-being*. Journal of Student Well-being.

Paroi, S. (n.d.). *LEVEL AND SOURCES OF STRESS: EVIDENCE FROM COLLEGE STUDENTS*. https://www.researchgate.net/publication/378306172

Patel, R. H., Foltz, E. A., Witkowski, A., & Ludzik, J. (2023). Analysis of Artificial Intelligence-Based Approaches Applied to Non-Invasive Imaging for Early Detection of Melanoma: A Systematic Review. In *Cancers* (Vol. 15, Issue 19). Multidisciplinary Digital Publishing Institute (MDPI). DOI: 10.3390/cancers15194694

Patil, D. P. (2013, January 1). Variability And Accuracy Of Sahli's Method In Estimation Of Haemoglobin Concentration. *National Journal of Integrated Research in Medicine*, 4, 38.

Patil, V., Singhal, D., Naik, N., Hameed, B., Shah, M., Ibrahim, S., Smriti, K., Chatterjee, G., Kale, A., Sharma, A., Paul, R., Chłosta, P., & Somani, B. (2022). Factors Affecting the Usage of Wearable Device Technology for Healthcare among Indian Adults: A Cross-Sectional Study. *Journal of Clinical Medicine*, 11(23), 7019. DOI: 10.3390/jcm11237019 PMID: 36498594

Patreon. (n.d.). *Functional Vision To Legally Blind Individuals - Interview With eSight's Director Of Product Marketing, Roland Mattern* Retrieved September 30, 2024, from https://www.patreon.com/posts/78398589

Peacock, J., Austin, A., Shapiro, M., Battista, A., & Samuel, A. (2023). Accelerating medical education with ChatGPT: An implementation guide. *MedEdPublish*, •••, 13. PMID: 38440148

Peksa, J., & Mamchur, D. (2023). State-of-the-art on brain-computer interface technology. *Sensors (Basel)*, 23(13), 6001. DOI: 10.3390/s23136001 PMID: 37447849

Pelletier, S. G. (2014). *Technology in academic medicine: Medicine takes a closer look at Google Glass*. AAMC Reporter.

Pennefather, P., & Krebs, C. (2019). Exploring the role of XR in visualisations for use in medical education. *Advances in Experimental Medicine and Biology*, 1171, 15–23. DOI: 10.1007/978-3-030-24281-7_2 PMID: 31823236

Pereira, A. S., Moreira, A. A., Chaló, P., Sancho, L., Varela, A., & Oliveira, C. (2016). Development challenges of a full integrated app in higher education. In *Gaming and Technology Addiction: Breakthroughs in Research and Practice* (Vol. 1, pp. 117–139). IGI Global. DOI: 10.4018/978-1-5225-0256-2.ch001

Perez, C., & Kumar, R. (2021). Real-time feedback in AI-enhanced education. *Education in Science*.

Philips. (n.d.). *DreamStation Go* Retrieved September 30, 2024, from https://www.usa.philips.com/c-e/hs/sleep-apnea-therapy/i-currently-use-sleep-apnea-therapy/dreamstation-go.html

Pillai, K. R., Upadhyaya, P., Prakash, A. V., Ramaprasad, B. S., Mukesh, H. V., & Pai, Y. (2021). End-user satisfaction of technology-enabled assessment in higher education: A coping theory perspective. *Education and Information Technologies*, 26(4), 3677–3698. Advance online publication. DOI: 10.1007/s10639-020-10401-2

Pollreisz, D., & TaheriNejad, N. (2017). A simple algorithm for emotion recognition, using physiological signals of a smart watch. *2017 39th Annual International Conference of the Ieee Engineering in Medicine and Biology Society (EMBC)*, 2353–2356.

Pope, Z. C., Barr-Anderson, D. J., Lewis, B. A., Pereira, M. A., & Gao, Z. (2019). Use of wearable technology and social media to improve physical activity and dietary behaviors among college students: A 12-week randomized pilot study. *International Journal of Environmental Research and Public Health*, 16(19), 3579. DOI: 10.3390/ijerph16193579 PMID: 31557812

Pratley, R. E., Kanapka, L. G., Rickels, M. R., Ahmann, A., Aleppo, G., Beck, R., Bhargava, A., Bode, B. W., Carlson, A., Chaytor, N. S., Fox, D. S., Goland, R., Hirsch, I. B., Kruger, D., Kudva, Y. C., Levy, C., McGill, J. B., Peters, A., Philipson, L., & Miller, K. M. (2020). Effect of Continuous Glucose Monitoring on Hypoglycemia in Older Adults With Type 1 Diabetes: A Randomized Clinical Trial. *Journal of the American Medical Association*, 323(23), 2397–2406. DOI: 10.1001/jama.2020.6928 PMID: 32543682

Qasim, S. H. (2024). Beyond the Classroom: Emerging Technologies to Enhance Learning.

Qureshi, M. I., Khan, N., Raza, H., Imran, A., & Ismail, F. (2021). Digital Technologies in Education 4.0. Does it Enhance the Effectiveness of Learning? *International Journal of Interactive Mobile Technologies*, 15(4), 31. Advance online publication. DOI: 10.3991/ijim.v15i04.20291

Qu, Z. (2022). Analysis of the Effect of Martial Arts on Students' Physical and Mental Health Based on Internet of Things Technology. *Security and Communication Networks*, 2022, 1–9. Advance online publication. DOI: 10.1155/2022/8663157

Rahiman, H. U., & Kodikal, R. (2024). Revolutionizing education: Artificial intelligence empowered learning in higher education. *Cogent Education*, 11(1), 2293431. Advance online publication. DOI: 10.1080/2331186X.2023.2293431

Rahiman, H. U., Kodikal, R., & Suresh, S. (2023). Game on: Can gamification enhance productivity? *F1000 Research*, 12, 818. DOI: 10.12688/f1000research.131579.2 PMID: 38434668

Ramaswamy, G., Vohra, K., Yadav, K., Kaur, R., Rai, T., Jaiswal, A., & Kant, S. (2021). Point-of-care testing using invasive and non-invasive hemoglobinometers: Reliable and valid method for estimation of hemoglobin among children 6-59 months. *Journal of Tropical Pediatrics*, 67(1), fmaa111. Advance online publication. DOI: 10.1093/tropej/fmaa111 PMID: 33367788

Ramirez, M., & Thomas, D. (2021). *Technical and financial challenges in AI for education*. Journal of Education and Technology.

Ramírez-Moreno, M. A., De, J., & Lozoya-Santos, J. (2024). Wearable Biosensor Technology in Education: A Systematic review. *Sensors (Basel)*, 24(8), 2437. DOI: 10.3390/s24082437 PMID: 38676053

Rana, M., & Bhushan, M. (2023). Machine learning and deep learning approach for medical image analysis: Diagnosis to detection. *Multimedia Tools and Applications*, 82(17), 26731–26769. DOI: 10.1007/s11042-022-14305-w PMID: 36588765

Rao, T. V. N., Gaddam, A., Kurni, M., & Saritha, K. (2022). Reliance on artificial intelligence, machine learning and deep learning in the era of industry 4.0. *Smart healthcare system design: security and privacy aspects*, 281-299.

RaviPrakash, H., & Anwar, S. M. (2023). Do It Yourself: Wearable Sensors and AI for Self-Assessment of Mental Health. *AI in Clinical Medicine: A Practical Guide for Healthcare Professionals*, 94-103.

Razdan, S., & Sharma, S. (2022). Internet of Medical Things (IoMT): Overview, Emerging Technologies, and Case Studies. In *IETE Technical Review (Institution of Electronics and Telecommunication Engineers, India)* (Vol. 39, Issue 4, pp. 775–788). Taylor and Francis Ltd. DOI: 10.1080/02564602.2021.1927863

REAL® System. (n.d.). *The REAL® System, Virtual Reality Rehabilitation Tool* Retrieved September 30, 2024, from https://www.realsystem.com/

Regona, M., Yigitcanlar, T., Xia, B., & Li, R. Y. M. (2022). Opportunities and Adoption Challenges of AI in the Construction Industry: A PRISMA Review. *Journal of Open Innovation*, 8(1), 45. DOI: 10.3390/joitmc8010045

Rehman, S. &. (2023). Towards a Cybersecurity Awareness Program for Non-Technical Audiences in Malaysia. 85-97.

Rehman, S. U. (2020). Towards sustainable IoT ecosystem. *In 2020 2nd International conference on industrial electrical and electronics (ICIEE) (pp. 135-138). IEEE.* (pp. 135-138). IEEE.

Rehman, S. U. (2023). Impact of DoS/DDoS attacks in IoT environment: A study. *AIP Conference Proceedings (Vol. 2760, No. 1).* AIP Publishing.

Rehman, S. U.-C. (2023). *Privacy calculus model for online social networks: a study of Facebook users in a Malaysian university*. Springer.

Rehman, S. U., & Manickam, S. (2024). Application of smart sensors for internet of things healthcare environment: study and prospects. In *Next-Generation Smart Biosensing* (pp. 287–305). Academic Press. DOI: 10.1016/B978-0-323-98805-6.00006-3

Rehman, S. U., Sadek, I., Huang, B., Manickam, S., & Mahmoud, L. N. (2024). IoT-based emergency cardiac death risk rescue alert system. *MethodsX*, 13, 102834. DOI: 10.1016/j.mex.2024.102834 PMID: 39071997

Resusci, A. (1971). *Advanced Cardiac Life Support Training guides*. American Heart Association.

Rich, E., Gray, S., Camacho Miñano, M. J., & MacIsaac, S. (2024). Digital health education: Critical perspectives. *Health Education Journal*, •••, 00178969241255438.

Rickards, E., Miller, S., Hampson, O., & Wat, D. (2021). Evaluating the accuracy of pulse oximetry: Finger probe versus earlobe probe in patients with confirmed or suspected COVID-19 within the community setting. *The European Respiratory Journal*, 58(suppl 65), PA456. Advance online publication. DOI: 10.1183/13993003.congress-2021.PA456

Rincon-Flores, E. G., & Santos-Guevara, B. N. (2021). Gamification during Covid-19: Promoting active learning and motivation in higher education. *Australasian Journal of Educational Technology*, 37(5), 43–60. DOI: 10.14742/ajet.7157

Rivera, E. S., & Garden, C. L. P. (2021). Gamification for student engagement: A framework. *Journal of Further and Higher Education*, 45(7), 999–1012. DOI: 10.1080/0309877X.2021.1875201

Roberto De Fazio, A.-R. A.-H. (n.d.). An Energy-Autonomous Smart Shirt Employing Wearable Sensors for Users' Safety and Protection in Hazardous Workplaces. *MDPI*, 2022.

Robinson, T., Condell, J., Ramsey, E., & Leavey, G. (2023). Self-Management of Subclinical Common Mental Health Disorders (Anxiety, Depression and Sleep Disorders) Using Wearable Devices. *International Journal of Environmental Research and Public Health*, 20(3), 2636. DOI: 10.3390/ijerph20032636 PMID: 36768002

Rodgers, M. M., Alon, G., Pai, V. M., & Conroy, R. S. (2019). Wearable technologies for active living and rehabilitation: Current research challenges and future opportunities. *Journal of Rehabilitation and Assistive Technologies Engineering*, 6, 205566831983960. DOI: 10.1177/2055668319839607 PMID: 31245033

Romanzini, M., Petroski, E. L., & Reichert, F. F. (2012). Accelerometers thresholds to estimate physical activity intensity in children and adolescents: A systematic review. *Revista Brasileira de Cineantropometria & Desempenho Humano*, 14, 101–113.

Roos, L. G., & Slavich, G. M. (2023). Wearable technologies for health research: Opportunities, limitations, and practical and conceptual considerations. *Brain, Behavior, and Immunity*, 113, 444–452. DOI: 10.1016/j.bbi.2023.08.008 PMID: 37557962

Ruiz, P., & Fusco, J. (2024, March 31). Glossary of Artificial Intelligence Terms for Educators – CIRCLS. Circls.org. https://circls.org/educatorcircls/ai-glossary

Ruotsalainen, P., & Blobel, B. (2020). Health information systems in the digital health ecosystem—Problems and solutions for ethics, trust and privacy. *International Journal of Environmental Research and Public Health*, 17(9), 3006. Advance online publication. DOI: 10.3390/ijerph17093006 PMID: 32357446

Russell, S., & Norvig, P. (2021). *Artificial Intelligence: A Modern Approach*. Pearson.

Saaida, M. B. (2023). AI-Driven transformations in higher education: Opportunities and challenges. *International Journal of Educational Research and Studies*, 5(1), 29–36.

Saddique, A., & Seng Choi, C. (2023). *Systematic Review Provides Evidence That Anxiety and Negative Thoughts are Prevalent among Medical Students in Malaysia*. https://doi.org/DOI: 10.6007/IJARBSS/v13-i7/16938

Sadek, I. C. (2019). *Memory snapshot dataset of a compromised host with malware using obfuscation evasion techniques*. Elsevier. DOI: 10.1016/j.dib.2019.104437

Sadek, I., Codjo, J., Rehman, S. U., & Abdulrazak, B. (2022). Security and privacy in the Internet of Things healthcare systems: Toward a robust solution in real-life deployment. *Computer Methods and Programs in Biomedicine Update*, 2, 100071. DOI: 10.1016/j.cmpbup.2022.100071

Sadek, I., Rehman, S. U., Codjo, J., & Abdulrazak, B. (2019). Privacy and security of IoT based healthcare systems: concerns, solutions, and recommendations. In *How AI Impacts Urban Living and Public Health: 17th International Conference, ICOST 2019, New York City, NY, USA, October 14-16, 2019* [Springer International Publishing.]. *Proceedings*, 17, 3–17.

Saghir, N., Aggarwal, A., Soneji, N., Valencia, V., Rodgers, G., & Kurian, T. (2020). A comparison of manual electrocardiographic interval and waveform analysis in lead 1 of 12-lead ECG and Apple Watch ECG: A validation study. *Cardiovascular Digital Health Journal*, 1(1), 30–36. DOI: 10.1016/j.cvdhj.2020.07.002 PMID: 35265871

Sahoo, K. C., Sinha, A., Sahoo, R. K., Suman, S. S., Bhattacharya, D., & Pati, S. (2024). Diagnostic validation and feasibility of a non-invasive haemoglobin screening device (EzeCheck) for'Anaemia Mukt Bharat'in India. *Cureus*, 16(1).

Sahoo, K., Sinha, A., Sahoo, R., Suman, S. S., Bhattacharya, D., & Pati, S. (2024, January 24). Diagnostic Validation and Feasibility of a Non-invasive Haemoglobin Screening Device (EzeCheck) for 'Anaemia Mukt Bharat' in India. *Cureus*, 16(1), e52877. DOI: 10.7759/cureus.52877 PMID: 38406104

Saini, A., Shukla, R., Joe, W., & Kapur, A. (2022). Improving nutrition budgeting in health sector plans: Evidence from India's anaemia control strategy. *Maternal and Child Nutrition*, 18(2), e13253. DOI: 10.1111/mcn.13253 PMID: 35231160

Sandall, B. K. (2016). Wearable technology and schools: Where are we and where do we go from here? Journal of Curriculum, Teaching. *Learning and Leadership in Education*, 1(1), 9.

Saputra, N., Satispi, E., & Herlina, M. G. (2021). Growing Learning Agility by Combining Learning Culture with Knowledge Management. *2021 International Conference on Advanced Computer Science and Information Systems, ICACSIS 2021*. DOI: 10.1109/ICACSIS53237.2021.9631346

Sarea, A., Subramanian, S., Alareeni, B., Shaikh, Z. H., Hawaldar, I. T., & Elshaker, A. H. (2021). Financial Disclosures by Using Machine Learning Analysis. In Alareeni, B., Hamdan, A., & Elgedawy, I. (Eds.), *The Importance of New Technologies and Entrepreneurship in Business Development: In The Context of Economic Diversity in Developing Countries. ICBT 2020. Lecture Notes in Networks and Systems* (Vol. 194). Springer., DOI: 10.1007/978-3-030-69221-6_5

Sarraf, D., Vasiliu, V., Imberman, B., & Lindeman, B. (2021). Use of artificial intelligence for gender bias analysis in letters of recommendation for general surgery residency candidates. *American Journal of Surgery*, 222(6), 1051–1059. DOI: 10.1016/j.amjsurg.2021.09.034 PMID: 34674847

Sato, Y., Wang, M., & Chen, L. (2023). Ethical standards in AI-driven wearable technologies for education: Addressing data privacy and transparency. *International Journal of Educational Ethics*, 12(3), 211–225.

Sayam, A., Rahman, M. M., Sayem, A. S. M., Ahmed, A. F., & Alimuzzaman, S. (2024). Carbon-Based Textile-Structured Triboelectric Nanogenerators for Smart Wearables. *Advanced Energy and Sustainability Research*, 5(11), 2400127.

Sazonov, E. (Ed.). (2020). *Wearable Sensors: Fundamentals, implementation and applications*. Academic Press.

Schoenenberger, C.-A., Korkut, S., Jaeger, J., & Dornberger, R. (2016). BioTourney: Gamifying a biology class by applying a content-independent learning game framework. In L. Boyle, L. Boyle, T. M. Connolly, & T. M. Connolly (Eds.), *Proceedings of the European Conference on Games-based Learning* (Vols. 2016-January, pp. 575–583). Dechema e.V. https://www.scopus.com/inward/record.uri?eid=2-s2.0-84997018044&partnerID=40&md5=6ab4bce210b027b2dc922d71f7df4702

Schroeder, N. L., Romine, W. L., & Kemp, S. E. (2023). A scoping review of wrist-worn wearables in education. *Computers and Education Open*, 5, 100154. Advance online publication. DOI: 10.1016/j.caeo.2023.100154

Seçkin, A. Ç., Ateş, B., & Seçkin, M. (2023). Review on Wearable Technology in Sports: Concepts, Challenges and Opportunities. *Applied Sciences 2023, Vol. 13, Page 10399, 13*(18), 10399. DOI: 10.3390/app131810399

Selwin, S., Saseendran, H., Vemavarapu, K. R., Shaikh, I., Shaikh, Z. H., & Harris, S. B. (2024). AI-Driven Employee Engagement: Transforming Banking Workforce Dynamics Using the BEULA Model. In Awwad, B. (Ed.), *The AI Revolution: Driving Business Innovation and Research. Studies in Systems, Decision and Control* (Vol. 525). Springer., DOI: 10.1007/978-3-031-54383-8_32

Selwyn, N. (2011). Is technology really transforming education? A critical perspective. *Educational Researcher*, 40(2), 75–80.

Seneviratne, S., Hu, Y., Nguyen, T., Lan, G., Khalifa, S., Thilakarathna, K., Hassan, M., & Seneviratne, A. (2017). A survey of wearable devices and challenges. *IEEE Communications Surveys and Tutorials*, 19(4), 2573–2620. DOI: 10.1109/COMST.2017.2731979

Seng, K. P., Ang, L.-M., Peter, E., & Mmonyi, A. (2023). Machine Learning and AI Technologies for Smart Wearables. *Electronics (Basel)*, 12(7), 1509. DOI: 10.3390/electronics12071509

Shahul, A., Hovan George, A. S., & George, A. S. (2023). *Partners Universal International Innovation Journal (PUIIJ) Wearable Sensors: A New Way to Track Health and Wellness*. DOI: 10.5281/zenodo.8260879

Shaikh, A. A., Kumar, A., Jani, K., Mitra, S., García-Tadeo, D. A., & Devarajan, A. (2022). The role of machine learning and artificial intelligence for making a digital classroom and its sustainable impact on education during COVID-19. *Materials Today: Proceedings*, 56, 3211–3215. DOI: 10.1016/j.matpr.2021.09.368 PMID: 35464152

Shaikh, Z. H., Irfan, M., Nomran, N. M., & Lawin, M. L. (2024). The Effect of Strategic Partnership on Innovation and Business: Performance of the Fintech Industry in Bahrain. In Irfan, M., Kadry, S., Sharif, M., & Khan, H. (Eds.), *Fintech Applications in Islamic Finance: AI, Machine Learning, and Blockchain Tech-niques* (pp. 236–254). IGI Global., DOI: 10.4018/979-8-3693-1038-0.ch016

Shaikh, Z. H., Irfan, M., Nomran, N. M., & Ratnakaram, S. P. (2024). *Unlocking the Potential of AI for Efficient Governance: Innovative Approaches of Bahrain., AI-Driven Decentralized Finance and the Future of Finance*. IGI Global., DOI: 10.4018/979-8-3693-6321-8.ch005

Shaikh, Z. H., Sarea, A., & Irfan, M. (2022). *Islamic Banking Strategies in the World of Fintech: Success Story of Bahrain* (Vol. 423). Springer., DOI: 10.1007/978-3-030-93464-4_10

Shaikh, Z., Irfan, M., Sarea, A., & Panigrahi, R. R. (2024). *The Emergence of Islamic Fintech and Bahrain: Prospect for Global Financial Sectors* (Vol. 503). Springer., DOI: 10.1007/978-3-031-43490-7_52

Shajari, S., Kuruvinashetti, K., Komeili, A., & Sundararaj, U. (2023). The Emergenceof AI-Based Wearable Sensors for Digital Health Technology: A Review. *Sensors (Basel)*, 23(23), 9498. DOI: 10.3390/s23239498 PMID: 38067871

Sharma, A., Arora, S., Kumar, S., & Bansal, A. (2023, January). Ai ML Enabled Wearable Smart Sensors Detecting Psychological Disorders. In *2023 13th International Conference on Cloud Computing, Data Science & Engineering (Confluence)* (pp. 651-656). IEEE. DOI: 10.1109/Confluence56041.2023.10048832

Sharma, R. (2024, October 30). Future of Education: AI Compliance with FERPA and GDPR - Hurix Digital. Digital Engineering & Technology | Elearning Solutions | Digital Content Solutions. https://www.hurix.com/data-privacy-in-education-through-ferpa-and-gdpr-adherence/

Shishakly, R., Almaiah, M. A., Lutfi, A., & Alrawad, M. (2024). The influence of using smart technologies for sustainable development in higher education institutions. *International Journal of Data and Network Science*, 8(1), 77–90. DOI: 10.5267/j.ijdns.2023.10.015

Shoikova, E., Nikolov, R., & Kovatcheva, E. (2017). Conceptualising of smart education. E+E, 52(3–4), 29–37.

Signori, G. G., De Guimarães, J. C. F., Severo, E. A., & Rotta, C. (2018). Gamification as an innovative method in the processes of learning in higher education institutions. *International Journal of Innovation and Learning*, 24(2), 115–137. DOI: 10.1504/IJIL.2018.094066

Siklander, P., & Harmoinen, S. (2021). Ice Age is Approaching: Triggering University Students' Interest and Engagement in Gamified Outdoor Playful Learning Activities. In *Science and Drama: Contemporary and Creative Approaches to Teaching and Learning* (pp. 125–143). Springer International Publishing. DOI: 10.1007/978-3-030-84401-1_8

Silva, L. M., Cibrian, F. L., Monteiro, E. M., Bhattacharya, A., & Beltran, J. A. (2023). *Clarisse Bonang, Epstein, D. A., Schuck, S., Lakes, K. D., & Hayes, G. R.* Unpacking the Lived Experiences of Smartwatch Mediated Self and Co-Regulation with ADHD Children., DOI: 10.1145/3544548.3581316

Simbeck, K. (2024). They shall be fair, transparent, and robust: Auditing learning analytics systems. *AI and Ethics*, 4(2), 555–571. Advance online publication. DOI: 10.1007/s43681-023-00292-7

Singh, P. (2022). Navigating the digital wave: Transformative strategies in modern business with the integration of artificial intelligence (ai) and machine learning (ml). *International Journal of Computing and Artificial Intelligence*, 3(2), 78–85. DOI: 10.33545/27076571.2022.v3.i2a.77

Singh, P. K. (2023). Digital transformation in supply chain management: Artificial Intelligence (AI) and Machine Learning (ML) as Catalysts for Value Creation. *International Journal of Supply Chain Management*, 12(6), 57–63. DOI: 10.59160/ijscm.v12i6.6216

Siren Care. (n.d.). *What is the Siren Care Diabetic Sock and Foot Monitoring System?* Retrieved September 30, 2024, from https://www.siren.care/blog/what-is-siren-diabetic-sock-system

Smith, T., & Jones, P. (2020). *The impact of personalized feedback on student engagement*. Learning Technologies Journal.

Sokolovskaia, S., Nezhkina, N., Orlova, E., Karpycheva, M., Bocharin, I., Kolokoltsev, M., Limarenko, O., Romanova, E., & Martusevich, A. (2022). Pedagogical online technology to increase the level of students' conscious attitude to physical and mental health. *Journal of Physical Education and Sport*, 22(10), 2380–2385. DOI: 10.7752/jpes.2022.10303

Sommer, I., Nußbaumer-Streit, B., & Gartlehner, G. (2021). WHO Guideline: Physical Activity, Sedentary Behavior and Sleep for Children under 5 Years of Age. In *Gesundheitswesen* (Vol. 83, Issue 7). DOI: 10.1055/a-1489-8049

Song, L. (2022). An Intelligent Method for Determining the Mental Health of University Students Based on IoT Technology. In Y. Pei, C. J.-W., & J. C. Hung (Eds.), *Lecture Notes in Electrical Engineering: Vol. 935 LNEE* (pp. 1283–1289). Springer Science and Business Media Deutschland GmbH. DOI: 10.1007/978-981-19-4132-0_175

Song, X., Zhang, Y., & Chang, X. (2024). Development of a Micro-course App for College Students' Mental Health Education Courses Based on Computer Software Technology. In Y. Zhang & N. Shah (Eds.), *Lecture Notes of the Institute for Computer Sciences, Social-Informatics and Telecommunications Engineering, LNICST: Vol. 583 LNICST* (pp. 572–581). Springer Science and Business Media Deutschland GmbH. DOI: 10.1007/978-3-031-63139-9_59

Sousa, A. C., Ferrinho, S. N., & Travassos, B. (2023). The use of wearable technologies in the assessment of physical activity in preschool-and school-age youth: Systematic review. *International Journal of Environmental Research and Public Health*, 20(4), 3402. DOI: 10.3390/ijerph20043402 PMID: 36834100

Sridhar, S., Saraswathi, S., Bharathi, S. V. S., & Devipriya, A. (2024). Gamification applications based on metaverse for healthcare industry. In Dhanaraj, R. K., Vashishtha, S., Sathyamoorthy, M., Balusamy, B., & Cengiz, K. (Eds.), *Metaverse Technologies in Healthcare* (pp. 233–244). Academic Press., https://doi.org/https://doi.org/10.1016/B978-0-443-13565-1.00011-7 DOI: 10.1016/B978-0-443-13565-1.00011-7

Srinivasa, K. G., Kurni, M., & Saritha, K. (2022). Harnessing the Power of AI to Education. In *Learning, teaching, and assessment methods for contemporary learners: pedagogy for the digital generation* (pp. 311–342). Springer Nature Singapore. DOI: 10.1007/978-981-19-6734-4_13

Stansbury, J. A., & Earnest, D. R. (2017). Meaningful Gamification in an Industrial/Organizational Psychology Course. *Teaching of Psychology*, 44(1), 38–45. DOI: 10.1177/0098628316677645

Steiner, D. F., Macdonald, R., Liu, Y., Truszkowski, P., Hipp, J. D., Gammage, C., Thng, F., Peng, L., & Stumpe, M. C. (2018). *Impact of Deep Learning Assistance on the Histopathologic Review of Lymph Nodes for Metastatic Breast Cancer*. www.ajsp.com

Stellarix. (n.d.). *Challenges in Wearable Technology In Healthcare* Retrieved October 2, 2024, from https://stellarix.com/wearable-health-technology-benefits-challenges-types/

Steve Jobs Was Real Inspiration Behind Creation Of Apple Watch To Improve Health Care. (n.d.). Retrieved October 2, 2024, from https://www.techtimes.com/articles/157379/20160510/steve-jobs-real-inspiration-behind-creation-apple-watch-improve-healthcare.htm

Stevens, G. A., Paciorek, C. J., Flores-Urrutia, M. C., Borghi, E., Namaste, S., Wirth, J. P., & Rogers, L. M. (2022). National, regional, and global estimates of anaemia by severity in women and children for 2000–19: A pooled analysis of population-representative data. *The Lancet. Global Health*, 10(5), e627–e639.

Stock, R., Gaarden, A. P., & Langørgen, E. (2024). The potential of wearable technology to support stroke survivors' motivation for home exercise–Focus group discussions with stroke survivors and physiotherapists. *Physiotherapy Theory and Practice*, 40(8), 1795–1806. DOI: 10.1080/09593985.2023.2217987 PMID: 37246716

Sturman, D., & Bower, M. (2015). What are the educational affordances of wearable technologies? *Computers & Education*, 88, 343–353. DOI: 10.1016/j.compedu.2015.07.013

Sulla-Torres, J., Calla Gamboa, A., Avendaño Llanque, C., Angulo Osorio, J., & Zúñiga Carnero, M. (2024). Classification of Motor Competence in Schoolchildren Using Wearable Technology and Machine Learning with Hyperparameter Optimization. *Applied Sciences (Basel, Switzerland)*, 14(2), 707. Advance online publication. DOI: 10.3390/app14020707

Sulphey, M. M. (2017). Game based learning as an aid for extenuating higher education sector issues-the case of Saudi Arabia. *International Journal of Simulation: Systems, Science and Technology, 18*(1), 6.1-6.10. DOI: 10.5013/IJSSST.a.18.01.06

Sultan, N. (2015). Reflective thoughts on the potential and challenges of wearable technology for healthcare provision and medical education. *International Journal of Information Management*, 35(5), 521–526. DOI: 10.1016/j.ijinfomgt.2015.04.010

Sunarya, P. A. (2022). Machine learning and artificial intelligence as educational games. *International Transactions on Artificial Intelligence*, 1(1), 129–138.

Szeto, K., Arnold, J., & Maher, C. (2024). The Wearable Activity Tracker Checklist for Healthcare (WATCH): A 12-point guide for the implementation of wearable activity trackers in healthcare. *The International Journal of Behavioral Nutrition and Physical Activity*, 21(1), 30. DOI: 10.1186/s12966-024-01567-w PMID: 38481238

Tamargo, J., Le Heuzey, J. Y., & Mabo, P. (2015). Narrow therapeutic index drugs: A clinical pharmacological consideration to flecainide. *European Journal of Clinical Pharmacology*, 71(5), 549–567. DOI: 10.1007/s00228-015-1832-0 PMID: 25870032

Tamkeen to empower 60 Bahraini nationals with Machine Learning and Artificial Intelligence training in Collaboration with AWS and BIBF. (2023, December 5). https://www.tamkeen.bh/en/tamkeen-to-empower-60-bahraini-nationals-with-machine-learning-and-artificial-intelligence-training-in-collaboration-with-aws-and-bibf/#:~:text=The%20program,%20in%20collaboration%20with%20Amazon

Tan, C. T. (2013). Towards a MOOC game. *ACM International Conference Proceeding Series*. DOI: 10.1145/2513002.2513040

Tang, K. H. D. (2024). Implications of Artificial Intelligence for Teaching and Learning. *Acta Pedagogia Asiana*, 3(2), 65–79. DOI: 10.53623/apga.v3i2.404

Taylor, E. W., & Cranton, P. (2017). TRANSFORMATIVE LEARNING THEORY. In *Mapping the Field of Adult and Continuing Education: an International Compendium* (Vol. 1). Adult Learners., DOI: 10.4324/9781003445913-9

Teja Nallamothu, P., & Bharadiya, J. P. (2023). Artificial Intelligence in Orthopedics: A Concise Review. In *Asian Journal of Orthopaedic Research* (Vol. 9, Issue 1). https://www.sdiarticle5.com/review-history/98595

Thakkar, H. K., Chowdhury, S. R., Bhoi, A. K., & Barsocchi, P. (2022). *Applications of wearable technologies in healthcare: An analytical study. In 5G IoT and Edge Computing for Smart Healthcare*. Academic Press.

Thomas, B. (2021). Artificial Intelligence: Review of Current and Future Applications in Medicine. *Federal Practitioner*, 38(11). Advance online publication. DOI: 10.12788/fp.0174 PMID: 35136337

Tian, T. (2016). *Proceedings, 2016 IEEE International Conference on Bioinformatics and Biomedicine :* Dec 15-18, 2016, Shenzhen, China. IEEE.

Tiger Medical. (n.d.). *Covidien Specialty SpO2 Sensors* Retrieved October 1, 2024, from https://tigermedical.com/products/specialty-non-adhesive-spo2-sensors-covsc-pr-oo?srsltid=AfmBOopn6zXnthlui9BQqG0z_E1eeyVivw71xSoHZOrQQcnHHnTdq3-o

Titus, S., & Ng'ambi, D. (2014). Exploring the use of digital gaming to improve student engagement at a resource poor institution in South Africa. In C. Busch (Ed.), *Proceedings of the European Conference on Games-based Learning* (Vol. 2, pp. 742–748). Dechema e.V. https://www.scopus.com/inward/record.uri?eid=2-s2.0-84923531536&partnerID=40&md5=790e0c18815d4118d44671efc7c203e4

Tiwari, R. (2023). The integration of AI and machine learning in education and its potential to personalize and improve student learning experiences. *INTERANTIONAL JOURNAL of SCIENTIFIC RESEARCH in ENGINEERING and MANAGEMENT*, 07(02). Advance online publication. DOI: 10.55041/IJSREM17645

Trabelsi, Z., Alnajjar, F., Parambil, M. M. A., Gochoo, M., & Ali, L. (2023). Real-time attention monitoring system for classroom: A deep learning approach for student's behavior recognition. *Big Data and Cognitive Computing*, 7(1), 48. DOI: 10.3390/bdcc7010048

Trotsko, A., Kharkiv, H., Olena, B., Kirilenko, G., Halyna, B., & Trush, O. (2019). TEACHERS' PROFESSIONAL SELF- IMPROVEMENT IN THE CONDITIONS OF DISTANCE LEARNING IMPLEMENTATION IN HIGHER EDUCATION INSTITUTIONS Liudmyla S. Rybalko. *Information Technologies and Learning Tools*, 72(4). Advance online publication. DOI: 10.33407/itlt.v72i4.3088

Tsay, C. H.-H., Kofinas, A., & Luo, J. (2018). Enhancing student learning experience with technology-mediated gamification: An empirical study. *Computers & Education*, 121, 1–17. DOI: 10.1016/j.compedu.2018.01.009

Tuomi, I. (2019). *The Impact of Artificial Intelligence on Learning, Teaching, and Education: Policies for the Future. JRC Science for Policy Report*. European Commission.

Turner Medical. (n.d.). *Ear-Clip Sensor Pulse Oximeters* Retrieved October 1, 2024, from https://www.turnermedical.com/EAR-PULSE-OXIMETER-s/74.htm

UAE Government. (n.d.). *The Emirati Genome Programme* Retrieved October 1, 2024, from https://u.ae/en/information-and-services/health-and-fitness/research-in-the-field-of-health/the-emirati-genome-programme

Ul Rehman, S., & Manickam, S. (2016). A study of smart home environment and its security threats. *Ul Rehman, S., & Manickam. International Journal of Reliability Quality and Safety Engineering*, 23(03), 1640005. DOI: 10.1142/S0218539316400052

Van Merriënboer, J. J. G., & Kirschner, P. A. (2017). *Ten steps to complex learning: A systematic approach to four-component instructional design*. Routledge. DOI: 10.4324/9781315113210

Vanduhe, V. Z., Nat, M., & Hasan, H. F. (2020). Continuance Intentions to Use Gamification for Training in Higher Education: Integrating the Technology Acceptance Model (TAM), Social Motivation, and Task Technology Fit (TTF). *IEEE Access : Practical Innovations, Open Solutions*, 8, 21473–21484. DOI: 10.1109/ACCESS.2020.2966179

Vator TV. (n.d.). *When Fitbit was young: the early years* Retrieved October 2, 2024, from https://vator.tv/news/2017-02-21-when-fitbit-was-young-the-early-years

Venter, M. (2020). Gamification in STEM programming courses: State of the art. In A. Cardoso, G. R. Alves, & T. Restivo (Eds.), *IEEE Global Engineering Education Conference, EDUCON* (Vols. 2020-April, pp. 859–866). IEEE Computer Society. DOI: 10.1109/EDUCON45650.2020.9125395

Vijayan, V., Connolly, J., Condell, J., McKelvey, N., & Gardiner, P. (2021). Review of wearable devices and data collection considerations for connected health. In *Sensors* (Vol. 21, Issue 16). MDPI AG. DOI: 10.3390/s21165589

Villagrasa, S., & Duran, J. (2013). Gamification for Learning 3D Computer Graphics Arts. *ACM International Conference Proceeding Series*, 429–433. DOI: 10.1145/2536536.2536602

Villagrasa, S., Fonseca, D., Redondo, E., & Duran, J. (2014). Teaching case of gamification and visual technologies for education. *Journal of Cases on Information Technology*, 16(4), 38–57. DOI: 10.4018/jcit.2014100104

Vincent-Lancrin, S. (2021). Smart Education Technology: How It Might Transform Teaching (and Learning). *New England Journal of Public Policy*.

Wang, C., He, Y., & Yang, Y. (2019). The role of artificial intelligence in education: A systematic review. *International Journal of Artificial Intelligence in Education*, 29(3), 397–422.

Wang, F. H. (2017). An exploration of online behaviour engagement and achievement in flipped classroom supported by learning management system. *Computers & Education*, 114, 79–91. DOI: 10.1016/j.compedu.2017.06.012

Wang, Q.-Z., & Gao, L. (2019). Neural network for processing privacy-protected data. *Journal of Cryptologic Research*, 6(2), 258–268. DOI: 10.13868/j.cnki.jcr.000300

Wang, S., Christensen, C., Cui, W., Tong, R., Yarnall, L., Shear, L., & Feng, M. (2023). When adaptive learning is effective learning: Comparison of an adaptive learning system to teacher-led instruction. *Interactive Learning Environments*, 31(2), 793–803. DOI: 10.1080/10494820.2020.1808794

Wang, X., & Jiang, J. (2021). The potential of wearable devices and AI for personalized learning. *International Journal of Educational Technology in Higher Education*, 18(1), 1–15.

Wang, Y. (2023). The research on the impact of distance learning on students' mental health. *Education and Information Technologies*, 28(10), 12527–12539. DOI: 10.1007/s10639-023-11693-w PMID: 37361750

Wang, Y., & Liu, J. (2023). Blockchain Technology in Education: A Review. *IEEE Access : Practical Innovations, Open Solutions*, 11, 11439–11453.

Wang, Z., Ong, C. L. J., & Fu, Z. (2022). AI Models to Assist Vancomycin Dosage Titration. *Frontiers in Pharmacology*, 13, 801928. Advance online publication. DOI: 10.3389/fphar.2022.801928 PMID: 35211014

Wei Gao, T. W. (2022, August 1). Toward a Systematic Survey on Wearable Computing for Education Applications. *IEEE Xplore*, 9, 15.

Wei, L., Li, Z., Dai, Z., Ding, L., Wei, H., Li, J., Zhu, H., & Zhu, Y. (2024). Wearable sweat management Technologies. *Advanced Materials Technologies*, 9(7), 2301812. Advance online publication. DOI: 10.1002/admt.202301812

Wellue. (n.d.). *O2Ring Smart Ring Pulse Oximeter. Overnight Tracking* Retrieved September 30, 2024, from https://getwellue.com/pages/o2ring-oxygen-monitor

Wen, J., Xu, B., Gao, Y., Li, M., & Fu, H. (2021). Wearable technologies enable high-performance textile supercapacitors with flexible, breathable and wearable characteristics for future energy storage. *Energy Storage Materials*, 37, 94–122. DOI: 10.1016/j.ensm.2021.02.002

Wiedermann, C. J., Barbieri, V., Plagg, B., Marino, P., Piccoliori, G., & Engl, A. (2023, May). Fortifying the foundations: A comprehensive approach to enhancing mental health support in educational policies amidst crises. [). MDPI.]. *Health Care*, 11(10), 1423. PMID: 37239709

Williams, D., & Carter, H. (2024). Immersive VR labs and their impact on student comprehension in science education. *Journal of Interactive Learning Environments*, 32(1), 87–102.

Wirani, Y., Nabarian, T., & Romadhon, M. S. (2022). Evaluation of continued use on Kahoot! as a gamification-based learning platform from the perspective of Indonesia students. *Procedia Computer Science*, 197, 545–556. https://doi.org/https://doi.org/10.1016/j.procs.2021.12.172. DOI: 10.1016/j.procs.2021.12.172

Wong, K. P., Tse, M. M. Y., & Qin, J. (2022). Effectiveness of Virtual Reality-Based Interventions for Managing Chronic Pain on Pain Reduction, Anxiety, Depression and Mood: A Systematic Review. In *Healthcare (Switzerland)* (Vol. 10, Issue 10). MDPI. DOI: 10.3390/healthcare10102047

Wu, Y., & Feng, R. (2023). Application of virtual reality technology in promoting mental health of college students. *Chinese Journal of School Health, 44*(7), 1095-1098and1102. DOI: 10.16835/j.cnki.1000-9817.2023.07.031

Wu, M., & Luo, J. (2019). Wearable technology applications in healthcare: A literature review. *On-Line Journal of Nursing Informatics*, 23(3).

Wu, T., Dameff, C., & Tully, J. (2014). Integrating Google Glass into simulation-based training: Experiences and future directions. *Journal of Biomedical Graphics and Computing*, 4(2), 49. DOI: 10.5430/jbgc.v4n2p49

Xia, Y., & Lin, Z. (2021). College Students' Mental Health Education System Based on Data Mining Technology. *Proceedings - 2021 International Symposium on Advances in Informatics, Electronics and Education, ISAIEE 2021*, 106–109. DOI: 10.1109/ISAIEE55071.2021.00033

Xie, L. (2021). Practice of Big Data Technology and Improved BP Neural Network in College Students' Mental Health Education. In V. Sugumaran, Z. Xu, & H. Zhou (Eds.), *Advances in Intelligent Systems and Computing: Vol. 1384 AIST* (pp. 831–837). Springer Science and Business Media Deutschland GmbH. DOI: 10.1007/978-3-030-74811-1_120

Xie, Y., Lu, L., Gao, F., He, S. J., Zhao, H. J., Fang, Y., Yang, J., An, Y., Ye, Z., & Dong, Z. (2021). Integration of artificial intelligence, blockchain, and wearable technology for chronic disease management: A new paradigm in smart healthcare. *Current Medical Science*, 41(6), 1123–1133. DOI: 10.1007/s11596-021-2485-0 PMID: 34950987

Xu, F., Sepúlveda, M.-J., Sepúlveda, S., Jiang, Z., Wang, H., Li, J., Liu, Z., & Yin, Y. (2020). Effect of an Artificial Intelligence Clinical Decision Support System on Treatment Decisions for Complex Breast Cancer. *JCO Clinical Cancer Informatics*, 4(4), 824–838. DOI: 10.1200/CCI.20.00018 PMID: 32970484

Yacef, K., Caillaud, C., & Galy, O. (2018). Supporting learning activities with wearable devices to develop life-long skills in a health education app. In Artificial Intelligence in Education: 19th International Conference, AIED 2018, London, UK, June 27–30, 2018 [Springer International Publishing.]. *Proceedings*, 19(Part II), 394–398.

Yamauchi, M., Moritsuchi, Y., & Hamada, E. (2024). Novel concept of personalized therapy with continuous positive airway pressure. In *Sleep and Breathing*. Springer Science and Business Media Deutschland GmbH., DOI: 10.1007/s11325-024-03095-0

Yanamandra, R. (2024). Developing an integrated framework of application of wearable devices in education. *International Journal of Business Excellence*, 32(1), 85–104. Advance online publication. DOI: 10.1504/IJBEX.2024.135923

Yang, M. (2018). Application of Virtual Reality Technology in College Students' Mental Health Education. *Journal of Physics: Conference Series*, 1087(6), 062056. Advance online publication. DOI: 10.1088/1742-6596/1087/6/062056

Yang, Y. (2024). Application of wearable devices based on artificial intelligence sensors in sports human health monitoring. *Measurement. Sensors*, 33, 101086. DOI: 10.1016/j.measen.2024.101086

Yaokumah, W. (2020). Predicting and explaining cyber ethics with ethical theories. *International Journal of Cyber Warfare & Terrorism*, 10(2), 46–63. DOI: 10.4018/IJCWT.2020040103

Yapici, M. K., & Alkhidir, T. E. (2017). Intelligent medical garments with graphene-functionalized smart-cloth ECG sensors. *Sensors (Basel)*, 17(4), 875. Advance online publication. DOI: 10.3390/s17040875 PMID: 28420158

Yasugaki, S., Okamura, H., Kaneko, A., & Hayashi, Y. (2023). Bidirectional relationship between sleep and depression. *Neuroscience Research*. Advance online publication. DOI: 10.1016/j.neures.2023.04.006 PMID: 37116584

Yee Chung, J. W., Fuk So, H. C., Tak Choi, M. M., Man Yan, V. C., & Shing Wong, T. K. (2021). Artificial Intelligence in education: Using heart rate variability (HRV) as a biomarker to assess emotions objectively. *Computers and Education: Artificial Intelligence*, 2, 100011. Advance online publication. DOI: 10.1016/j.caeai.2021.100011

Yllaconza, J. E. C., Elescano, R. A. V., Cuevas, T. L. V., Orellana, M. V., Campos, J. H. T., Montenegro, J. S., Livia, D. P. R., & Loayza, E. F. P. (2024). Use of wearable technologies in health promotion in Human Medicine students. *EAI Endorsed Transactions on Pervasive Health and Technology*, 10. Advance online publication. DOI: 10.4108/eetpht.10.5701

Yoo, J. H., & Kim, J. H. (2023). Advances in Continuous Glucose Monitoring and Integrated Devices for Management of Diabetes with Insulin-Based Therapy: Improvement in Glycemic Control. In *Diabetes and Metabolism Journal* (Vol. 47, Issue 1, pp. 27–41). Korean Diabetes Association. DOI: 10.4093/dmj.2022.0271

Yoshida, S., Kihara, K., Takeshita, H., & Fujii, Y. (2014a). A head-mounted display-based personal integrated-image monitoring system for transurethral resection of the prostate. *Wideochirurgia i Inne Techniki Malo Inwazyjne*, 9(4), 644–649. DOI: 10.5114/wiitm.2014.43040 PMID: 25562008

Yoshida, S., Kihara, K., Takeshita, H., & Fujii, Y. (2014b). Instructive head-mounted display system: Pointing device using a vision-based finger tracking technique applied to surgical education. *Wideochirurgia i Inne Techniki Malo Inwazyjne*, 9(3), 449–452. DOI: 10.5114/wiitm.2014.44132 PMID: 25337172

Yoshida, Y., & Yuda, E. (2023). Workout detection by wearable device data using machine learning. *Applied Sciences (Basel, Switzerland)*, 13(7), 4280. DOI: 10.3390/app13074280

Young, M. F., Raines, K., Jameel, F., Sidi, M., Oliveira-Streiff, S., Nwajei, P., McGlamry, K., Ou, J., Oladele, A., & Suchdev, P. S. (2021). Non-invasive hemoglobin measurement devices require refinement to match diagnostic performance with their high level of usability and acceptability. *PLoS One*, 16(7), e0254629. DOI: 10.1371/journal.pone.0254629 PMID: 34270591

Zafari, M., Bazargani, J. S., Sadeghi-Niaraki, A., & Choi, S. M. (2022). Artificial intelligence applications in K-12 education: A systematic literature review. *IEEE Access : Practical Innovations, Open Solutions*, 10, 61905–61921. DOI: 10.1109/ACCESS.2022.3179356

Zamani, A. S., Hashim, A. H. A., Akhtar, M. M., Samdani, F., Siddiqui, A. T., Alluhayb, A., Hamza, M. A., & Ahmad, N. (2023). The prediction of sleep quality using wearable-assisted smart health monitoring systems based on statistical data. *Journal of King Saud University. Science*, 35(9), 102927. Advance online publication. DOI: 10.1016/j.jksus.2023.102927

Zarella, M. D., Bowman, D., Aeffner, F., Farahani, N., Xthona, A., Absar, S. F., Parwani, A., Bui, M., & Hartman, D. J. (2019). A practical guide to whole slide imaging a white paper from the digital pathology association. In *Archives of Pathology and Laboratory Medicine* (Vol. 143, Issue 2, pp. 222–234). College of American Pathologists. DOI: 10.5858/arpa.2018-0343-RA

Zeng, R., Johnson, E., & Patel, M. (2023). Building smart classrooms: Leveraging AI and wearables for adaptive learning environments. *Smart Learning Environments*, 9(4), 334–349.

Zhang, M., Luo, M., Nie, R., & Zhang, Y. (2017). Technical attributes, health attribute, consumer attributes and their roles in adoption intention of healthcare wearable technology. *International Journal of Medical Informatics*, 108, 97–109. DOI: 10.1016/j.ijmedinf.2017.09.016 PMID: 29132639

Zhang, P., Han, W., & Liu, Q. (2024). Research on Predicting the Mental Health of College Students with Prediction Models based on Big Data Technology. *IEIE Transactions on Smart Processing and Computing*, 13(4), 393–401. DOI: 10.5573/IEIESPC.2024.13.4.393

Zhang, X., Ding, Y., Huang, X., Li, W., Long, L., & Ding, S. (2024). Smart classrooms: How sensors and ai are shaping educational paradigms. *Sensors (Basel)*, 24(17), 5487. DOI: 10.3390/s24175487 PMID: 39275398

Zhao, A. P., Li, S., Cao, Z., Hu, P. J. H., Wang, J., Xiang, Y., Xie, D., & Lu, X. (2024). AI for science: Predicting infectious diseases. In *Journal of Safety Science and Resilience* (Vol. 5, Issue 2, pp. 130–146). KeAi Communications Co. DOI: 10.1016/j.jnlssr.2024.02.002

Zhao, Y., & You, Y. (2021). Design and data analysis of wearable sports posture measurement system based on Internet of Things. *Alexandria Engineering Journal*, 60(1), 691–701. DOI: 10.1016/j.aej.2020.10.001

Zheng, H. (2021). *Assisting Young Adults with Neurodiversity to Engage in Learning Activities: Smartwatch Applications in Inclusive Education* (Doctoral dissertation, George Mason University).

Zhou, Z. X., Tam, V., Lui, K. S., Lam, E. Y., Yuen, A., Hu, X., & Law, N. (2019, November). Applying deep learning and wearable devices for educational data analytics. In 2019 IEEE 31st International Conference on Tools with Artificial Intelligence (ICTAI) (pp. 871-878). IEEE.

Ziccardi, G. (2020, may 29). Wearable Technologies and Smart Clothes in the Fashion Business: Some Issues Concerning Cybersecurity and Data Protection. *MDPI*.

About the Contributors

Shafiq Ul Rehman is an experienced educator and professional consultant with a background in curriculum development and educational technology. With over 10 years in the field, Dr. Shafiq has worked with various educational institutions to implement sustainable practices and innovative teaching strategies. Dr. Shafiq received his Ph.D. degree from Universiti Sains Malaysia (USM), Malaysia, in 2017. He was a Postdoctoral Research Fellow at Singapore University of Technology and Design (SUTD), Singapore, from 2017 to 2020. He is currently Acting Dean, College of Information Technology and Chairman Department of Computer Science, Kingdom University, Bahrain. He is also an Assistant Professor with research interests in cybersecurity, artificial intelligence, and the latest emerging IT technologies. He lectures in various Computer Science and IT courses, including developing curriculum and courseware in tandem with current technology trends. He has authored and coauthored more than 50 peer reviewed publications, published in journals, conference proceedings, and book chapters. He supervises Ph.D., postgraduate, and undergraduate students. He also gives talks and training on cybersecurity, artificial intelligence, and the latest emerging IT technologies at various organizations, conferences, seminars, and workshops.

***

Joji Abey serves as an Assistant Professor in Finance & Accounting Department at College of Business Administration, Kingdom University, Kingdom of Bahrain. With nineteen years of teaching experience at higher education level, in which Seventeen years in the Kingdom of Bahrain. Dr.Abey holds the prestigious Chartered Manager designation from the Chartered Management Institute (CMI), UK. This designation represents professional recognition and is the highest status that can be achieved in the Management and leadership profession. She achieved the status of Fellow of Higher Education Academy (HEA) in recognition of attainment against the UK Professional Standards Framework for teaching and learning support in higher education. She published articles in national, Scopus-indexed journals and presented papers and participated in various international seminars and conferences.

Mohammad Irfan is presently working as an Associate Professor at NSB Academy Business School, Bangalore. He was associated with the School of Business, AURO University, Surat, Gujarat for five years. He is MBA (Finance) and M.Com (Account and Law). Dr. Irfan has done this Ph.D. from the Central University of Haryana. He has qualified UGC-JRF/SRF/NET in Management and Commerce. Dr. Irfan certified NSEs (NCFM) and BSEs certification. He has experience of sixteen years in the area of SAPM, Artificial Intelligence, Machine Learning, Blockchain, Cryptocurrency, Financial Engineering, Fintech, Green Finance, and Alternative Finance. He has to his credit more than 30 Scopus Indexed articles, includes The Journal of Economic Cooperation and Development (Q2), International Journal of Business Excellence (IJBEX), International Journal of Economics and Management (IJEM), Montenegrin Journal of Economics (Q2), Cogent Business & Management (Taylor & Francis) (Q2), Indian Journal of Finance (IJF) and Journal of Islamic Monetary Economics and Finance (JIMF). His citations reached 200+ along with 7 H-index. Dr. Irfan has published 6 books in Springer, IGI Global Publication (Scopus indexed) and national publisher. He is an editorial board member/reviewer in several national and international journals.

Rashmi Kodikal is an experienced and innovative professor with 21 years of teaching expertise at the postgraduate level. She has contributed significantly to research, presenting 112 papers at international and national conferences and publishing 54 articles in renowned journals indexed under Scopus, Web of Science, and ABDC. Her research focuses on organisational behaviour, service quality, and sustainability. She has served on various academic boards and panels and is currently the Chief Editor of two national-level management journals. She is also a skilled freelance trainer in behavioural sciences, conducting programs on self-development, behavioural diagnostic tools, transactional analysis, and emotional intelligence. Currently, she serves as a Professor of Management Studies at Graphic Era Deemed to be a University in Dehradun, India.

Durga Madhab Mahapatra, is currently working as an Associate Professor in Department of Commerce, North-Eastern Hill University, Meghalaya, India. He has 9 years of UG teaching experience and 19 years of PG teaching experience apart from 9 years of research experience. With his rich academic and research experience, he has published 9 books, 74 research papers and 18 book chapters

Devi Manikeswari is a dynamic educational professional with over nine years of work experience in both teaching and industry. With a natural flair for teaching, she has worked in prominent educational institutions in more than two countries and is currently employed as an Assistant Professor at the School of Business in Bahrain Polytechnic. She has made several presentations in seminars and published research articles in both national and international journals including Scopus indexed journals. She is a Certified Psychometric Test User (BPS Level 1 & 2 Certified Occupational Test User-Ability & Personality).

Naji Mansour Nomran, PhD, is currently Assistant Professor in Finance & Accounting Department, College of Business Administration, Kingdom University, Kingdom of Bahrain. His research interests include Islamic banking and finance, corporate governance, risk management, financial performance, and financial markets. His attention has recently shifted to a variety of topics, including the role that green finance plays in combating climate change and promoting sustainable development, examining issues related to cryptocurrencies, blockchain technology in finance, besides issues related to artificial intelligence and FinTech in finance. He has published several publications in international peer-reviewed journals as well as edited books.

Nilamadhab Mishra is an Associate Professor in the School of Computer Science and Engineering, AI-ML Division, at VIT Bhopal University, India. He received his Doctor of Philosophy (Ph.D.) in Computer Science & Information Engineering with a specialization in Data Science & Machine Learning from Chang Gung University, Taiwan. He has more than 22 years of national and international involvement in academic teaching and research at recognized Indian, Taiwanese, and African universities. He has over 50 publications in SCI/SCIE and SCOPUS-indexed journals, ISBN books and chapters, Indian and Australian patents, IEEE conference proceedings, and others. He has served as a reviewer, associate editor, and editorial board member for SCI/SCIE-indexed journals and conferences. Dr. Mishra worked on multiple funded research projects from the MOST, NSC, and CGU Memorial Hospital, Taiwan, during his Ph.D. and has been involved with several professional bodies. He is currently an Associate Editor for two SCOPUS-indexed journals, and his research interests span the areas of AI, Data Science, Machine Learning, and Cognitive Analytics & Applications.

Bhabani Shankar Mohanty is currently working as an Assistant Professor in the Department of Statistics and Applied Mathematics, School of Mathematics and Computer Sciences, Central University of Tamil Nadu, Tamil Nadu. He is an M.Phil, and a Doctorate (Ph.D.) in Statistics from the P.G. Department of Statistics,

Utkal University, Bhubaneswar. He has completed a Master's Degree in Statistics from Utkal University and worked for one year as Project-JRF at IIT-Bhubaneswar. He has presented research papers at National/ International conferences in India and has published papers in indexed international journals of repute. He has over 10 years of teaching and research experience. His research interest is Inventory Control, Fuzzy Logic, Optimization Techniques, and Sustainable Supply Chain. He has also command over analyzing high dimension data generated from different real-life phenomena.

Arul Kumar Natarajan currently serves as an Assistant Professor in the Department of Computer Science at the Samarkand International University of Technology in Uzbekistan. He earned his Doctor of Philosophy degree from Bharathidasan University, India, in 2017. Concurrently, he is engaged in postdoctoral research in Generative AI for Cybersecurity at the Singapore Institute of Technology, Singapore. Throughout his 14-year teaching career, Dr. Arul has held esteemed positions at various institutions, including Christ University, Bishop Heber College in India, and Debre Berhan University in Ethiopia. Dr. Arul has made significant contributions to academia, specializing in cybersecurity and artificial intelligence, evidenced by his portfolio of scholarly works. He has authored 32 international publications indexed in Scopus and 03 international publications and has delivered 34 conference presentations. Additionally, he has edited and published three books with IGI Global Publisher, USA, focusing on Python data structures, algorithms, and geospatial application development. In addition to his academic pursuits, Dr. Arul is a prolific innovator, holding 17 patents in India and 01 granted patent in the United Kingdom across diverse domains, including communication and computer science. He demonstrates proficiency in networking and cybersecurity, having completed the CCNA Routing and Switching Exam from CISCO and the Networking Fundamentals exam from Microsoft. He maintains a sincere interest in GenAI for Cybersecurity.

Habeeb Ur Rahman is an innovative academician with over a decade of expertise in higher education. He holds a PhD in Management and an MBA from VTU India and has achieved the prestigious 'Fellow' status from Advance HE, UK, in recognition of his significant contributions to global higher education. In addition to his academic qualifications, he has received certifications in FinTech from the University of Cambridge, Artificial Intelligence from the Massachusetts Institute of Technology (MIT), US, Transactional Analysis from The Berne Institute, England, and ISO Lead Auditor. These certifications highlight his commitment to continuous learning and staying updated with the latest advancements in his field. He is an active researcher in the field of social science and humanities, with his work published in reputable journals indexed under Scopus, ABDC, and Web of

Science. He possesses a deep understanding of behavioural sciences and is a skilled freelance trainer, conducting programs on self-development, behavioural diagnostic tools, transactional analysis, and emotional intelligence. Currently, he is serving as an Assistant Professor and the Head of the Staff Development Unit at Kingdom University, Bahrain.

Satya Pavan Kumar Ratnakaram is an accomplished academician currently employed as an Assistant Professor in the Faculty of Business and Logistics, Bahrain Polytechnic since September 2022. Prior to this, he has worked as a Program Manager and IQA member in the Commercial Studies Division, Bahrain Training Institute, Kingdom of Bahrain for more than 16 years. With over two decades of teaching/training/industry experience across three countries, presented and published more than 25 articles in national and international conferences and journals in the fields of banking, insurance, accounting, CSR, ESG, Psychometrics, Higher education etc.

Saroja Kumar Rout (Senior Member, IEEE) received M.Tech (CS) from Utkal University in the year 2007 and a Ph.D. in Computer Science in the year 2018 from Siksha 'O' Anusandhan University for work in the field of Wireless Sensor Networks. Currently working as Associate Professor in the Department of Information Technology, Vardhaman College of Engineering (Autonomous), Hyderabad, India. Contributed more than 60 research-level papers to many National and International journals and conferences. Also having 8 Indian patents. Research interests include Sensor Networks, Machine Learning, Cloud computing, and Health Informatics.

Zakir Shaikh is an Assistant Professor, Department of Accounting and Finance, School of Business Administration, Kingdom University, Bahrain. Prior to this, he worked more than 28 years in commercial and academic industry. He has published numerous articles in referred journals and presented many papers in various conferences, both local and abroad in the different area of Accounting, management and Islamic Banking and Finance. He has also participated in a variety of seminars, forums, workshops and international conferences. He obtained his Ph.D. (Islamic Banking and Finance) from India in 2018. His doctoral thesis explored the entrepreneurial phenomenon from an Islamic perspective and argued for profit and loss sharing (PLS) contracts as viable alternatives to conventional interest-based financing instruments. He holds Master degree in Commerce, Business Administration & Finance and Control. He is a member of AAA, Chartered Institute of Islamic Finance Professionals (CIIF), Malaysia; Indian Accounting Association; Orissa Commerce Association, India and associated with Accounting & Auditing Organization for Islamic Financial Institution (AAOIFI), Bahrain

Anand Soni, with 19 years of teaching experience and over 4 years of industry exposure, currently working as a faculty in the school of business at Bahrain Polytechnic (Ministry of Higher Education) apart from regular teaching he holds positions as a Co-supervisor for a PhD Scholar for some Indian University and as a reviewer for book chapters at Springer IGI. Noteworthy accomplishments adorn Dr. Soni's career, including the authorship of the book "Business Analytics and Data Science for Decision Making," comprising three chapters, a patent, and multiple publications in esteemed national and international journals and conferences. The breadth of Dr. Soni's engagements extends beyond research. They have actively participated in various capacities in academic and non-academic realms, contributing to event organization, delivering keynote speeches, and engaging in presentation opportunities at seminars and conferences. In the realm of education, Dr. Soni's teaching prowess shines through a diverse array of subjects across specializations such as accounting, human resources, and management, catering to undergraduate and master's level students. Their multifaceted expertise and dedication to both teaching and scholarly pursuits underscore a rich and impactful academic career.

Mohammad Taqi is presently working as an Assistant Professor of Accounting and Finance in Bahrain Polytechnic, Bahrain. He has around seven years teaching and research experience at UG and PG level. Dr. Taqi has published 22 research papers and 05 book chapters. Dr. Taqi has participated in international conferences in France, UK, Uzbekistan, Bahrain, Bangladesh, Thailand, Bhutan and Nepal for the presentation of his research papers.

Index

A

C

D

E

G

H

I

M

P

S

T

V

W

www.ingramcontent.com/pod-product-compliance
Lightning Source LLC
LaVergne TN
LVHW080346011026
849161LV00009B/2201
* 9 7 9 8 3 6 9 3 7 8 1 7 5 *